D1000618

Maintenance Management

Maintenance Management

Revised Edition

Lawrence Mann, Jr.
Louisiana State University

LexingtonBooks
D.C. Heath and Company
Lexington, Massachusetts
Toronto

Library of Congress Cataloging in Publication Data

Mann, Lawrence.
Maintenance Management.

Bibliography: p.
Includes index.
1. Plant maintenance. I. Title
TS192.M38 1982 658.2'02 81-47628
ISBN 0-669-04715-5

Published simultaneously in Canada

Printed in the United States of America

International Standard Book Number: 0-669-04715-5

Library of Congress Catalog Card Number: 81-47628

To Suzanne,
Chip, and Nancy

Contents

Figures

Tables

Preface

It would be more accurate to describe this revised edition of *Maintenance Management* as a revised *addition*. Very little of the first edition has been taken out. Some corrections have been made and some data have been updated—and much has been added. During the seven years since the original edition, maintenance costs have risen, on the average, 15 percent a year—a considerably higher rate than that of the Consumer Price Index.

In chapter 1, a section about the utilization of engineers in maintenance has been added. A number of typical work-order forms have been added to chapter 2. In chapter 3, a section on controlling maintenance backlogs has been added. Chapter 4 now includes coverage on establishing and justifying maintenance standards. Additions to chapter 5 involve establishing a preventive-maintenance (PM) inspection schedule, measuring and analyzing vibration, infrared thermography, and a discussion of designing for maintenance. Chapter 6 now includes a number of specific productivity indicators and a section on a suggested composite index.

Chapter 7 now has a section that depicts patterns of maintenance spares usage. Chapter 8 has been enlarged to include projection of maintenance costs by factored budgeting and an extensive discussion of forecasting maintenance-manpower requirements. Chapter 9 is augmented by discussions on justifying a computer for maintenance management and on deciding whether to buy, share, or rent a computer. Chapter 10 includes a discussion of performance-based training and training of maintenance supervisors. Chapter 11 now defines the factors to consider when evaluating contract maintenance. Chapter 12 now has a section on motivating by new shift schedules.

Chapter 13, which is entirely new, discusses statistical and operations-research applications in maintenance and touches on the theoretical bases of maintenance. If the book is used as a text in an engineering program, it might be advisable to amplify these topics or to cover these and more theoretical considerations as outside assignments.

The two appendix topics are also new. The maintenance-management inventory form in appendix A is for the do-it-yourselfer. It provides a means of accumulating data from which decisions can be made regarding the design or redesign of the maintenance-management information system (MMIS). Appendix B is a maintenance game or simulator. For those who use the book for an academic text, the game provides practical, hands-on experience in managing a maintenance program. It can be as superficial or as complex as the instructor desires. Finally, the new bibliography lists some maintenance source material.

Preface to the First Edition

This book results from a need that was made known to me by a group of maintenance managers. The lack of qualified individuals to manage maintenance operations led to a series of seminars and has resulted in an academic course encompassing these matters.

I have had experience in many plants and have found that, although plant management is aware of the need for maintenance management, few programs are operating as they were expected to operate. The material in this book is intended not only to assist maintenance managers in implementing the more modern aspects of their jobs but also to assist those managers in designing up-to-date programs if that is their desire.

For years, industrial-engineering programs have been concerned with the design and operation of production-control systems. In process-plant operations, the ratio of maintenance craftsmen to operators is greater than is the traditional hardgoods industry. This fact leads to the realization that the efficient management of these resources would yield more benefits to a firm than a belief that maintenance is not so important. Furthermore, the dependence of the process industry on preceding and succeeding steps indicates that maintenance must be more responsive to the needs of operations than it is where redundant systems exist. Therefore, there is no reason to assume that these operations cannot be as formalized as production-control operations.

In the production-control situation, the malfunction of the system is immediately apparent, usually resulting in the failure to meet a delivery date for material. The benchmarks and indexes that indicate how well the maintenance-management system is operating are not so apparent, and there is a real opportunity for inefficiencies and system failure to be present but not readily apparent. These situations indicate that more attention should be given to the maintenance-control system. Many plants profess to have maintenance-operating systems that are failing to live up to the expectations of management.

Throughout this book, emphasis is placed on computerizing the maintenance-management information system. Few plants do not have computers or do not have access to a computer with which to obtain information needed for decisions regarding the maintenance-management program.

Chapter 1 deals with organizing maintenance operations. This chapter considers such problems as area or centralized maintenance facilities and describes the organization of the maintenance function. Chapter 2 discusses the basic maintenance-control system from the paperwork standpoint. The work-order system is emphasized, as are time-reporting procedures. Chapter 3 considers job planning and scheduling, with discussion of techniques of planning and scheduling and job-planning functions. Chapter 4 is concerned with maintenance work measurement and standards. The various methods of measuring indirect labor and the creation of standards from those measurements are covered in this chapter. Chapter 5 describes various aspects of the preventive-maintenance system. Equipment inspection and its relationship to preventive maintenance are also presented. Chapter 6 considered the problems and solutions involved in measuring and appraising maintenance performance. The performance of the labor force and the maintenance-management system are covered, and various indexes for monitoring these systems are discussed. Maintenance-stores control is considered in chapter 7, and the relationship between plant maintenance and stores is explored. Advantages and disadvantages of various types of stores organizations are discussed, as well as the principles of the maintenance-stores operation. Chapter 8 considers maintenance budgets. This chapter considers long-range planning, forecasting, estimating return on investment, and equipment-replacement problems. The use of the computer in the maintenance-management information system is the basis for chapter 9. The various subroutines necessary to a maintenance-management information system are discussed in detail. Chapter 10 is concerned with maintenance training, and various aspects of training are discussed. Chapter 11 is devoted to consideration of contract maintenance, and the various advantages and disadvantages of contract maintenance are presented. Chapter 12 is concerned with motivating the maintenance foreman and maintenance craftsman.

Acknowledgments

I am indebted to the following, who have granted me permission to quote from their publications.

Thanks are due to Clapp and Poliak, Inc., for permission to quote from various articles from their publication, *Techniques of Plant Engineering and Maintenance.* Appreciation is also due to Morgan-Grampian, Inc., for permission to quote from *Factory* magazine. *Hydrocarbon Processing* magazine granted me permission to use various sources that appeared in their publication. The McGraw-Hill Book Company granted permission to quote from sections of their publications, which are noted in the appropriate place, as are other quoted references. The American Management Associations gave their permission to use a figure from their Report 77. Other abstracts and illustrations appear with permission of Technical Publishing Company from various issues of *Plant Engineering* magazine, copyright by Technical Publishing Company, 1301 South Grove Avenue, Barrington, Ill. 60010.

1 Organizing for Maintenance Operations

The management of maintenance activities has been a long-neglected field. For over fifty years, industrial engineers have been applying work-measurement and methods-improvement techniques to repetitive tasks in the operations areas. It was felt that, since maintenance activities were not repetitive in the same manner as operations tasks, maintenance did not lend itself to systemization. Recent developments in the field of indirect work measurement have convinced management that maintenance is amenable to the same sort of analyses as is operations.

It has become increasingly evident that maintenance can no longer be ignored; it must be engineered, as other plant functions are. The fact that the cost of maintenance labor is increasing faster than the index of total plant operating costs dictates increased prudence in spending for maintenance activities. In addition, the increasing scarcity of capable, well-trained craftsmen has meant that management must use those that it has more effectively .

Maintenance can be defined as the activities required to keep a facility in as-built condition, continuing to have its original productive capacity. For purposes of discussion, maintenance usually is categorized on the basis of when the work must be done: *emergency maintenance* means that the work must be done in the immediate future; *routine maintenance* normally means that the work must be done in the finite, foreseeable future; and *preventive maintenance* denotes maintenance that is carried out in accordance with a planned schedule.

During the past decade, the general field that includes industrial engineering, management science, and systems engineering has evolved to the point that it has become useful in solving many of the maintenance manager's traditional problems. Organizations in the forefront of technology have applied management information systems, inventory and materials controls systems, scheduling algorithms, work standards, indirect work measurement, and replacement theory to the everyday problems of maintenance. Other organizations are in the process of implementing these systems, although some have not yet made use of them. These procedures, which primarily have quantified items that previously have not been quantified, have made possible the elevation of maintenance management from an art to a science.

An adequate maintenance system should adhere to four principles:

1. Labor is only as efficient as management had planned it to be.
2. Good organization is essential for good performance.
3. For greatest efficiency, the functions of maintenance and production should be separated.
4. The greatest productivity results when each worker is given a definite task, to be performed in a definite timeframe and in a definite manner.

The key features of such a system are

1. organization of the function,
2. a work-order system,
3. planning and scheduling,
4. labor standards,
5. methods improvement,
6. material control
7. performance indexes,
8. preventive maintenance,
9. training of the work force.

We are concerned here mainly with the maintenance-management program or maintenance-management information system (MMIS). Some of the overall objectives of this program are to provide an MMIS that will

1. maximize the response of a facility to the wishes of the operator;
2. minimize maintenance cost in view of the overall plant objective of maximizing profits;
3. accumulate data to be used in analysis and performance measurement of the system;
4. include methods of informing all levels of management of the performance of the system so that corrective action can be taken by all levels of supervision; and
5. make the greatest use of standards in view of the economic decision not to standardize every operation.

Before considering each of these objectives, it is necessary to define the step-by-step approach that should be taken in designing a system like the MMIS:

1. Define the objectives of the system.
2. Establish responsibilities and authority.
3. Spell out the actions to be taken.
4. Determine the personnel required.

5. Establish the financial-control procedures.
6. Provide for feedback and performance indexes to monitor the system.

Designing the Maintenance-Management Information System

Define Objectives

The objectives of an MMIS can be many. In general, most organizations want a system that uses the work order as an input device and that supplies printout (1) showing the work orders to be worked on; (2) assigning men, equipment, and material to the various jobs; and (3) indicating the degree of performance of both the maintenance-management group and the direct labor group. The system can also include variance reports that indicate appreciable discrepancies between actual and estimated funds or man-hours; overtime reports; and reports on the number of work orders closed out per time period. Other reports will be mentioned in later discussions.

Establish Responsibilities and Authority

Once the system has been evolved and flow diagrams and other graphic representations have been completed, the maintenance functions must be packaged into positions. Work-measurement methods must be used to determine whether management is expecting too much or too little of the individuals who have been assigned to new positions in the plant. These positions usually carry such titles as planner, scheduler, coordinator, expediter, area-maintenance foreman, and maintenance manager.

Spell Out Actions

In the design of the MMIS, it is necessary to define all action steps in the system. This includes such activities as initiating the work order, estimating the resources required for the work order, ordering material that is not on hand, and reserving equipment for the work order. Other action steps include completing the work order after the job has been finished, closing out the work order, and returning unused material to the storehouse.

Determine Required Personnel

A job-description format is recommended for defining the many tasks involved in the functions of planning, scheduling, expediting, coordinating, and supervising operation of the MMIS. Careful coordination and frequent training sessions are recommended when the program is initiated. Most plants find that it is wise to promote a craft or area foreman to the maintenance position that deals with planning, scheduling, and estimating.

Establish Financial-Control Procedures

Maintenance-management systems require two financial and accounting subsystems. The first of these is a costing system, in which all aspects of the work order are divided into codes that account for all expenditures, including labor, material and equipment considerations, capital versus expense, geographical location, identifiable equipment, preventive or emergency maintenance, and other expenditures necessary to fulfill the needs of both the accounting and the maintenance functions.

The second financial system is used for budgeting. Maintenance departments must account for expenditures and must function within a budget. A short-range budget covers the day-to-day or week-to-week operations of the department. An intermediate budget encompasses expenditures projected for the immediate future, usually thirty days to six months. A long-range budget is designed to inform management of the resources maintenance will need for one to five years in the future.

Provide for Feedback and Performance Indexes

To evaluate the effectiveness of any system, it is necessary to use feedback information and performance indexes. The MMIS is no exception; those who estimate for the work order must have continual feedback to ascertain whether or not their estimates are valid. Many maintenance estimators base their estimates on knowledge gained when they were in the field; but procedures, methods, and equipment change, and such estimates do not truly represent what is currently taking place in the field.

Performance indexes are essential to an evaluation of the system. These indexes should reflect how well maintenance manages the work force and how well the work force performs their particular tasks.

The Functions of the System

Top management's primary responsibility in relation to MMIS is to establish the tempo of the program. Managers must express their desires for the program in sufficient detail so that everyone involved in the system understands what management expects of them and of the system. Top management is responsible for such major decisions as whether or not to contract all or portions of maintenance. Exercise of this responsibility depends on information received from all organizational levels. Attitude is also a top-management responsibility; a positive attitude must start with top management. The example set by top management will be reflected by all levels and will dictate, to a great extent, the seriousness with which the organization takes the MMIS. Evolution of the MMIS should be a joint effort by all levels of management.

Input from the supervisor must be considered in the design of the system. The supervisor's responsibility and sense of authority are very important. Although modern maintenance-mangement information systems require that the maintenance craftsman come in contact with a number of supervisors, there should be no question regarding to whom he is responsible. Another top-management decision involves supervisor density—how often individual workers see the supervisor and to what degree they are on their own. When work-sampling studies reveal that craftsmen are often waiting for instructions or material, supervision may be spread too thin. Management then must provide more supervisors or reallocate the duties of existing supervisors so that they come in contact with the workers more often.

This gives rise to another problem that must be decided very early in the creation of the MMIS—that is, whether area supervision or central supervision is more desirable. This will be discussed in greater detail later.

Perhaps the greatest innovation that has taken place in maintenance in the past twenty-five years is the acceptance of the multicraft maintenance mechanic as a labor category. Use of multicraft mechanics generally involves a collective-bargaining agreement, and management should proceed with caution in instituting such a program. The large number of plants that are changing the structure of their maintenance personnel away from traditional craft lines to include at least some maintenance mechanics indicates that this trend probably will continue and accelerate in the future.

The process by which a plant changes from the traditional craft breakdown to use of the multicraft mechanic is interesting. If the collective-bargaining agreement prohibits this sort of organizational change, it is necessary for management to bargain with the union, granting sufficient items that the union wants as well as a substantial wage increase in order to bargain for the multicraft organization. The large number of plants that have changed over in the recent past attests to the ability of management to make these changes.

The change usually starts on a rather modest scale, involving the joining together of all crafts that are involved in pipe crews. This means that the craftsmen would be trained in and asked to perform such varied tasks as insulating, burning, welding, pipe fitting, and possibly some minor instrumentation work. Concrete crews could be joined by training the craftsmen in carpentry work for the forms, boilermaker work for cutting and tying rerod, and concrete finishing. These groupings are but a step toward developing the true multicraft mechanic, who is trained to perform all tasks and who can perform those tasks if necessary. In practice, a craftsman who has been well trained in his primary craft seldom is asked to vary from his usual tasks, but management can use him in other categories should the need arise. A good foreman will use a craftsman in the craft for which he is most experienced.

A number of functions in the MMIS can be grouped into what might be called a work-order program. Every system must have some work-requesting document, which, for simplicity, we will call the work order, although it has other names. This form, which will be discussed in greater detail in chapter 2, includes such items as a narrative description of the work requested, estimates of the man-hours required, estimates of material and equipment required, authorization by the responsible authority, priority considerations, and codes that appropriately describe the work and the apportioning of cost.

Information from the work order is used by the scheduler to make short- and long-range schedules and by the planner who must consider all the details necessary to accomplish a particular job—that is, the needed personnel, the estimated man-hours, material, equipment, and costs; and the procedure to be followed. The work order should be completed in sufficient detail to provide all the information necessary for job planning.

One of the planner's tasks is to list all material required for the job, as specified in the work order; and the system should include some procedure by which it can be ascertained that the material is on hand. Frequently, material is set aside or reserved for a specific work order to ensure that it is not used before that job is scheduled.

The system should also include a consideration of standards. Standards specify the material and equipment necessary to do a job and the procedure to be used in order to accomplish it in the safest manner. Standards also include the time it should take to do the job, thereby providing measurement of results by comparison with the actual time used. The more repetitive the task, the more beneficial the standard is.

The final function to be considered is financial accounting, which include apportionment of the costs of a particular work order to the geographical areas of the plant; identifiable equipment; and capital versus expense (if capital projects are encompassed in the maintenance system). The work orders for maintenance jobs not yet completed provide an indication of how budgeted funds for maintenance will be used. For forecasting purposes, these budgets normally are translated into man-hours.

If there is an existing maintenance system, it is necessary to analyze it and to decide which elements of that system, if any, are applicable or transferable to the newly designed MMIS. As a practical matter, for instance, the work-order form may be redesigned to accumulate the data necessary for the new system while retaining many of the features with which the field forces are familiar. The existing system also must be evaluated as to its measure of success. Such evaluation should include such considerations as whether the system is responsive to management decisions; whether the system accumulates the necessary data for decision making; whether the productivity of the work force is satisfactory; whether idle time or overtime is excessive; and whether the cost of maintenance is commensurate with the production rate. Answering these questions will emphasize points that must be considered in the redesign of the system.

Managing the Program

If the MMIS is to be managed properly, it is necessary that those administering the program be entirely familiar with the requirements for such a system. To this end, what is expected of the system should be described in a narrative way, and an organizational chart should be constructed to assist in appreciation of the tasks at all levels of organization. Next, the communicator should be considered. (In most maintenance programs, a work order is the primary communicator for the system.) It is then necessary to establish some criteria for task standards and performance indexes.

Another consideration is records, which obtain their information from the work order and become the inputs to various reports that enable management to control the system.

The interface between the maintenance and stores departments is a vital facet of any maintenance system. No amount of planning and scheduling will suffice if the proper material and tools are not available when the job is initiated. Therefore, it is necessary for the maintenance system to have a clear and positive relationship with the stores department.

From the organizational standpoint, there is little rationale for the stores department to be under any function other than maintenance. In the process plant, the raw materials for the plant are rarely the items that normally would be found in a storehouse. The raw materials usually arrive by pipeline, ship, rail car, or tank truck. These materials are stored in tanks or other vessels and do not resemble the items that normally are required for maintenance. Therefore, the stores department exists merely to serve maintenance, and good organizational policy dictates that, if responsibility for maintenance is placed within the maintenance department, then the maintenance department should be provided with the freedom to use all resources as it sees fit. Accordingly, most successful maintenance organizations place the storehouse under the authority of the maintenance manager.

If one considers that the purchasing department is but an extension of the stores department, one realizes that there must be a clear and positive link between maintenance and purchasing. The purchasing department must be aware of the priorities of what it is buying. It must be aware that the maintenance organization is awaiting material to complete a work order; otherwise, the purchasing agents are merely ordering material to replace stock. The maintenance organization must have good communications with the purchasing department to assure a steady flow of items that previously were not stocked or that are out of stock in the storehouse.

An adequate preventive-maintenance program requires a group of people often known as equipment inspectors. This group performs various nondestructive tests, such as vibration tests, to predict the need for maintenance.

Also, if productivity, performance, and expenses are to be monitored, maintenance management must find some way to quantify the level of maintenance required. The degree to which any facility is maintained is usually dictated by the operator of that facility. Operators have different ideas about what constitutes necessary maintenance. Therefore, the amount of funds spent on a specific facility, in many cases, may depend entirely on the operator, not on what is actually necessary.

Finally, maintenance management must be aware of the necessity for training programs that maintain or enhance the productivity of the maintenance labor force.

Plant Maintenance Needs an Annual Report[1]

For years, plant maintenance has been at the mercy of the capital-versus-expense controversy. Traditionally, all monies spent for maintenance have been in the expense category, and no one has felt it necessary to justify their allocation. This arrangement does not permit management to judge the ability of maintenance to program improvements. Furthermore, maintenance does not have an opportunity to participate in capital programs that will increase its productivity. Part of this problem lies in the difficulty of expressing maintenance productivity as compared to the ease with which the production side of the firm can express its productivity. The dynamics of the firm, in part, prevent maintenance from directly relating its efforts to reduced costs—that is, the relationship of maintenance costs to hours of operation, pounds of product, and so forth, is not so clear-cut as the relationship of production costs to those factors.

Mention must be made here of the dynamics of a process plant. Very few plants remain the same from one time period to the other. The factors that cause plants to be dynamic include different mixes and different quantities of products—that is, the degree of full capacity to which each production unit in the plant has been run during the time period, the degree of contract maintenance, the level of maintenance that has been decided by management, the productivity of the work force, and various degrees of automation through materials handling and power equipment. Whenever an index is compared from one time period to the other, these dynamic situations must be realized and factored into the index.

Indexes

The most significant portion of a maintenance report should be the indexes that are designed to give plant management a quantitative assessment of how well maintenance tasks are being performed. Maintenance-performance indexes should be divided into three classes: top management—to indicate how well all levels of maintenance management are spending the resources that have been granted to them for maintaining the plant; maintenance control—to indicate how well maintenance management, planning, and scheduling perform their tasks; and labor productivity—to indicate how well the maintenance crews perform their jobs.

The most common labor-productivity indexes compare a predetermined standard time for work orders to the actual time expended on those work orders. This comparison can be done either weekly or monthly.

Maintenance-control indexes could include any or all of the following. Work-load indicators include current backlog of work, total backlog, man-hours spent on preventive maintenance, number of open work orders, and number of hours of assigned or area work. Planning indicators include jobs completed on schedule, forecasting effectiveness, maintenance hours planned versus those actually worked, emergency work, overtime, and downtime. Performance indicators include productively engaged maintenance manpower, maintenance effectiveness while working, maintenance-labor costs compared to maintenance-material costs, maintenance cost per unit of production, and the number of maintenance workers compared to the total number of plant personnel.

Return-on-Investment Concept

In the past, maintenance has ignored the return-on-investment concept that is so popular for evaluating production projects. Maintenance has a need for and can justify the investment of capital funds. Nevertheless, most maintenance organizations have not wished to compete with production for these funds. I do not wish to convey the impression that maintenance and production are equal; however, from the standpoint of the organizational structure of the firm maintenance is and always will be a service organization that is an adjunct to production.

Technological Forecasting

Necessary to any long- or short-range planning for maintenance is the integration of maintenance planning with technological forecasting. Just as production needs to know the future plans of the firm, maintenance must know and integrate the effects of those plans into their own planning. Several basic ideas about prediction, judgment, and the nature of data inputs to be used form the principles of forecasting practice. These principles, which have to do with the manner with which time series can be extended into the future, pertain to all technological forecasting techniques, including implicit or intuitive forecasting, explicit or formal forecasting, forecasting by extrapolation of existing trends, forecasting by analysis of precursive events, forecasting by analysis of primary trends, and dynamic forecasting. Examples of such forecasting that might affect maintenance are retirement or addition of productive facilities in the plant, consideration of contract maintenance, and the percentage of utilization anticipated for facilities that will remain active. Long-range planning might also include forecasts of increased or decreased productivity; the effects of training programs; and the effects of large-scale technological developments, such as computer control or drastic improvement in material, that might have a large impact on maintenance requirements. This is covered in detail in chapter 8.

Capital Projects

Maintenance management must also have a plan for capital projects that are intended to improve the ability of maintenance to perform the work required of it. These capital projects should include such efforts as a program to increase the utilization of power tools: alterations in the materials-handling system; acquisition of equipment, such as cranes, hoists, or prefabricated scaffolding; and acquisition of electronic data-processing equipment that would increase the effectiveness of maintenance planning. Any training program, such as craft training or maintenance-management training, would improve management's ability to perform and would be classified as an investment program.

The Annual Report

The annual report is a communicator from maintenance management to plant management indicating the manner in which maintenance management is using the resources given to them; it should cover

1. intermediate and long-range plans,
2. forecast criteria and bases,
3. comparison of contract versus own forces,
4. capital-investment program,
5. training programs, and
6. cost evaluation.

Immediate and long-range plans should be broken down into plans for manpower, equipment, and material resources. This is the traditional line-item budgeting process. In addition to this traditional breakdown, a program-budgeting process should be followed. Program budgeting requires that the total cost of maintenance be separated into various programs, such as a painting program for a specific portion of the plant, an insulation program, the normal maintenance required for a process unit, and other specific groups of tasks. In this way, management can be aware of where money is being spent; and, if the budget must be curtailed, management can choose the programs that they wish to curtail.

Manpower planning should include craftsmen—broken down into crafts and maintenance-mechanic categories—and administrators, such as maintenance schedulers, planners, and methods analysts. The need for equipment should be presented so that it can be integrated into the organization's capital-investment program. Material resources are reported according to the dollar value of spare parts carried in stock. Frequently, the turnover of the storehouse is of interest to management, as is the number of transactions that occurred during a particular time period.

The forecast criteria and bases for intermediate and long-range plans should be derived from the most up-to-date information available from the production and design divisions, so that the latest decisions are integrated into the annual report.

In many maintenance situations, contract maintenance has become an increasingly attractive alternative to a maintenance-management system. Before any long-range plan can be devised, decisions must be made to determine to what extent, if any, the company plans to contract maintenance forces.

The capital-investment programs considered in the annual report should include not only capital equipment, such as power tools and materials-handling equipment, but also participation in training programs and any other capital-investment program that would have an effect on maintenance performance.

Cost indicators could include the ratio of direct and general maintenance cost to total labor cost, indirect maintenance cost as a portion of total maintenance cost, the percentage of maintenance payroll, actual maintenance cost compared with budgeted maintenance cost, and maintenance administrative cost as a percentage of total maintenance cost.

The annual report not only informs plant management of how well maintenance is doing its job; it also informs maintenance itself of the areas of strength and weakness in the program. The fact that an annual report is prepared indicates to management that plant maintenance does plan and that the requested capital funds are part of a long-range, comprehensive program. Finally, the annual report raises maintenance to a more professional level of activity.

What Could an Engineer Do for You in the Maintenance Game?[2]

The maintenance engineer plays an important role in today's maintenance organization. Specific maintenance responsibilities are discussed throughout this book; this section merely recommends that those who are responsible for establishing the maintenance organization consider assigning the maintenance functions to a maintenance engineer.

In most well-organized maintenance departments, those functions have been assigned and, presumably, are being adequately covered. This section attempts to describe the full range of responsibilities that probably can best be handled by an engineer. Figure 1-1 provides job description of the maintenance engineer's responsibilities, and figure 1-2 indicates his place in the organization. Many of the topics mentioned in this section are covered in more detail later. (If definitions of terms and techniques are needed, consult the index.) The maintenance engineer coordinates the activities of maintenance personnel and is responsible for the work-order information system, the standards program, the preventive-maintenance (PM) planning and scheduling system, and the use of modern diagnostic techniques.

In short, the maintenance department operates as it should because the maintenance engineer brings continuity to all the essential elements of a maintenance-management effort (see figure 1-1).

Normally, in a maintenance work force of fewer than fifty persons, the administrative and technical aspects of maintenance management are fragmented among key people. Such an arrangement lacks some of the consis-

Scope—The maintenance engineer is responsible for ensuring that equipment and facilities are properly installed, correctly maintained, monitored for performance, and modified, if needed.

Specific responsibilities include assessment of newly installed equipment to ensure maintainability and adequacy of parts, maintenance instructions, prints, design, and installation procedures. He monitors ongoing work to ensure that sound practices are followed.

He reviews the information system to ensure its adequacy and uses it to review the workload and backlog with the view of recommending changes in the workforce level.

He observes the adequacy of spare parts and arranges the necessary actions to ensure the stocking of correct parts in the proper quantity. Further, he monitors the quality of parts used, to ensure good performance and initiates action to correct inadequacies.

He reviews repair history and costs to determine repair, rebuild, overhaul, or corrective maintenance needs.

He reviews the conduct of PM services to ensure they are conducted on time and according to standard. He evaluates deficiencies uncovered, and helps develop standards for individual major jobs, procedures for performing work, cost levels, and quantity of resources used.

Periodically, he reviews costs and benefits to verify make or buy actions. He also develops recommendations for training. As required, he prescribes methods for nondestructive testing.

Control—The actions of the maintenance engineer are prescribed by the maintenance superintendent. Usually, the maintenance engineer reports directly to the superintendent.

Measuring Performance—The maintenance engineer's performance is generally measured by the type of problems he uncovers and the quality of the solutions prescribed. However, in the broad scope, the most successful maintenance engineer is one who can develop the most practical, realistic technical solutions. A good maintenance engineer provides the superintendent with reliable, technical support solutions.

Benefits—The maintenance engineer helps bring continuity and better overall performance to a maintenance organization.

Qualifications—The maintenance engineer should be an experienced maintenance supervisor with an engineering background. Ideally, he should have a degree in engineering and have had field experience as a foreman or general foreman. He is usually identified as being next in line for the maintenance superintendent's job.

Source: Paul D. Tomlinson, "The Maintenance Engineer," *Plant Engineering,* July 10, 1980, p. 75.

Figure 1–1. The Maintenance Engineer's Responsibilities

tency that would be possible under a single supervisory head. In a maintenance work force of more than fifty persons, however, the need to ensure continuity is more serious (see figure 1–2, which shows the relationship of the maintenance engineer to other engineering functions). If these administrative and technical activities were attempted without the assistance of a maintenance engineer, several problems could arise, including the following:

1. Work orders might be used improperly, and standards might not be observed.
2. Application of the PM program might be marginal, and sophisticated diagnostic techniques often might not be employed.
3. Essential data might not be used properly, and repair histories often might be missing.
4. Control of spare-parts quantity and quality might be erratic.

Such problems frequently develop because these essential responsibilities are added to the normal duties of foremen, planners, or superintendents, with the result that some of these activities are performed poorly, incompletely, or not at all.

The effectiveness of each element of the work force can be assured by placing a maintenance engineer in charge to see that all functions are admin-

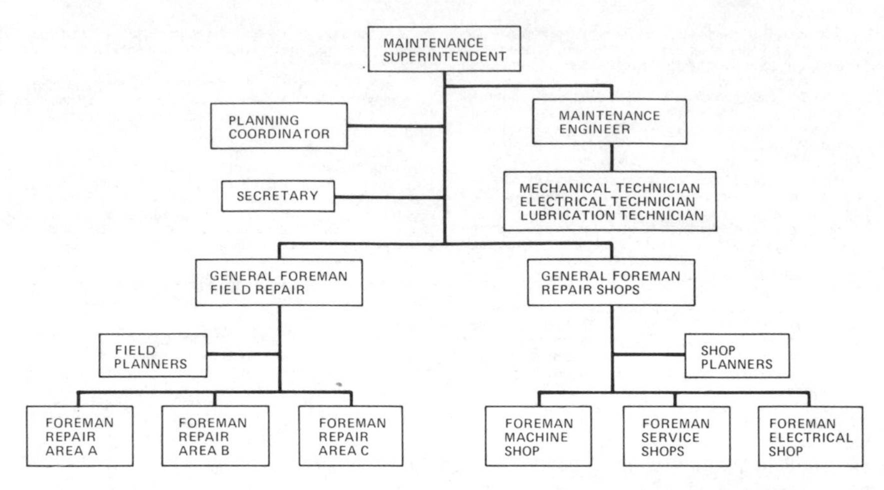

Source: Paul D. Tomlinson, "The Maintenance Engineer," *Plant Engineering,* July 10, 1980, p. 75.

Figure 1-2. The Maintenance Engineer's Place in the Organization

istered properly. The following discussion indicates how useful such a staff member can be to the maintenance organization.

Preventive Maintenance

Development of the basic PM program—including service frequencies, checklists, and use of diagnostic techniques—and continual monitoring of the program are the principal functions of a maintenance department. The maintenance engineer is involved not only in what maintenance is performed, when it is performed, and how it is performed, but also in the results. He monitors the PM program to observe both manpower use and the overall effectiveness of the program. He reviews the manpower required for PM services and recommends changes in staffing levels. He makes certain that data on equipment inspections are available as an important source of information on machine performance. The maintenance engineer and his staff also monitor the lubrication program to ensure proper lubrication, oil sampling, resale of used oil, and proper replenishment.

Predictive Maintenance

The use of diagnostic devices to monitor equipment performance is becoming more important as machines increase in complexity. Selection of the proper devices, correct use of them, and interpretation of results are essential components of a good predictive maintenance program. The maintenance engineer usually is behind the effort to see that these devices are specified and used effectively.

Information

The work order provides the primary information needed for good maintenance management, and it should be designed carefully. The maintenance engineer is the ideal person to design the information system because he understands its operation, has the time to develop it, and is in the best position to monitor its use.

Because he is the principal user of maintenance information, he constantly reviews repetitive repairs, machine performance, and cost trends so that he can recommend repair or overhaul actions.

Once in use, the work order must be monitored to detect inadequate data or errors and to make modifications; seldom does a superintendent, planner, or foreman have sufficient time to deal with these matters. The maintenance department and all its operations benefit from placing this responsibility with the maintenance engineer.

Work Force

A effective work force must be composed of the right number of maintenance personnel and the proper balance of craftspeople. Ineffective maintenance results often can be traced to an improper work-force mix. Data concerning the work load should be available to the maintenance engineer; with such information, he can periodically check the work load and recommend work-force adjustments.

How maintenance manpower is used is of vital interest to the maintenance manager. Too little preventive maintenance results in excessive emergency work, and ineffective use of crews leads to excessive unscheduled repairs. Inadequate planning and information reduce the opportunity for scheduled major maintenance work and improvements in productivity. By accepting more construction, installation, or modification work than it can handle, a maintenance department risks undermining its capability to perform basic maintenance.

All these actions should be carefully supervised, and the maintenance engineer, with this knowledge of the work load, is in the best position to monitor manpower use.

Use of Standards

The exact way in which a major maintenance job should be performed is predicated on standards. The maintenance engineer should be responsible for specifying what work requires standards and what those standards should be. If overhauls are to be performed, for example, the maintenance engineer should specify when and how they are to be done, their cost limitations, and specifications for the completed overhaul.

The maintenance engineer should constantly observe work procedures and repair techniques. He is in the best position to clarify necessary actions of foremen and crew members. His observations often result in the development and use of improved standards and, subsequently, in better performance.

Repair History

The repair history reveals the pattern of chronic equipment problems that call for corrective measures before damage is done is a machine. The repair history is also the source of information for determining the need for replacing equipment. The maintenance engineer can be instrumental in setting up the repair-history program. Repair-history data also provide vital information that he can use in prescribing repair and maintenance actions.

Repair Materials and Spare Parts

A common area of disagreement between maintenance and purchasing is the requisitioning, inventorying, and storage of maintenance spare parts. The maintenance engineer can alleviate this problem by monitoring spare-parts procedures.

The major effort of the maintenance engineer in controlling these storage items is through quality control and standardization. Information on failure analysis and the quality of parts and materials will help establish specifications that will ensure improved performance. Through standardization, fewer parts will be needed, and their quality and durability can be improved.

Make-or-Buy Decisions

A frequently necessary decision is whether a replacement item should be purchased or fabricated in the shop. If it is to be manufactured in the shop, it is necessary to know how many must be made to ensure a competitive price. There is often a compulsion to manufacture items just because the shop has the capability. The make-or-buy decision can be a natural extension of the maintenance engineer's duties. He has the repair-history records and all the data collected from various sources to enable him to develop guidelines for arriving at a reliable decision.

Overhauls and Rebuilds

Most equipment eventually deteriorates to such a point that routine servicing and component replacements are insufficient to ensure good performance. At that point, major components must be replaced or rebuilt, and often the entire unit must be overhauled. Decisions to overhaul must be based on factual data that are properly analyzed and interpreted. These data include the repair history and the cost of replacement. Because overhauls are costly and time-consuming, specific procedures are necessary for determining when to overhaul. Development of the necessary records can be handled easily by the maintenance engineer.

New Installations

Installation of new equipment must meet exacting specifications. In addition to supervising the installation of a new machine and seeing that it operates properly, someone must be responsible for obtaining operating and maintenance instructions, prints, and spare-parts listings from the manufacturer. Because this equipment becomes the responsibility of maintenance, the maintenance engineer should see to it that all pertinent data about the new equipment are supplied and filed for future use.

Quality Control

The foreman normally check the quality of the work performed. The maintenance engineer should also inspect the quality of major jobs and should verify adherence to standards. He also could be called upon to check the quality of work performed by contractors.

Cost Trends

Information relating to costs is available to the maintenance engineer from all the records he maintains. Such data will enable him to interpret cost trends.

A plant engineer who is aware of flaws in the present operation of his maintenance department should seriously consider the addition of a maintenance engineer to his staff. The maintenance engineer's careful control of all maintenance resources will help the department operate more efficiently.

The Maintenance-Management Manual

A necessary part of the MMIS is documentation of that system so that continuity can be maintained. The necessity of reducing any system to written form can also lead to a better understanding of that system, allowing corrections to be made early in the life of a system. The maintenance-management manual should include narrative descriptions, flow sheets, job descriptions, sample forms, continuing productivity-measurement programs, contracting, indexes, and annual reports.

The manual should be distributed to all those who have any connection with the program, including the stores and purchasing departments, the operating departments, and plant management. It should answer procedural questions related to such matters as disposition of work orders, material lists, stockouts, and the many other problems normally encountered in day-to-day maintenance operations.

The manual should start with a brief narrative description of the purpose and workings of the maintenance-management system, followed by a flow sheet showing the work forms, the manner in which information is recorded, and the uses to which that information is put. The flow sheet should also indicate the ultimate disposition of the various copies of forms. (See figure 2–4 in chapter 2 for a typical flow diagram.)

Job descriptions are an integral part of the manual. They can be used to locate personnel with the qualifications necessary for various positions in the maintenance-management function and to convey to the position holder the limits and responsibilities of his position.

A sample of all the forms that are necessary for the system should be included in the manual, together with instructions for completing each form. How and why each form is used should be explained, so that the interested parties will view accurate completion of the form in a responsible manner. The disposition of completed forms should also be included in this section.

An important part of the manual is a description of the various programs used to monitor and improve maintenance productivity. These programs might include work-sampling programs, productivity-assessment indexes, training programs and methods, and a continual review of maintenance operations from the standpoint of minimizing costs.

If the plant uses a contracting alternative, the information necessary to decide whether or not to contract work should be included in the manual.

Finally, the various indexes and the format for the annual report should be included. A uniform report format simplifies comparison of data from year to year.

Three Basic Plans

Any listing of the advantages and disadvantages of the central, area, and departmental (or unit) maintenance organizations may be debatable. Many of the same pros and cons are claimed or disavowed by the supporters of each of the three plans. This possible disagreement should be appreciated when considering an individual plan.

Central Maintenance

In a central-maintenance organization, maintenance mechanics of craftsmen are assigned to work in any or all areas of the plant but report to the same maintenance head. The number of workers may vary, from one to more than a thousand in a large operation such as a crude-oil refinery. In industries or plants that normally require large crews to check equipment, a central-maintenance system is generally used.

The advantages of a central-maintenance system include the following:

1. Sufficient men are available to handle the work requirements of the plant.
2. Considerable flexibility is available in assigning members of the maintenance crew with different craft skills to the various jobs.
3. Emergency jobs, breakdowns, and new work are handled quickly.
4. The total number of maintenance personnel can be held reasonably level, thus minimizing hirings and layoffs.
5. Specialists can be used more effectively.
6. Special maintenance equipment is used effectively.
7. One individual is responsible for all maintenance.
8. Accounting for all maintenance costs is centralized.
9. More control is obtained over capital or new work.
10. Workers are better trained in their craft skills.
11. A central group can more readily justify the need for trained engineers.

The disadvantages may include the following:

1. Members of the maintenance crew are scattered all over the plant and are not properly supervised.

2. Time is lost in traveling to a job, obtaining tools, and receiving instructions.
3. Coordination or scheduling of several crafts to a job becomes difficult.
4. More administrative controls are necessary for effective maintenance performance.
5. Different individuals are assigned to the same equipment; consequently, no one really becomes proficient in its repair.
6. The interval between initial job request and completion of routine work is too long.
7. Scheduling priorities are given to the various production jobs by a maintenance man, not by a production man.
8. The capital expense of pickup trucks, bicycles, and motor scooters is necessary to reduce travel time.
9. Time-consuming planning sessions with the production departments and plant manager are often needed to agree on the order of work. At the very least, these should be held weekly.

Maintenance supervisors are now being recognized by industries as key persons on the management team. Special courses are offered by colleges, universities, and organizations to improve the skill of plant engineers, and these engineers are rapidly developing new concepts, techniques, and practices that provide improved maintenance service at lower costs. The central-maintenance organization should attract people of this caliber.

Area Maintenance

If maintenance is organized on an area basis, maintenance crews are assigned to specific areas in the plant, and all report to the same maintenance head. The areas may be defined geographically, by product or production grouping, or by service function (utilities, research, and so on).

The advantages of area maintenance include the following:

1. Maintenance personnel are readily accessible to production personnel. Thus, scheduling is simplified and the competition for service is reduced to one or two departments. (Work normally labeled "rush" under a central-maintenance plan can be scheduled for the day after tomorrow.)
2. Time spent traveling to a job and obtaining tools is reduced.
3. Time lag between the issuance of the work order and its completion is minimized.
4. The same individuals service and repair the equipment and come to know its peculiar characteristics. Equipment is repaired more rapidly

by a maintenance group that is familiar with its assembly and spare-parts stock.
5. Maintenance crews are better supervised.
6. Production-line or process changeover is faster.
7. There is greater continuity from one shift to another.
8. Maintenance supervisors and workers become more familiar with production schedules, problems, special jobs, and so forth.

The disadvantages may include the following:

1. There is a tendency to overstaff the area.
2. Major repairs or servicing jobs are difficult to handle.
3. There are more personnel problems and regulations pertaining to transfer, hirings, and overtime.
4. Special equipment is difficult to justify because usage may be limited.
5. There is a duplication of service equipment, and multiple area-maintenance shops use more plant space.
6. More clerical help is needed if the area groups are large.
7. It is difficult to use specialists effectively.

Proper area staffing is a major problem, which becomes more difficult in a plant that observes strict craft lines. The work does not occur uniformly; it varies daily, even hourly. Four pipe fitters may be necessary one week but only two the next week. Consequently, there is a tendency to select the number of craftsmen on the high side of average. If the low side of average is used, more workers must be obtained for some jobs. This transfer of personnel can become difficult and complex.

Departmental or Unit Maintenance

In a departmental- or unit-maintenance organization, maintenance mechanics are assigned to a definite department, unit, or function and report to a production supervisor. (Sometimes they report to a maintenance foreman, who in turn reports to a production supervisor.)

The advantages of the unit-maintenance group are similar to those of the area-maintenance group. The major difference, claimed by many to be the greatest advantage, is that the maintenance mechanics report to the production supervisor, who does the scheduling. There is no problem of conflicting priorities with other groups.

The disadvantages are also similar to those of area maintenance, but may also include the following:

1. Production supervisors are not qualified to direct a maintenance job.
2. Production supervisors cannot give technical assistance to a mechanic.
3. Production supervisors may neglect maintenance in order to meet schedules.
4. The maintenance responsibility of the plant is divided.
5. Maintenance costs are harder to determine and control.
6. Personnel problems are more pronounced than they are with area maintenance.

The disadvantages of a unit-maintenance group that are universally cited by maintenance men are that production supervisors are neither qualified nor primarily interested in their equipment. The equipment may be repaired improperly, may be operated with minimum maintenance, and, finally, may need major overhauling. Also, production supervisors may agree to minor equipment changes, modifications, or new work that does not follow approved procedure, and, the results of inadequate or technically improper maintenance may be costly.

Management-labor grievances are sure to arise when, without central control, the scope of the work for the same craft varies from department to department. Another personnel problem is created when minor adjustment or servicing of equipment—such as cleaning, which is normally done by operators—is gradually assigned to maintenance mechanics because they are available, their work load is low, and there is a natural desire to keep them busy. In one firm, the mechanics were performing so much of the operator's work that, after a job audit, they were reclassified as operators.

A Combination Plan

Many plants attempt to resolve the problem of balancing service and maintenance costs by combining a central-maintenance group with an area or departmental (unit) group. The variations on such combinations are unlimited. The advantages and disadvantages of the basic systems exist in proportion to their contribution to any combination maintenance system.

In planning a combination maintenance system, the basic factors and responsibilities that represent the maintenence function must be reviewed and modified as necessary. The maintenance responsibility of process lines, service mains, equipment, and buildings, for example, must be clearly assigned to either the central or the noncentral group. Thus, preventive maintenance may become the basic responsibility of an area group, whereas the installation of new processes and equipment may be done only by the central group. The success of the combination plan frequently is dependent on this allocation of basic factors.

Small plants generally use a central-maintenance plan. Large plants that manufacture one product or only closely related products (steel plants, for example) generally use a central-maintenance group of a thousand or more men. Plants with several different major products or service functions (such as research) most often use a combination maintenance plan.

The selection of the best maintenance plan requires careful analysis of the maintenance function. A study that considers only a few of the factors may result in an organization that leaves many procedures and responsibilities unresolved.

Which Maintenance Plan?[3]

In many industries, the maintenance operation has become a major factor influencing and determining costs. The ratio of maintenance mechanics to production workers is steadily increasing as additional labor-saving production equipment and controls are installed. Because the impact of automation in many industries has resulted in more maintenance workers than production workers on the payroll, the spotlight has inevitably focused on the maintenance function and its organizational structure.

No magic formula can be found for selecting the best maintenance organization plan for a plant. Before any contemplated change is made, several general factors common to all maintenance functions must be evaluated carefully. In addition, the advantages and disadvantages of the central, area, and departmental plans must be reviewed and considered.

Specially appointed management committees, corporate-staff personnel, and consulting firms have spent considerable time analyzing maintenance operations. Their recommendations have included organizational changes and installation of modern maintenance practices, such as work-order controls, equipment records, standardization, and preventive maintenance. The long-accepted production techniques—work measurement, time study, method and work simplification, and scheduling work—are also being modified and used by maintenance departments.

The results of these changes vary widely. Often, after several years of operations, management is not completely satisfied with the new system. Frequently, a second study of the maintenance function is conducted, and the only significant change recommended is in organizational structure. In one firm, for example, the maintenance department was reorganized from central maintenance to a combination of central and area maintenance, and the latest maintenance techniques and practices were installed. The initial results were good; after several years, however, management still was not completely satisfied. Another study was made, and the only significant recommendation was the reassignment of many of the area-maintenance workers to production departments.

In another case, the initial study resulted in maintenance mechanics being assigned to production departments. After a few years, the maintenance costs still were not satisfactory, and a second study completed the cycle with the recommendation that the mechanics be reassigned back to central maintenance.

Admittedly, achieving a balance between service and costs may be difficult. Service can always be improved if costs are ignored, and, conversely, lower costs can be obtained if service is minimized. The key or fulcrum is often the type of maintenance-organization plan. More consideration is therefore being given to the general factors that influence the selection of a particular plan, as well as to its advantages and disadvantages.

The significance of each of these general factors will vary from plant to plant. The important point is to recognize that each factor has an effect on the selection of the maintenance plan. A maintenance-organization change based on only one or two factors will not be satisfactory.

General Factors for Evaluation before Reorganization

A new and revised maintenance plan must be compatible with top management's organizational philosophy. Today, many companies advocate functional responsibility rather than line-staff responsibility for all phases of operation. When a change from line-staff to functional responsibility has been made, maintenance mechanics have been assigned to such functions as production, material handling, and utilities.

In planning a change in maintenance organization, existing maintenance policies and procedures such as the following must be understood:

1. structure of the maintenance organization;
2. maintenance practices—preventive maintenance, work scheduling, delivery of materials, standards;
3. provisions for approval of work—capital and maintenance;
4. accounting procedures—accumulation and distribution of maintenance charges from outside suppliers; and
5. training programs.

The organizational structure should be charted and the responsibilities of each group established. This may disclose such flaws as overlap of responsibilities, inadequate utilization of staff and clerical groups, wide variations in the number of workers assigned to each supervisor, and maintenance crews too large to be supervised effectively.

Study the Techniques

Existing maintenance techniques and the extent of their application and effectiveness should be analyzed. Sometimes, a particular technique becomes fashionable and receives considerable publicity. Too often, however, maintenance departments, under the guise of keeping up to date, adopt the new procedure in name only. Many firms, for example, claim to have preventive-maintenance plans; but, in practice, some are not effective because they are too broad, the clerical staff is inadequate, no systematic review is made of the records, or the program has been oversimplified to nothing more than a lubrication system. The present efficiency of maintenance mechanics, in particular, should be measured. Of the various methods used, work sampling is proving to be most successful. Maintenance supervisors are often surprised to learn that only 30 to 40 percent of the mechanic's time is spent working with his tools. Such a study also reveals the amount of time workers spend traveling to a job, obtaining material, getting tools, waiting for crafts, or waiting for production equipment to shut down. The results of this type of study frequently have been the basis for a change in the maintenance structure.

The origin, type, and cost of maintenance work are obvious factors for consideration. A representative period of work should be analyzed to determine the departments or areas in which work is being done, the type of work (routine, new, emergency), the crafts involved, and the labor and material costs.

Union Contracts

Union contractual agreements, formal and informal, are extremely important. In companies where union contract and practice follow strict craft lines, a high degree of coordination is necessary if maintenance is to perform with a minimum of lost time. Occasionally, this is the basic reason why an area- or departmental-maintenance group has not been established. In recent years, however, many companies have been fortunate in establishing a utility or area maintenance-mechanic classification. Use of these multicraft workers makes it possible to reduce the size of area crews and prevents time being lost because of a missing craft on a multicraft job.

Interdepartmental Communications

The maintenance-production communication system has an important effect on the way production rates maintenance service. If the system permits production to give work-order requests quickly to maintenance, production generally is satisfied. Delays in locating a mechanic or receiving service, however, may prompt a production-worker to request that maintenance workers be assigned to his area directly or indirectly. Communication is a continuing problem. Although autocalls, dispatchers, autowriters, and pocket radios, are being used to locate maintenance workers quickly, there is nothing more satisfying to a production worker with equipment failure on a key line than to convey the repair order to a maintenance worker personally.

Perhaps the most important communicator is the meeting held periodically between maintenance and production to determine which work orders receive what priority and which equipment the operations people can let the maintenance people have. Most plants have a short daily meeting, usually lasting 30 to 45 minutes, to accomplish this task. Occasionally, a plant will have a weekly operations meeting to determine these matters. An indicator of the degree of control maintenance management has over the work force is the time at which this meeting is held. If the meeting is held early in the morning to plan the day's activities, that is an indication that there is sufficient control over the work force to know which work orders will be closed out on that day. If the meeting is relegated to the late afternoon, that is an indication that maintenance does not have sufficient control to know which resources will be available for new work orders the next morning.

General Services

Some general plant services are not maintenance functions but may be the responsibility of the maintenance manager. These include

1. maintenance stores,
2. safety,
3. plant production,
4. porter service,
5. interplant mail and messenger service,
6. telephone operators,
7. utility operations,
8. waste disposal,
9. plant engineering, and
10. interplant trucking.

When maintenance systems are reorganized, these services often are assigned to other functions. Careful consideration must be given to this reassignment to avoid having a service become an appendage to another department and to insure that a service receives maximum benefit from the maintenance change. This is particularly true for maintenance-stores operation when the location or the method of charging out material may require modification. Occasionally, an area- or unit-maintenance group is established and no change is made in the store functions. Consequently, the time saved by locating workers closer to the work is lost in the travel time necessary to obtain materials.

Physical factors—plant size, number of major products manufactured, and number of employees—influence the type of maintenance organization. Generally, when there is more than one product or production center, or when there are large service centers, the problem of giving good maintenance service to everyone becomes difficult, and the problem is exacerbated by the fact that few departments appreciate the priority system.

The extent of new construction work, process, and equipment installation to be done by the maintenance group should be established. Any consideration of the use of outside contractors must be approached in terms of the possible impact on management-labor relations.

Other Considerations

Centralized or Decentralized Shops

A question that frequently arises is whether all maintenance-shop areas should be concentrated in one building or location. Some years ago, there was a trend to aggregate all maintenance shops in one area. This trend has continued, and certain advantages have resulted, including minimum required supervision, minimum disruptions because of weather, the ability to congregate tools and equipment, and the ability to place the stores area immediately adjacent to the shop.

Lately, however, certain disadvantages of the centralized shops have become apparent. These disadvantages could be grouped under the term *production-control problems,* which means that material that flows through the centralized-shops areas frequently becomes lost or strayed. It is not infrequent, for example, that an area supervisor sends a pump to the shop, inquires two days later about when he will get his pump, and finds that no one knows about it. What has happened in such situations is that no one has designed the production-control system so that all items that enter the central shops are tracked and followed along each step required for the production or reconditioning of the material.

Rotating Operations and Maintenance Personnel

In many organizations, there is a very clear-cut barrier between maintenance personnel and operations personnel. In such organizations, there are ample opportunities for problems to arise, since the operations people often consider the maintenance people second-class citizens. This can be avoided by rotating individuals periodically from maintenance to production so that they become aware of the problems in each of the areas. This prevents the build up of problems that are specific to each group. Furthermore, it enables technical people, such as engineering personnel, to be rotated between the two groups without feeling that they are permanently relegated either to production or to maintenance. This is a particular problem for engineering personnel, since engineers who are assigned to maintenance might feel that they have been removed from the mainstream of the promotion process and might therefore begin to seek employment elsewhere.

Plant Management

The trend today is to rotate plant managers frequently in order to increase their training opportunities. The effect of this on maintenance is that many plant managers are maintenance-oriented while others are not. When the plant manager is maintenance-oriented, the maintenance organization probably receives its share of the resources necessary to accomplish its work, and maintenance management has a sympathetic ear available to listen to the procedures that they would like installed to improve maintenance. When the plant manager is not maintenance-oriented, his attitude is more likely to be "don't do anything to the equipment until it stops running" or "we don't have time to perform maintenance now."

Degree of Contracting

The maintenance organization is affected by the degree to which the organization contracts maintenance activities. If maintenance activities are heavily contracted, the organization of the maintenance work force and the maintenance management are considerably different from those in a plant in which all or nearly all of the work is done by the plant's own work force. Accordingly, the degree to which the plant plans on contracting in the future must be considered when establishing the maintenance organization.

Notes

1. This section is adapted from Lawrence Mann, Jr., "Plant Maintenance Needs an Annual Report," *American Society of Mechanical Engineers Report* No. 75-PEM-4, 1975, pp. 1-4.

2. See, Paul D. Tomlinson, "The Maintenance Engineer," *Plant Engineering,* July 10, 1980, p. 74.

3. This section is adapted from "When Maintenance Plan for You," by Hermann F. Bottcher, Director, Manufacturing and Operations Services Management Consulting Div., Ebasco Services, Inc., New York, N.Y., in *Plant Engineering Library,* Maintenance, Part A, p. 5.

2 Paperwork Control

A set of forms is necessary in any maintenance-control system. Information flow should parallel the operational design of the organization. Naturally, the paperwork system should be as simple as possible while collecting all the information necessary to accomplish and monitor maintenance work. All too often, maintenance forms are adapted from other plants and therefore ask for information that is not really needed. As a rule, forms that ask for unnecessary information are not completed adequately by individuals who are aware of their irrelevancies.

The request for maintenance—the work order—is the heart of any paperwork system for maintenance control. The work order describes the project in sufficient detail so that cost estimates and material estimates can be made. Although it may be initiated by foremen, craftsmen, or safety personnel, the work order usually is initiated by someone at the supervisory level. The work order should be approved, if not initiated, by the member of operating personnel who is financially responsible for operation of and maintenance expenditures for the equipment concerned. (This is particularly true in plants operating under the cost-center concept.) The work order becomes part of the historical records used for planning, monitoring, and evaluating plant activities. Work-order files may supply information on the amount of downtime and may serve as the basis for formulating engineering standards and preventive maintenance (PM) programs. A copy of the work order goes into the equipment file so that the history of individual, identifiable items of equipment can be traced. The accounting department also receives a copy.

The Work-Order System

The format of the work order varies from company to company. Plants with a minimal maintenance-management information system (MMIS) or no system usually use a very brief work-order form, which describes the job in a narrative format and has spaces for estimated man-hours, material, and labor. Plants with an MMIS use the work order to collect the information necessary as input to that system. Consequently, a complete work-order form might include material, labor, and equipment costs; accounting-function charges; management information system input; PM program information; priority of work orders; and information necessary for equipment history. Figures 2–1 and 2–2 illustrate a complete work order form. The appendix to this chapter shows other typical work-order forms used for both minor work requests and work orders.

Material cost can be entered on the work-order form by the initiator of the work order, or it can be costed out, usually by computer, in the accounting department. If the stores-catalog number of the material is entered on the work order, the cost can be retrieved from the computer when the form is processed. The labor cost entered on the work order generally is an estimate derived from established standards or from planning. Equipment cost generally is shown in hours on the work order and may be retrieved from the computer system.

Accounting classifications are listed on the work order as charge numbers—usually nine- to twelve-digit numbers, each digit or combination of digits indicating information such as physical location of the work to be done, whether the work is capital or expense, whether the work is labor or

MAINTENANCE WORK ORDER

No. 18724

To Be Completed by Requester

Department ______ Equip. Desc. ______

Charge No. ______ Equip. No. ______

PRIORITY: EMERGENCY | PROD. RED. | SAFETY | TURNAROUND | PM | ROUTINE | Comp. Date | S.T. | O.T.

DESCRIPTION OF WORK:

Date Issued: Requested By: Dept. Approval: Record into Kardex | Yes | No

To Be Completed by Maintenance

Date Received FOREMAN ASS'D: 1. 2. 3. AREA ENGR.:

MATERIAL LOCATION: Whse. | Area | None Req'd. | Drawings/Sketches | Order Req'd. | Req. No./Date

Special Safety Precautions: | Gas Free Certificate | Acid Protection | Other

REMARKS:

INDUSTRIAL MAINTENANCE MECHANICS						OPERATORS		CONSTRUCTION	CONTRACT
QTY. DESC.	MECH. m/h est.	QTY. DESC.	E/I m/h est.	QTY. DESC.	C/C m/h est.	DESC.	EQUIP. est. hrs.	M/H est.	C.M.I.
___Eng. Mech.		___Elec.		___Carp.		500 T		Design m/h est.	Laborers m/h est.
___Mach.		___Electron.		___Cement		44 S. C.			
___Pipe Ftr.		___Instr.		___Insul.		D-7			
___Sht. Mtl.		___Refrig.		___Painter		A. W.			
___Welder						J. D. 300		Total est. hrs.	Equip. est hrs.
						Hydra Boom			
						Forklift H.D.			
						Operator			

Date Scheduled: Date Re-Scheduled:

Released Safe By: Date: Accepted By: Date:

REMARKS:

1

Distribution: Pink Copy–PLANNING, White Copies–FOREMEN, Yellow Copy–REQUESTER

Figure 2–1. Front of Work Order

IMM	Qty.	S.T.	O.T.	Total
Eng. Mech.				
Mach.				
Pipe Ftr.				
Sht. Mtl.				
Welder				
Elec.				
Electron.				
Instr.				
Refrig.				
Carp.				
Cement				
Insul.				
Painter				

Equipment	S.T.	O.T.	Total
500 T			
44 S.C.			
D-7			
A. W.			
J. D. 300			
Hydra-Boom			
Fork Lift H.D.			
Operator			

C.M.I.	S.T.	O.T.	Total

C.M.I. Equip.	S.T.	O.T.	Total
Total			

QTY.	MATERIAL DESCRIPTION	Warehouse Code

Work Performed:

IMM:

Date:

Equip. No. ______________ Foreman ______________

Figure 2–2. Back of Work Order

material, and various other data for appropriate apportioning of the charges. When the work order includes both estimated cost (material, labor, or equipment) and actual cost and provides for a variance between these two figures, then it can be used by maintenance management to assess the effectiveness of the maintenance program.

The work order includes some priority, either alphanumeric or according to date wanted, so that a job sequence can be scheduled.

Preventive-maintenance information includes a definitive recording of when, where, and what was done in specific geographical locations or to identifiable items of equipment. Repeated cycles of a certain type of job enable maintenance management to establish appropriate inspection intervals and to preschedule work that must be done periodically. It is a responsibility of the production department to provide timely and complete information about the anticipated necessity for maintenance so that it may be properly scheduled. When production cannot or does not foresee maintenance needs, every job becomes an emergency.

As shown in figures 2–1 and 2–2, the work order may be a rather comprehensive document. Many plants have determined that the cost of completing such a document does not justify its use in minor jobs. Accordingly, these organizations have resorted to a minor work order or maintenance request, such as that shown in figure 2-3. The plant formulates decision rules, which might include the maximum cost or maximum man-hours for jobs that can be done on a maintenance request and might specify that the work order be used for larger jobs. A typical decision rule would be that the maintenance request be used for all maintenance jobs requiring less than four man-hours.

It would be unrealistic to assume that all work done by the maintenance forces eventually will appear in either a work order or a maintenance request. There is always going to be a certain amount of maintenance activity that does not appear on one of these two documents. Most organizations employ the open-work-order concept to accumulate all these miscellaneous man-hours and costs. This procedure operates very satisfactorily if it is not abused. Normally, plants should become concerned when more than 7 or 8 percent of the man-hours are relegated to the open work order. What is actually happening, then, is that information is being thrown away because records of maintenance that is accomplished are being lost in the open-work-order number.

To combat this problem, it is wise to maintain a maintenance log, similar to an operating log, on each production unit. The maintenance log records all activities of a particular cost center and accumulates the equipment history, which would be lost if a record of minor activities were not

MINOR WORK ORDER 540331 DATE JOB NO.

CRAFT √
BMKR.
BURN.
WELD.
CARP.
PNTR.
ELEC.
GEN. L.
MACH.
MAS.
CNCR.
INSU.
M. & I.
RIG.
PFR
TIN.
MO. EQ.
TOTAL

PRIORITY ☐ EMERGENCY ☐ ONE ☐ TWO
DATE NEEDED
REASON FOR EMERGENCY
PLANT OR UNIT
WORK DESCRIPTION
ORIGINATOR
AREA NO.
TEL. NO.
COMPLETED BY
DATE
№ 67

Figure 2-3. Typical Work-Order Form for Minor Jobs

kept. The maintenance log also provides continuity from shift to shift or from day to day regarding the work that currently is being done on a particular item of equipment. Maintenance logs also record maintenance work that is done on all shifts, so that the PM crew can know what has taken place in their absence.

There does not seem to be unanimity about the appropriate number or distribution of work orders. Certainly, the information should reach all functions of the plant that have a use for it. A suggested distribution is that the original copy go to maintenance planning. One copy must go to accounting, and that copy could also serve the MMIS. One copy must go to the craft or area-maintenance supervisor involved; in plants with multicraft maintenance mechanics, the area supervisor would receive this copy. Another copy would go to the operating unit involved. Note that no copy is sent to the stores department. It is assumed that the maintenance-planning copy would initiate determination of whether or not the material is on hand.

The distribution described here is a minimum one; certainly many plans distribute more than an original and three copies. Also, with judicious planning, it is possible, for one copy to serve more than one purpose; for instance, a copy should go into the equipment-history file when identifiable items of equipment are involved. Frequently, one of the copies is sent to that function after its primary purpose has been fulfilled.

Figure 2–4 shows a maintenance procedure in which four copies of the work order are used:

1. The request originates by telephone, memorandum, or a personal visit to the operating unit.
2. A work-order clerk prepares the work orders and, if more than one craft is involved, the suborders. Copy 1 is retained by the work-order clerk in an open-work-order file. The clerk enters standardized times on the work order.
3. The craft foreman places copies 2 and 3 in his backlog file, which he uses to make job assignments. The worker performs the necessary work and notes the time on copy 2 of the work order. He also prepares an indirect-labor timecard, which goes to the accounting department after it has been approved. The work-order clerk is responsible for posting times on the reverse side of the first copy. Upon completion of the work, copy 3 goes to the originator as a completion notice. It is then forwarded to plant maintenance for PM review. Copy 2 goes to the industrial-engineering department for incorporation in the standards program and preparation of an earned-hour control report to evaluate department or craft performance.

The items in the following list appear on many work-order forms; each of them should be considered when designing a system or a work-order format:

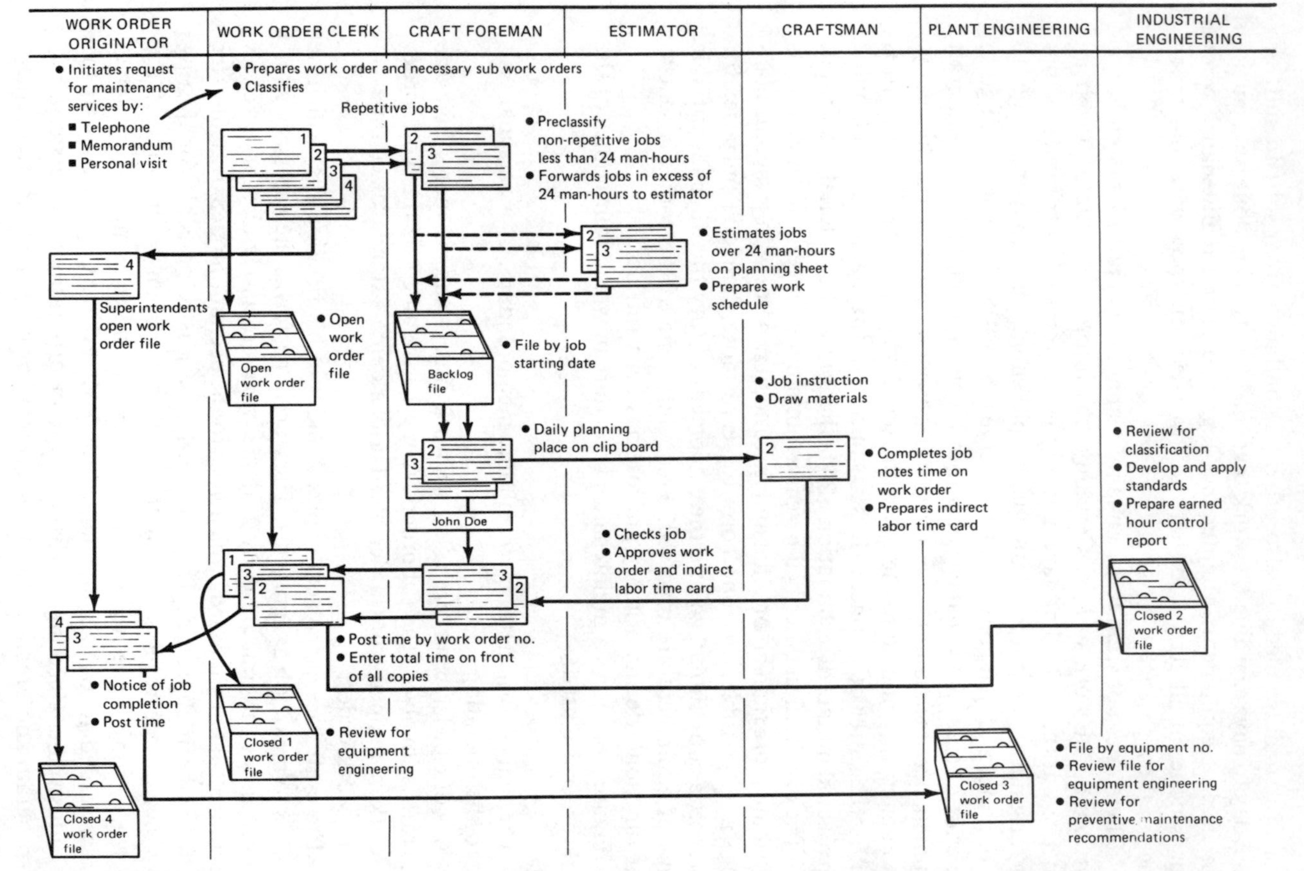

Source: Reproduced by permission from American Management Association Management, "Modern Maintenance Management," Report No. 77 (1963), pp. 56, 57.

Figure 2-4. Maintenance Work Order Procedure

1. date requested
2. date needed
3. accounting code for funds allocation
4. equipment number or plant location
5. priority
6. work-order number
7. originator
8. approved by
9. crafts required
10. equipment required
11. material required
12. estimated hours
13. actual hours
14. job description
15. maintenance foreman or supervisor

Priority

A priority sequence is necessary in any maintenance system. The following is a suggested system; details of its operation would vary from plant to plant.

The originator of the work order assigns a date needed to each work request, and a coordinator or scheduler assigns a priority. Priorities and any imperative starting or completion dates determine when work will be done. In some instances, priorities are determined by the importance of a job. In other instances, priorities show only when the job can be done. Possible priority definitions are as follows:

Number 1 priority designates emergency or safety work. Emergency work is defined as work necessary to stop a serious loss or to prevent a pollution violation. This priority covers any work that must start within twenty-four hours after the need for it becomes apparent. Such a job may be scheduled or unscheduled. This priority carries automatic approval of overtime, and work will go on until it is completed (on at least a two-shift basis).

Number 2 priority designates any work, except shutdown work, that must start on a specific future date. This priority is assigned only to jobs that must be started at a particular time to prevent product or quality loss.

Number 3 priority designates any work, other than emergencies, that is required to prevent or stop a loss. This priority can also cover additions and alterations necessary to improve the quality or quantity of the product. Jobs with this priority are started within seven days of the time the work order is written unless the priority is renegotiated with the originator of the work order.

Number 4 priority designates all work not covered by any other priority. Jobs in this classification will be started within thirty days of the time the work order is written, unless the originator of the work order is notified to the contrary.

Number 5 priority designates work that can be postponed until a substantial portion of the producing unit is shut down and production is being lost because of the shutdown. It covers work that can be done only when all or a large part of a unit is shut down.

Although the foregoing priority system lists priorities from 1 to 5, it is suggested that priority levels be given names rather than numbers or letters. Priority systems that are denoted by names are more likely to be used in an unbiased manner.

Furthermore, any priority system needs to be reinforced constantly by meetings, similar to safety meetings, with those who initiate work orders. At these meetings, the definitions of the various levels of priority are confirmed and discussed. Thus, if one of the priority levels is "emergency," the initiator of a work order knows, by assigning that priority to the work order, that he is authorizing overtime and that he is authorizing that the particular job be started on the same shift where the fault was discovered.

The Coordinator

A maintenance clerk or coordinator is suggested as part of an MMIS. The coordinator's primary tasks would be

1. to receive, check, and classify work orders according to priority and to note man-hour, equipment, and materials requirements;
2. to send closed-out work orders to accounting;
3. to file duplicate copies of work orders in the equipment file; and
4. to analyze data to pinpoint needs for methods study, contracting work, or preventive maintenance.

Time-Reporting Procedure

Accurate accounting of time is essential for cost control and accountability. The time-reporting system should be compatible with both the accounting and the maintenance-management systems. Most plants have a time-report form as well as a work order. The time-report form lists maintenance personnel by name and number, identifies the work orders on which each employee has worked, and records the number of hours each employee has spent on each work order.

The individual maintenance worker's name does not appear on most work orders. Nevertheless, during any time period, the total man-hours attributed to the completed work orders should equal the total man-hours on the time report. This is a good cross-checking device.

It is possible to eliminate the time report if the worker's name appears on the work order and if the foremen conscientiously record all work on a work order. The open work order might be useful here for charging miscellaneous jobs of short duration. Care should be exercised to prevent too much work being charged to the open work order, lest accountability be lost. Most systems limit such work to 5 percent of the total man-hours.

If the design of the system is such that operations perform certain routine and simple maintenance jobs, it is usually difficult to obtain a record of all maintenance work that is done. Indeed, it is not desirable to record every small maintenance effort, whether performed by operators or by maintenance workers.

Appendix 2A: Typical Work-Order Forms

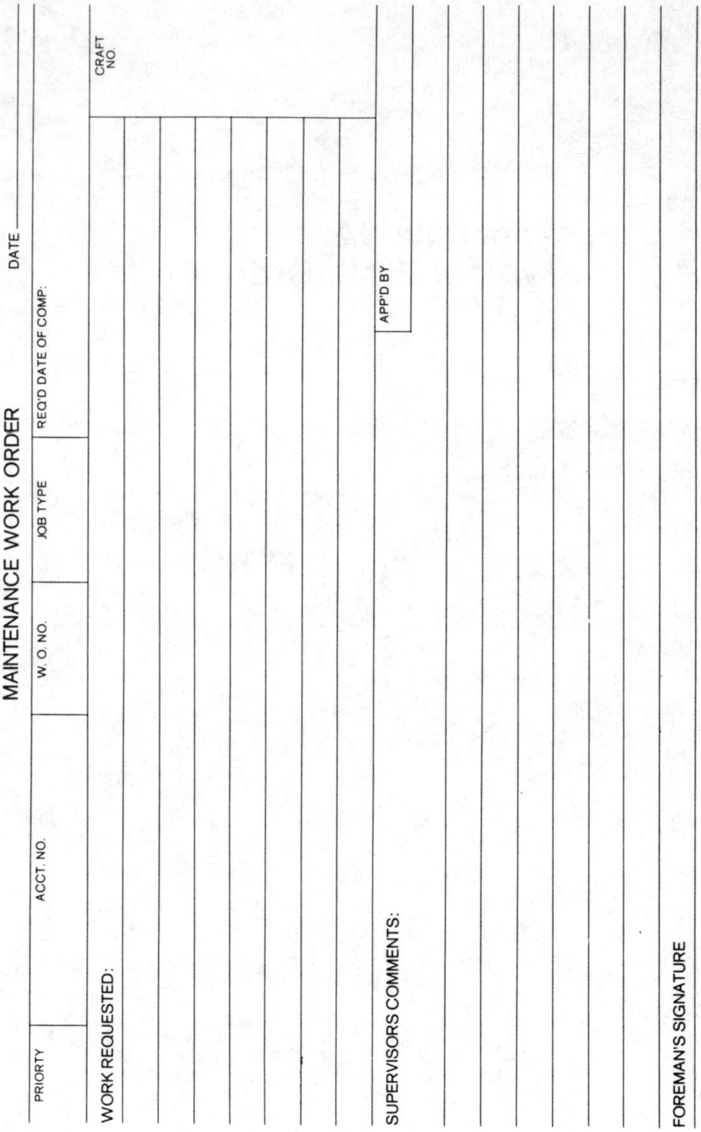

MAINTENANCE WORK ORDER

DATE ______

PRIORTY	ACCT. NO.	W. O. NO.	JOB TYPE	REQ'D DATE OF COMP:

WORK REQUESTED: CRAFT NO.

SUPERVISORS COMMENTS: APP'D BY

FOREMAN'S SIGNATURE

Figure 2A–1.

MAINT. USE ONLY

MAINTENANCE WORK REQUEST		Originator's Work Request Number:	☐ SAFETY ☐ OSHA ☐ ENERGY CONSERVATION ☐	Equip. No.
Date Wanted	Date and Time Equip. Available			Acct. No.
☐ True Maintenance—Keep Facilities In Condition to Function.		☐ Project Work		Work Site
Guestimate of Cost of Project Work: $		Cost Must Not Exceed: $		
Requested By and Date:	Dept. Manager Approval for Project Work and Date:			Job Completed By
Statement of Urgency:				Date Completed
What is Wanted:				

CR 493 (R 1/77)

Figure 2A-2.

WORK ORDER REQUEST

Requested By	Department	Date Issued

APPROVALS	ESTIMATE		ACCOUNTING		
Scheduler	Estimated By	Date	Special Project Field No.		
Maintenance General Foreman	Material -------- $		Work Order No.		
Engr. Supt./Plant Engr.	Labor --------- $		Dept.	Class	Zone
Responsible Dept. Supt.	Burden --------- $		Account No.		
Factory Manager	TOTAL $		A. R. No.		
Factory Accounting Mgr.	Date Approved		Unit No.		

PROJECT DESCRIPTION

RECORD OF MATERIAL USED

Purchase Order Number	Date	Material or Equipment	Cost
		TOTAL	

LABOR RECORD

Date	Hrs. Used	Balance	Date	Hrs. Used	Balance
	TOTAL EST'D.				

WORK ORDER
** COMPLETED **

Date

Supervisor

Figure 2A–3.

WORK ORDER

No. 59743

DATE	ITEM NO.	EQUIPMENT NAME	PRIORITY	ACCOUNT NO.	AREA

WORK TO BE PERFORMED ☐ MAINT. ☐ CHANGE & ADDITION DUE COMPL. DATE

ORIGINATOR ______ APPROVED ______

SAFETY REQUIREMENTS

☐ HOT WORK PERMIT ☐ VESSEL ENTRY PERMIT ☐ FRESH AIR EQUIP. ☐ PROTECTIVE CLOTHING
☐ LOCK OUT ☐ MASTER TAG ☐ FIREWATCH ☐ OTHER

PLANNING

SEQ.	JOB OPERATIONS	SKILL	NO. MEN	HRS.	TOTAL MAN HRS.	ACTUAL MAN HRS.

SUPERVISORS COMPLETION COMMENTS

MAINT. SUPER. ______ OPER. SUPER. ______ DATE COMPLETED ______

EQUIPMENT	SPECIAL TOOLS	MATERIAL = ☐ STORES ☐ SPECIAL PURCH.	
☐ SEE ATTACHED LIST ______		QUANT.	DESCRIPTION
☐ CRANE ______			
☐ CHERRY PICKER ______			
☐ WELD. MACH; NO. ______			
☐ ACET. WELD OR CUT ______			
☐ AIR COMP. ______			
☐ OTHER ______			
______			

MAINTENANCE FOREMAN

Figure 2A-4.

WORK REQUEST

ROUTING: (USE "X")

	1	2	3
PLANNING OFFICE			
ENGINEERING			
CONTRACTOR SUPERVISOR			
ORIGINATOR			
OS. SUPERVISOR—22-23-24-25 26-27-28-29			

NO. 59416

PRIORITY

EQUIPMENT INVOLVED ☐ OVERTIME APPROVED SECT. SUPR.

REQUESTED COMPLETION DATE ☐ BEGIN ASAP ☐ ROUTINE MAINTENANCE ☐ SHUTDOWN REQUIRED

JOB DESCRIPTION

CAUSE OF EQUIPMENT FAILURE

REQUESTED BY DATE

APPROVED BY

SUPERVISOR DATE

ENGINEERING DATE

MAIN	SUB	DETAIL	ITEM	JOB	MISC.

22 23 24 25 26 27

☐ ACTIVE 28 29 ☐

☐ BACKLOG

☐ SHUT DOWN OF EQUIPMENT:

MANPOWER ESTIMATE/ACTUAL

	PF	ELECT	MW	CARPT	PAINT	LABOR	OPER.	TEAM	INSL	OSC	OSE
EST.MANHOURS											
TOTAL MANHOURS EST.											
ACT.MANHOURS											

DESCRIPTION OF WORK PERFORMED /REASON FOR INCOMPLETE | DATE COMPLETED | MATERIAL PURCHASE ORDER NUMBERS

TECHNICIAN | SUPERVISOR

PLANNING OFFICE

Figure 2A–5.

WORK ORDER — W.O. No.

TO BE COMPLETED BY W.O. ORIGINATOR

W.O. ORIGINATOR (FULL SIGNATURE)	W.O. ISSUE DATE			COMPLETION REQD. BY:			WORK ORDER PRIORITY (x)			
	YR.	MO.	DAY	YR.	MO.	DAY	Emergency E	Urgent A	Future B	Normal C

TYPE OF WORK (X) AS APPROPRIATE				P.C.N. or P.A.R. No.	EQUIPMENT NUMBER	ACCOUNT NUMBER
Safety 1	Shut-down 2	Regular Maintenance 3	P.C.N. or P.A.R. 4			

PROBLEM

WORK REQUESTED

TO BE COMPLETED BY MAINTENANCE

AREA SUPVSR.	REVIEW DATE		EXPECTED WORK BY:		EXPECTED PLANNING EFFORT:		
	MO.	DAY	P	Contract C	Planned P	Estimated E	Unplanned U

JOB PLAN SUMMARY

PLANNING SUMMARY			
Planner	Plan Date MO. DAY		
	CODE	Job Plan Man-hours	Actual Man-hours Summary
Instr.	1		
Elect.	2		
Pipftr.	3		
Blrmkr.	4		
Mlwrt.	5		
Welder	6		
Gen. Labor	7		
JOB TOTALS			

SAFETY PERMIT No.	TYPE OF SAFETY PERMIT REQUIRED (X)				
	Cold Work	Hot Work	Excavation	Confined Space / Entry	Type Entry (x) A B C D

SPCL. TOOLS AND EQPT.

TO BE COMPLETED AFTER JOB IS DONE

PLAN CHANGES MADE ON JOB (Use reverse side if needed)

DATE JOB COMPLETED			JOB CHECKED BY: (Foreman)	FOLLOW-UP ACTION NEEDED?	OPERATIONS APPROVAL (Job complete, site clear)	ENGINEERING APPROVAL (For P.C.N., P.A.R. or special)
YR.	MO.	DAY		NO / YES (See reverse)		

W.O. No. — JOB TITLE (Standard Short Form)

Source: D.L. Lawson, "Maintenance Information Reports," *Plant Engineering*, June 22, 1978, p. 122.

Figure 2A-6.

WORK ORDER

1 ☐ PWO	2 ☐ CWO	3 ☐ CMWO	4 ☐ PMWO	5 ☐ SCHED REPAIR	6 ☐ UNSCHED REPAIR	7 ☐ SERV	8 ☐ OTHER	REQUIRED FOR ALL W.O.'s	CONT 1 (1)	STANDARD NO. (3–7)	WORK ORDER NO. (8–13)

PRINT (14) | QUANTITY (15–17) | WORK ORDER DESCRIPTION (18–42) | ☐ PET NOT REQ'D ☐ PET REQUIRED

CLASS (43) | COST CODE (44–47) | EQUIP. TYPE & ITEM (48–52, 53–56) | MAT'L $ (BELOW) (57–62) | APPR (63) | REOP (64) | DATE ISSUED | P. O. NO. | PET NO.

REQUIRED FOR: PWO'S, CWO'S & CMWO'S

CONT 2 (1) | W. O. (ABOVE) (8–13) | PROJECT NO. (14–18) | PROJECT AREA (19–24) | PROJECT ACCT. (25–29) | LABOR $ (BELOW) (30–35)

I. C. (36) | PRF (37) | ACCT | CODE | SUFFIX (48) | DESCRIPTION OF WORK DESIRED (BE AS SPECIFIC AS POSSIBLE)

LABOR EST. | CONT 3 (1) | W. O. (ABOVE) (8–13)

CRAFT | MAN-HOURS

14–20, 21–27, 28–34, 35–41, 42–48, 49–55, 56–62, 63–69, 70–72–75

CLOSE-OUT DATE | BLDG NO. | LIFE

WORK APPROVAL SIGN.	LEVEL $	WORK APPROVAL SIGN.	LEVEL $	WORK APPROVAL SIGN.	LEVEL $		
FOREMAN	1 / 500	SUPT. OR ASST. SUPT.	3 / 2000	RES. MGR. OR ASST.	5 / OVER 5000	LABOR $	
						MAT'L $	
CHIEF SUPVR. OR SUPVR.	2 / 1000	MGR. OR ASST. MGR.	4 / 5000	PLANNER		TOTAL $	TOTAL MAN-HOURS

CONT: 21, 22, 23 = OPEN; 31, 32, 33 = REVISE ALL FIELDS; 41, 42 = REVISE SELECTED FIELDS.

Figure 2A–7.

WORK ORDER

EQUIPMENT AVAILABILITY | DATE REQUIRED / / | WORK ORDER NO. 50750

DATE & TIME OR HOURS NOTICE

COST CENTER | AFE NO. | LT MP PRIORITY | REG. | TAG NUMBER

PERFORM THE FOLLOWING ON | EQUIPMENT COMMON NUMBER | EQUIPMENT COMMON NAME

ORIGINATOR | $ EST. REQD. ☐ APPROVED BY | DATE / /

SAFETY PRECAUTIONS OR REMARKS | SAFE WORK PERMIT REQ'D. Yes No | REASON | Safety Dept. Service Req'd. Yes No

Engineering Required Yes No

DRAWINGS REQUIRED

JOB ID | Type | Cause | Zone | Unit | Est. Manhours

JOB SEQ.	SKILL	DESCRIPTION	MANHOURS ESTIMATE	SCHEDULE DAY	SCHEDULE START	COMPLETED BY
			X			
			X			
			X			
			X			
			X			
			X			
			X			
			X			
			X			
			X			

MATERIALS

LABOR SUMMARY

Deliver Materials To ______

EXPEDITING INSTRUCTIONS
☐ Order By Phone
☐ Authorize Shop Overtime
☐ Air Freight
☐ Call Del. Estimate To Sched.

______ X ______ = $ ______
TOTAL M.H. RATE

Material Source	STOCK CODE	Quantity	Unit	DESCRIPTION	Date Available	Unit Cost	TOTAL

SPECIAL EQUIPMENT REQUIRED	ESTIMATED HOURS

MATERIALS ESTIMATED BY: | MATERIAL COST $

JOB ESTIMATED BY: | EST. JOB COST $

DATE / / | COMPLETED BY | ACCEPTED BY

Figure 2A–8.

3 Maintenance-Job Planning and Scheduling

Although adequate planning and scheduling are essential to a maintenance-management information system (MMIS), it is important to guard against overplanning and overscheduling. The maintenance planning and scheduling effort is directed toward minimizing the idle time of maintenance forces; maximizing the efficient use of work time, materials, and equipment; and maintaining the operating equipment at a level that is responsive to the needs of production. Although planning and scheduling are different functions, it is difficult to separate them, and the same individual is usually responsible for both.

As noted in the preceding chapter, maintenance jobs are introduced into the system in a number of ways. Regardless of how the work order gets into the system, however, its implementation must be planned to some extent. Naturally, in emergency work, it is not always possible to plan adequately, and inefficiencies inevitably occur. Some plants maintain an emergency manual that includes preplanned work orders for some of the emergencies that might occur in the plant.

Once initiated, work orders must be screened from several points of view, the first of which is priority. Priorities are best assigned by operating people and then coded for computer input. Work orders must also be screened with regard to material, equipment, and manpower needs. If material is not available, it must be ordered; if equipment is not available, it must be assigned. Work orders are also screened to determine whether the work can or should be done in the shop rather than in the field. Emergency work usually performed is on the spot, and the work order, if any, is written after the fact.

Engineering the Work Order

Maintenance-job planning includes the inseparable functions of job estimating and job engineering. Estimating a project is, in effect, engineering that projects; if the project supervisor does not follow the plan specified by the estimate, not only will the estimate have no relationship to what is actually done, but there is little likelihood of remaining within the financial budget for that job. Accordingly, the term *estimating,* as used here, will include engineering of the work order.

The degree to which the work order is engineered is determined by the complexity of the work-order form, which must supply sufficient information for the planner and scheduler to determine the resources required for the project. Although projects such as "replace the relief valve," "tighten the seals on the pump," or "replace the tube bundle in a heat exchanger" could all be planned and scheduled without additional information, more complicated jobs must be estimated to convey to the planner and scheduler the magnitude of the effort required for each project.

Who Plans and Schedules the Work Orders?

The training and attributes of the maintenance planner and scheduler are worthy of consideration here. This position requires an individual who is fully familiar with the production methods used throughout the plant or in the area or department (unit) to which he is assigned. He must have had sufficient experience to enable him to estimate labor, material, and equipment for the work orders he will process. Although it is preferable to have individuals in these positions who have had technical education, many plants promote area or craft supervisors to maintenance planning and scheduling positions. Care must be taken to assure that the planner continues to be familiar with field operations. Too frequently, the planner is found to be estimating work orders on the basis of the way things were done when he was in the field, whereas new procedures, methods, and equipment are always being introduced in the plant. The planner must be made familiar with new operations if his estimates are to continue to be realistic. It is absolutely essential for the planner to visit the field periodically and to discuss work orders with the field forces. This might be accomplished by requiring that the planner review all work orders that exceed some tolerance in both man-hours and dollars. A continuing surveillance program will help assure that the planner remains current with the methods as they are being performed in the field.

Individuals who work with maintenance planning and scheduling must be able to live with change. People who delight in creating a plan or a schedule but become upset when the program does not follow that plan or schedule are ill qualified for positions in maintenance planning and scheduling. Seldom does a planner come to work in the morning and find that he is able to follow completely the schedule he created the previous afternoon. A realization that this is part of the job is an essential attribute for the planner and scheduler. Furthermore, he must be able to converse easily with the field forces about the current status of jobs and methods of performing the work order.

It appears to be wise to locate the planner and scheduler in an area that is frequently visited by the area and craft foremen so that they can maintain an informal, continuing discussion. A good plan is to locate the planner and scheduler's desk in the same vicinity as the desk of the craft or other maintenance foreman. If the system requires that the planner and scheduler leave his work place to estimate every job, it often results in his working inefficiently. His physical remoteness from ongoing maintenance work also leads to difficulties when the scope of a work order becomes greatly enlarged after work has started, creating a need for additional material, equipment, or personnel.

Standards

Preplanned work orders are based on a set of maintenance standards. A continuing topic of discussion for those who are interested in maintenance management is what percentage of maintenance man-hours or work orders should be devoted to standards.

A maintenance standard has four parts: (1) the sequence of events required to perform the job; (2) a listing of the crafts and the man-hours required for each of those crafts; (3) a list of the material and equipment required; (4) a list of any special considerations that must be given to the job. Special considerations may include needs for special lock-out devices, a gas-free atmosphere, protective clothing, or having an individual stand outside an enclosed vessel when maintenance employees are working inside the vessel.

In many cases, the store's stock number is included in a standard to ensure withdrawal of the correct material. Standards facilitate estimation, assure that the job will be done in the most efficient manner, and contribute to training by indicating to the craftsmen the way the project should be performed.

There are different views about how to maintain a standards system. Some plants assign several industrial engineers to work continually with the standards system. Under this mode, the group examines each work order and writes a standard for that work order, in cooperation with the craft or area supervisor. The standard is reviewed after the work is completed as a check on its practicality. It is then filed until the next time that project is required and then can simply be drawn from the file, eliminating the necessity for further maintenance planning. The standards also provide the data necessary for scheduling craftsmen and equipment on the project. Other plants take a less complete approach—analyzing the work being done and choosing to prepare standards only for projects that are more or less repetitive. Plants that lack a full complement of trained craftsmen use standards primarily as procedural guides to assure that work gets done in a desired manner. In such cases, the standards may or may not include a specified job time or equipment and material lists.

Each plant must decide the extent to which it will design and use standards. It is unrealistic for a plant to claim that it does not use standards; even if they are not written down, they are used intuitively by drawing upon the knowledge that the field supervisors have built up over a period of time. This will be covered in more detail in chapter 4.

Materials, Equipment, and Locale

No work order can be executed without the necessary material and spare parts. The planner must be able to list all the material and equipment necessary to the work order, and the system must provide some means of locating and reserving material and equipment for the time at which the work will be done. Various procedures, such as red-tagging, use of holding pens, and delivery of material to the site, are designed to ensure that material is available when it is needed. The planner must have a direct avenue to the purchasing department so that material that is not in the plant can be obtained quickly.

Normally, maintenance equipment includes such items as welding machines; air compressors; scaffolding; hoists, cranes, and various other materials-handling equipment; and portable power tools. In most plants, the central garage comes within the scope of the maintenance management, and control over this equipment is absolute. The planner looks upon equipment problems in the same manner as he looks upon problems of personnel handling. Just as there is the possibility that a scheduled maintenance worker will not be present at some planned time in the future because of illness, absenteeism, and the like, equipment might be inoperative because of maintenance problems of malfunctions or may still be in use on some previously scheduled order that has gone over schedule.

The maintenance shops—whether central, area, or departmental—come within the scope of responsibilities of the maintenance department, as do the garages. The maintenance planner is responsible for the decision of whether a maintenance task is best done in the field or in the shop. Of course, there are some activities that can be done only in the shop and other activities that can be done only in the field. Between these extremes are a large number of work orders that can be worked to some degree in either of these locations. The maintenance planner must have sufficient experience to evaluate all of the variables in each situation in deciding where the work will be done. In general, the maintenance planner attempts to maximize the amount of work done in the shop, because that is usually the minimum-cost alternative.

Keeping in Touch

Indispensable to any maintenance-planning system is the feedback mechanism whereby the planner is constantly being made aware of discrepancies between planning and what actually takes place in the field. The most common method entails feedback of the total cost of a work order compared to the estimated cost and the total man-hours spent compared to the estimated man-hours. This feedback alerts the planner to changes in operational methods and procedures in the field, permits analysis of why discrepancies have occurred, and keeps maintenance estimates realistic.

The functions of a planner include periodical meetings with the maintenance supervisor and operating personnel to determine what work will be done and when it will be done. Most plants hold a short daily planning meeting. If each day's meeting is held early in the morning, the system can be responsive to the needs of management, in that maintenance can predict what will take place during the rest of that day. In addition, the scheduler can ask which work orders will be closed out that day and can reassign the resources then being used. If the planning meeting is held late in the afternoon, the system does not reflect adequately controlled maintenance management.

This meeting also provides a review of intermediate-range planning—that is, planning for the next two weeks. Operating personnel can either reaffirm their willingness to release equipment for maintenance or can change their minds. Maintenance management can keep operations aware of the effect of schedule changes on the efficiency of the maintenance system, and operations, in turn, can make maintenance aware of the factors that necessitate a delay in releasing equipment and operating units for maintenance. These maintenance-planning meetings are usually short, taking between thirty and forty-five minutes. In the MMIS, it has been assumed that one planner and scheduler can determine the needs for the complete maintenance system. There are many instances, however, when the maintenance work force is large enough to require more than one planner and scheduler. In such cases, the function usually is segregated by geographical areas of the plant, and it is necessary to have a coordinating supervisor over the planners and schedulers. All the advantages and disadvantages of centralization and decentralization discussed in chapter 1 apply to the situation in which there must be more than one planner and scheduler in the maintenance system.

One method used to determine whether a planner is aware of how work is done in the field entails statistical sampling of certain work orders—comparing estimated versus actual man-hours and dollars and requiring the planner to explain any discrepancies. This procedure is not intended to criticize the planner but rather to build into his routine a method by which he is continually aware of the way operations are conducted in the field.

Maintenance Scheduling

Maintenance scheduling is primarily concerned with arranging the sequence in which the work orders that have already been written will be worked. This scheduling must consider the job priority, the availability of material and equipment, and, of course, the availability of craftsmen or maintenance mechanics. Superimposed on these constraints are the desires of the operating personnel, which might change from day to day.

A number of considerations in maintenance scheduling warrant discussion here. The first is that schedules should be based realistically on what is likely to happen rather than on what the planner and scheduler would like to happen.

As already mentioned, the scheduler should not have a disposition such that he would become upset if the schedule were changed. Changes result from many circumstances—malfunctioning equipment, withdrawal of the wrong material from stores, worker absenteeism, and operations personnel changing their minds about when they can release a unit. A competent scheduler will arrange the job requests so that productive work can be assigned immediately to crews that are unexpectedly made idle by a last-minute schedule change.

A primary rule of maintenance scheduling dictates that a work order not be scheduled until all material needed for that work order is in the plant. Many a maintenance project has had to be canceled because it was scheduled for the expected arrival of material that was still in transit.

Much has been said here about cooperation between maintenance management and operations. No system can be effective without understanding by both parties about the results of delayed or postponed maintenance. Often this communication is accomplished by rotating personnel from operations into the maintenance-management system and from maintenance management into operations. In this way, each becomes aware of the other's point of view. Maintenance should inform operations of major expenditures, such as renting a large crane, that will be wasted if the work is not accomplished on schedule. Operations also should be kept fully informed about conditions existing outside the control of maintenance management that cause schedules not to be met.

It should be emphasized that poor scheduling and missed completion dates destroy the confidence of the operating personnel in the maintenance-management system.

The final rule of good scheduling is always to be realistic about the resources available and the demand for maintenance. It is certainly unrealistic to schedule 100 percent of the maintenance work force when there is a normal absenteeism rate of 5 to 7 percent or when experience has shown that the standards are so tight that schedules are seldom met. Some plants schedule 100 percent of the work force knowing that some percentage of the schedule will not be worked. Other plants schedule 75 percent of the work force confident in the knowledge that the balance of the force will be assigned to emergency work orders that occur during the off shifts. This figure may vary according to how predictable maintenance needs are in various situations.

Priority

One of the most important aspects of the scheduling function is the priority system. The priority system is established primarily to ensure that the most-needed work orders are scheduled first. Constant vigilance must be exercised so that the priority system is not abused. Experienced maintenance managers have some idea of the normal maintenance requirements for each cost center within the plant, and, by a continuing comparison of these norms with the type and amount of requested maintenance, they can assure an equitable flow of maintenance services.

Priority systems usually include from four to ten levels of priority; most plants find that four levels are adequate. In some plants, priority is assigned by the individual who initiates the work order; in other plants, a coordinator, such as the centralized planner and scheduler, does so. A four-level priority system would classify emergency work, normal maintenance, preventive maintenance, and other maintenance. A system that uses adjectives to describe priorities seems to be better than those that use numbers or letters. People are less likely to abuse the word *emergency* and more likely to overuse such designations as A or 1, particularly when the corresponding priorities are ill defined. In some plants, the initiator of the work order declares the date on which the work is wanted or the date on which the equipment must return to service. This date then determines the priority.

Occasionally a plant will claim that it does not find it necessary to have a priority system because there are enough resources to meet all the needs of maintenance. This is an indication that the plant is being overmaintained, since most maintenance managers agree that a reasonable backlog of jobs is a healthy situation.

Difficulties can arise when a plant has more than one planner and scheduler. The criteria by which these individuals classify work orders may be different, often leading to situations in which one area of the plant is in need of resources for high-priority work while other areas of the plant are using those resources for lower-priority work. Care must be taken to see that good schedulers and planners are not penalized by being deprived of resources that have been assigned to areas where the scheduling and planning have not been done well and where, therefore, emergencies frequently arise. Morale will inevitably suffer if such circumstances prevail. Training or discussion sessions concerned with priorities should be used to establish some uniformity of practice among the schedulers and planners and to improve the judgment of operating personnel who are responsible for assigning work-order priorities.

The Backlog

No maintenance-management system can operate effectively without a realistic backlog of work orders. This backlog provides the scheduler with alternative assignments when scheduled work orders cannot be worked. A backlog also informs management about future needs for maintenance. If the backlog is growing and if management has the willingness and capability, then contract maintenance might be a feasible alternative. An ever-growing backlog might be an indication that the maintenance work force is undermanned, and steps should be taken to acquire additional personnel.

The MMIS should include a frequent review of the backlog. One reason for this is that the scheduler does not always know when small low-priority work is completed at unscheduled times, usually as a fill-in after maintenance workers have finished a primary job. Continually carrying a number of small work orders on the backlog creates an erroneous impression of maintenance needs. Also, a review should include determination of whether operations still desires backlogged work orders to be performed. Some plants automatically eliminate all work orders that have not been worked after thirty days; if the job is still desired, the initator of the work order is asked to rewrite the work order.

Standards

The role of standards is as important in scheduling as it is in planning. A maintenance standard should include the approximate number of man-hours required to perform the task, categorized by crafts. The scheduler uses that estimate in scheduling standardized tasks and as a guide in scheduling similar tasks. Most large plants have standards that list the specific elements of a task in greater detail than the work order does. Data exist, for example, about how long doing a four-inch weld on a standard-weight pipe should take, according to the location of the pipe; there is a standard figure for man-hours per square foot for paving areas of processing units; and there is a standard number of man-hours required to lay drain pipe, according to the size and type of pipe and the depth.

Controlling Maintenance Backlog[1]

Maintenance-work backlog can be defined as an accumulation of unperformed tasks. These tasks are the essential maintenance and repair jobs that should be done but have not been attended to because of lack of resources—money and manpower. Maintenance jobs are considered essential when sound engineering judgment indicates that they should be done if the equipment involved is to be used effectively for its designated purposes.

Programmed work consists of jobs that have been approved for funding within existing resources. Unprogrammed work consists of jobs that have so little priority that money is not available to do them. Maintenance managers must employ some method of backlog control for programmed work to be able to compare estimated hours of workload with actual available man-hours. The purpose of such evaluation and control is to maintain minimum personnel requirements—covering crafts, area, or minor construction groups—and still ensure optimum maintenance-work quality and meet production requirements.

Figure 3–1 shows the format for a labor-control or backlog report covering all maintenance forces—shops, craft, area, and minor construction groups. This report can be used

1. to indicate the extent of current and forthcoming authorized work in relation to the ability of on-board personnel to absorb the load;
2. to indicate the need to rearrange, increase, or decrease field forces;
3. to indicate required personnel staffing for a balanced organizational pattern to achieve an effective maintenance-management system;
4. to serve as an aid in determining whether an adequate work-input-control procedure is in effect;
5. to indicate the extent to which the maintenance-control group is managing the backlog;
6. to indicate the capability and flexibility of field personnel to work on the backlog; and
7. to indicate the progress being made in raising facilities to the proper level of maintenance.

COST CENTER BACKLOG																
COST CENTER							BACKLOG OF CONTROLLED JOBS (IN SHOP DAYS)					BACKLOG OF ESSENTIAL MATERIAL FOR QUARTER ENDING:				
							In Field		Held in Maintenance Control Group							
No.	Name	NUMBER OF SUPERVISORS	TOTAL NUMBER OF PERSONNEL	OVERHEAD	NUMBER OF MEN ASSIGNED TO OTHER THAN CONTROLLED JOBS	NUMBER OF MEN AVAILABLE FOR CONTROLLED JOBS	Material Available	Material Not Available	Ready for Issue	Not Planned and Estimated	Total Backlog of Controlled Jobs	30 June	30 Sept	31 Dec	31 Mar	30 June
(1)	(2)	(3)	(4)	(5)	(6)	(7)	(8)	(9)	(10)	(11)	(12)	(13)				

Source: Bernard T. Lewis, "Controlling Maintenance Backlog," *Plant Engineering*, February 22, 1973, p. 203.

Figure 3–1. Basic Format for Cost-Center Backlog Report

The basic function of the report is as a tool to help maintenance managers manage the backlog effectively. The report format shown in figure 3-1 gives maintenance managers an idea of the maintenance work load that is tentatively scheduled for the field forces. All maintenance-department cost centers are listed, and the size of the work load or backlog is noted, expressed in field days.

Field-day figures are used in the report columns so that comparisons can be made by cost center without reference to the number of employees in the cost center; for example, fourteen field days has the same meaning whether there are fifty men or ten men assigned to the field. Field days are computed by relating the number of men available to work on the cost center's backlog to the total man-hour estimates of the jobs that must be done by that cost center.

If the backlog report is to be employed effectively as an aid to work-load management, it is necessary to understand the concepts involved in the term *managing the backlog*. The term may be defined as the art of balancing force availability with size of the backlog to obtain maximum effectiveness in real-property maintenance.

Managing the backlog begins when a maintenance deficiency is found and reported. Once the total backlog has been determined, maintenance managers must plan the time sequences and methods to be used in doing the jobs. In general, the decision to complete a job depends on the availability of funds—not necessarily when the decision is made, but at the time work is scheduled to start. Assuming that funds are available, the work can be done either by contract or by in-house personnel. If the job is to be done in-house, it will appear in the backlog report. Thus, to use the report properly, it is important to remember (1) that the backlog shown on the report is only that part of the total backlog for which funds are available and (2) that the work shown is to be done by in-house forces. These jobs must be carefully controlled through the various stages of planning, estimating, authorization, and on through completion.

A key factor in job control is the rate at which the work is introduced into the field. Effective control is assured only if the rate of work input allows the field personnel sufficient flexibility to schedule and accomplish jobs in an orderly, logical sequence.

The information given in columns 3 through 7 in the form shown in figure 3-1 is the maintenance-department staffing arranged by cost center. The desired objective is to have as many workers as possible available for controlled jobs. The data on the report are used to determine staffing relationships and to rearrange, increase, or decrease field loads in relation to the backlog.

Columns 8 and 9 in the report form show the in-field backlog. The figures in these columns are the field-day equivalents of all controlled jobs in the field. Controlled jobs with material available are in column 8, and those without available material are indicated in column 9. Data in these columns are used by maintenance managers to verify the adequacy of existing work-input-control procedures.

Backlog held in the maintenance-control group is reported in columns 10 and 11. The figures placed in these columns are the field-day equivalents of all the controlled jobs that will be done in-house and that have funds available but have not been forwarded to the field. Column 10—Ready for Issue—is for the field-day equivalents of work for which final estimates have been prepared. Column 11—Not Planned and Estimated—is the field-day equivalents of the remaining work for which funds are available and which will be done by in-house forces. When the data in columns 8, 9, 10, and 11 are balanced with available personnel (column 7), maintenance management's ability to manage the backlog can be determined.

Data contained in the column 13—Backlog of Essential Maintenance—are useful in observing trends in the dollar value of the backlog of essential maintenance. A downward trend can indicate an improvement in the level of maintenance; an upward trend can mean the opposite.

The backlog report gives the plant engineer a summary of operations from an organizational point of view—that is, in terms of cost centers and personnel as related to backlog reduction. With this report, he can make sure that only a minimum number of maintenance personnel are used for purposes other than backlog reduction. In addition, detailed plans can be developed by the maintenance-control group so that the field does not have to create work. With the backlog report, the plant engineer will be able to maintain a proper balance of jobs between the maintenance-control group and the in-field portion of the backlog. If the backlog is high, it indicates either insufficient work being sent to the field or understaffing in the field. Conversely, little or no backlog may indicate overstaffing, poor management of the backlog, or inefficient inspection practices.

Proper management of the backlog requires that the work be distributed so that jobs can be moved at a rate permitting maximum work-scheduling effectiveness. Work requests that would upset orderly work accomplishment can be identified. The report helps to avoid—or at least decrease—the number of rush, emergency jobs. It gives the maintenance engineer factual evidence, rather than opinions, to support his position.

Some jobs, though important, are not considered crucial enough to warrant immediate attention and thus are not included in the budget drawn up for maintenance activities. All such unfunded maintenance and repair work should be reported to management periodically, and efforts should be made to reduce the backlog figure gradually.

When the maintenance-management system is working properly, it gives the plant engineer the capability to maintain close control over both maintenance service and maintenance costs. Such control can be achieved by making certain that the number of men within each craft has a reasonable relation to the backlog of requested work. Too little backlog means poor scheduling and excessive craft waiting time; too much backlog means poor service. Maintenance managers generally agree that an acceptable backlog consists of enough work on hand to keep each craft busy from two to four weeks. This implies of course, a priority system that relegates certain jobs to a waiting list in order to allow some jobs to be handled at once.

An important element of control is the regular (usually weekly), systematic measurement of maintenance-work backlog. With an efficient planning and scheduling procedure, this measurement becomes easy, requiring only a small amount of adding-machine tabulation (see figure 3-2).

As big jobs are authorized, there will be fluctuations in the reported backlog. The reports will have more meaning if the work is pinned down to a definite period of time—a week, for example. The most significant factor about backlog, however, is the backlog trend. This trend should be charted for each craft and compared to a control line that designates the optimum backlog level (see figure 3-3).

When the backlog trend for a craft continues to rise, it may be possible to increase the amount of work being handled on a contracted basis, or

MAINTENANCE WORK BACKLOG REPORT

	Available to Be Scheduled (man-days)	Work Order Backlog which Could Be Scheduled in Week	
		Current Wk. (man-days)	4-Wk. Avg. (man-days)
Electricians	175	92	130
Millwrights	160	304	290
Machine Repair	160	220	250
Pipefitters	105	290	206
Laborers	85	250	240
Sheetmetal	40	85	60
Carpenters	30	109	150
Painters	10	50	40
Total Maintenance	765	1400	1356

Source: Bernard T. Lewis, "Controlling Maintenance Backlog," *Plant Engineering,* February 22, 1973, p. 203.

Figure 3-2. Maintenance-Work Backlog Report

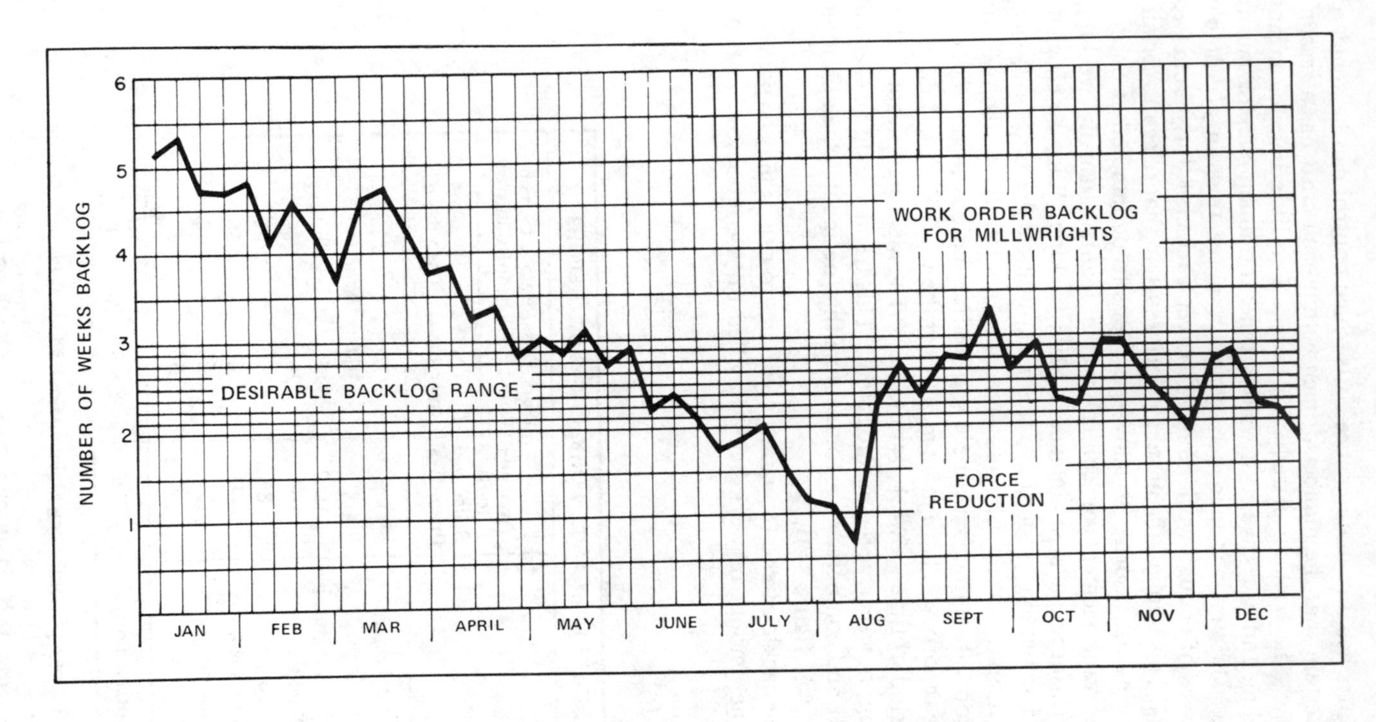

Source: Bernard T. Lewis, "Controlling Maintenance Backlog," *Plant Engineering*, February 22, 1973, p. 203.

Figure 3–3. Charting Backlog Data from Weekly Reports

transferring craftsmen from another craft or from a helper pool may be practical. Such a trend may indicate the need to increase the maintenance force, although this should be avoided unless continuation of the work load is anticipated.

If the backlog trend is downward and well below the optimum level, prompt action is necessary to avoid a reduced level of completed work. Transfers between crafts or a reduction in contracted work may be the solution. Forced reduction may be essential, despite the need to avoid frequent fluctuations in hard-to-train craft manpower.

It is a sound idea to keep the regular maintenance force small enough so that it is fully occupied in handling the minimum annual work load. Normally, outside contracting can be relied on to handle any unusual jobs or work beyond the expected work level. This practice permits the size of the maintenance force to be kept fairly constant during the year.

Several performance indicators can be useful in evaluating backlog control. These indicators are not necessarily applicable in every maintenance situation, and they can vary—even between maintenance organizations within the same company. The following questions are presented as guidelines for the maintenance manager in developing performance indexes (chapter 6 covers indexes in detail):

1. Is there weekly or monthly backlog measurement by craft?
2. Are there trend data to show any need for craft increases or cuts?
3. Are there data to predict any need for scheduled overtime by crafts and subcontracting?
4. Are there sufficient man-hour data to aid a management decision about what jobs must be delayed if new jobs are inserted?

A certain amount of backlog is necessary to plan and schedule maintenance work properly, but too great a backlog (more than two weeks' backlog, exclusive of construction) may result in emergencies, loss of production, or downtime. Too small a backlog (less than two weeks' backlog in a craft) can result in lost time, poor scheduling, and the like. Another way to view backlog is to examine equipment separately. Experience indicates that the proper range for equipment backlog is roughly one to five weeks. The most important aspect of backlog, however, is the monitoring of trends to determine corrective action.

Reports and Forms Required for Planning and Scheduling

The first report of the day used by maintenance management is the *maintenance schedule*. It is necessary that a maintenance scheduler have timely information about the resources available to him. Although he might schedule work orders the day before they are to be worked, he must know, the morning of the day of performance, whether the individuals he scheduled are in fact available. If he finds that they are not available, he must take the necessary action to assure that the most urgent jobs get manned and that manpower is shifted from jobs with lower priority.

The same is true for maintenance equipment. Although equipment might be available and operable the day before it is needed for a job, it might become inoperable during the off shifts or in the early hours of the shift for which it is scheduled. The scheduler must have an alternative plan to employ workers effectively when necessary equipment is unavailable.

The second report necessary for maintenance scheduling is the *backlog report*. The backlog report should include the backlog of maintenance man-hours, categorized according to craft (or classification, if craft classifications have not been eliminated). It usually lists work orders that are ready to be worked and that presumably have lower priority than the work orders being processed during that day. The backlog report may list the work orders sequentially by order number (as shown in table 3–1) or by priority.

Another report necessary for decision making in maintenance scheduling is the *production schedule*. This report keeps the maintenance scheduler informed about what operating equipment is in use and when that equipment can be released for maintenance. It is very important that the production schedule be reported accurately, since production from a particular processing unit might be vital one day but of minor importance some other day.

The *maintenance-planning work sheet,* shown in figure 3–4, is used in planning a work order. It may be used as an aid in writing the work order or as a document to accompany the work order. It is divided into four sections: (1) the man-hours required of each craft needed on the job are listed for each specific workday; (2) each item of required material is listed by name and storehouse stock number; (3) the necessary tools and equipment are listed and their estimated hours of use are shown (for large-scale, expensive equipment, this entry permits the scheduler to make provisions for using such equipment on other work orders while it is available); and (4) allowance is made for entry of special safety requirements and unusual working conditions, such as "acid area," "working in hot furnace," or "chlorine (or ammonia) environment."

Figures 3–5 and 3–6 show the front and back of the *maintenance and repair service ticket* used by some plants. The service ticket essentially replaces the work sheet; it may be used in place of the work order or as an aid in writing the work order.

Table 3–1
Backlog Report (Week Ending May 1, 197x)

Work Order No.	*Craft Code*	*Est. M–H*	*Priority*	*Description*	*Date Submitted*	*Date Due*	*Action Taken*
18325	30	21	2	Replace flare stack tip	4-2-7x	4-28-7x	Schedule
	23	32					
	1	128					
18326	10	15	1	Repair crane tread	4-2-7x	4-10-7x	Schedule
	21	18					
18331	30	221	1	Replace catalyst riser	4-3-7x	4-12-7x	Schedule
	23	306					
18333	28	42	3	Install control valve	4-5-7x	5-10-7x	Schedule
18334	15	56	3	Repair air compressor	4-8-7x	5-10-7x	Matl. Ordered
18336	14	10	2	Replace P-24A Foundation	4-8-7x	5-2-7x	Schedule
	25	12					
	26	9					
18338	4	158	3	Reset compressor	4-8-7x	5-1-7x	Schedule
	19	102					
	20	120					
18339	14	76	1	Clean drainage ditches	4-9-7x	4-15-7x	Matl. Ordered
18340	5	21	2	Repair control house door	4-9-7x	4-25-7x	Matl. Ordered
	7	30					
18341	20	142	1	Repack pump seal	4-11-7x	4-16-7x	Schedule
18342	31	130	1	Insulate line	4-11-7x	4-17-7x	Schedule
18343	7	72	3	Paint walkway	4-11-7x	5-20-7x	Schedule
18344	30	68	3	Replace dock loading line	4-11-7x	5-16-7x	Schedule
	33	70					

MAINTENANCE PLANNING WORK SHEET

Work Order No. ______________

Date: ______________ Planner: ______________

MANHOURS REQUIRED

Craft	Date								

MATERIAL REQUIRED

Stores Stock No.	Item

TOOLS AND EQUIPMENT REQUIRED

Item	Hours Needed

SAFETY REQUIREMENTS ______________

UNUSUAL WORKING CONDITIONS ______________

Figure 3–4. Maintenance-Planning Work Sheet

M & R SERVICE TICKET

CHARGE NUMBER	DEPT.	TICKET	U.C.	DEPT. TITLE

BUDGET REQUEST NUMBER:

DATE WRITTEN: REQU. COMP. DATE:

MACH TITLE:

MACH LOCATION:

WORK REQUESTED:

ORIGINATOR: APPROVED:

M & R FOREMAN'S INSTRUCTIONS:

JOB COST ESTIMATE: MAN HRS: MATERIAL: $

JOB COMPLETIONS COMMENTS:

ACTUAL JOB COST: MAN HRS. MATERIAL: $

TOTAL COST (TIME & MATERIALS): $

JOB COMPLETETION APPROVAL SIGNATURES	DATE
M & R FOREMAN:	
ORIGINATOR:	

TICKET ROUTING

WHITE → M & R DEPT. → M & R SUPT. → M & R FILE
YELLOW → M & R DEPT. → ORIGINATOR → PROD. SUPT.
PINK → M & R DEPT. → MECHANIC → CLERK
GREEN → M & R DEPT. → ELECTRICIAN → CLERK
ORANGE → ORIGINATOR'S FILE

M & R FORM 3

Figure 3–5. Front of Maintenance and Repair Service Ticket

COST RECORD

SERVICE TICKET NO:

DATE	CLOCK NO.	HRS.	MATERIAL USED	MAT. COST
TOTAL HRS.			TOTAL MATERIAL COST	

MAN-HOUR ESTIMATE SUMMARY

MECHANIC		WELDER	
MACHINIST		SHEETMETAL	
PIPE FITTER		ELECTRICIAN	

Figure 3–6. Back of Maintenance and Repair Service Ticket

MAINTENANCE PERSONNEL ASSIGNMENT SHEET FOR June 7, 19
date

CRAFTSMEN

NAME	NUMBER	for June 7	for June 8
BREAUX, E.	16802	1028, 1031	1030, 1036
COMO, P.	5126	1028, 1041	1044
DeBLIEUX, P.	13128	1043, 1046	1051
ERSKIN, X.	7826	1050, 1053	1056, 1059
FULSOM, D.	9438	1028, 1060	1044
KILROY, I.	12185	1043, 1062	1060
PLACID, A.	10867	1043, 1046	1051
SOTO, B.	10246	1057	1060
THOMS, C.	9782	1057	1061
WILLS, F.	8953	1043, 1062	1061
ZARCO, H.	9168	1028, 1063, 1064	1039, 1040

Figure 3–7. Maintenance Personnel Assignment Sheet

Scheduling Techniques

For many years, schedulers have been using the Gantt chart, developed by Henry L. Gantt during World War I for use in production scheduling. Later it became known as the bar graph and became a standard tool in maintenance, project, and production scheduling. The simplicity of this technique is primarily responsible for its widespread and continuing use.

The Gantt chart can be used for short-range, intermediate-range, and long-range scheduling. Figure 3–7 shows a Gantt chart for short-range scheduling—that is, for the present day and the day after. Various refinements have been added to the Gantt chart since its inception. It is possible to use a bar for scheduling and an additional bar in the same block to indicate progress; that is, if a certain work order is actually scheduled for the part of the morning and that work order is actually worked, then an additional bar is shown immediately below the scheduled one to indicate that the plan has been followed. The bar chart is a useful tool when no more than about twenty craftsmen are to be scheduled. When the size of the work force or the service load becomes so large that the scheduler has difficulty accurately anticipating the results of work-schedule changes—in terms of manpower and equipment schedules—it is time to consider using a computer. Seldom, if ever, can installation of a computer be justified solely on the basis of scheduling, but most plants that already possess electronic data-processing equipment can use it economically for scheduling purposes.

From the standpoint of maintenance scheduling, the computer is merely a data-processing instrument. The computer can quickly inform the scheduler about the effects of schedule alterations, and thus can minimize conflicts and idle time. A scheduling algorithm is an important subroutine in any MMIS. Use of a computer in the MMIS will be discussed in detail in chapter 9.

Network-scheduling techniques—also known as CPM, PERT, and critical-path scheduling—have been touted as significant improvements in scheduling operations. In maintenance, network scheduling is used primarily for the turnaround. Common to most large-scale chemical plants and refineries, the turnaround is a periodic, planned major overhaul or preventive maintenance service of process units. It usually requires these units to be out of service for three to thirty days. Because of the disruption in the production process, it is advantageous to minimize the downtime of these units; therefore, turnaround operations are usually conducted on a three-shift, seven-day-a-week basis. Since the turnaround is compressed into the shortest possible period of time, it is advantageous to plan and chart the work to ascertain the critical paths and to make the best use of resources. An illustration of this technique is shown in the next section. The critical-path technique is expensive and time-consuming and does not lend itself to planning for day-to-day operations.

The origin of the term *turnaround* is interesting. In the early days of oil-refining operations, it was common to boil off gasoline in large containers that resemble modern pressure vessels. In order to do this, a fire was built under the container. Repeated firings caused the shell of the container to become thin. The container was therefore periodically rotated, or turned around, so that its use could be extended. The term *turnaround* arose from this process.

The Use of the Critical-Path Method (CPM)[2]

Use of the Critical-Path Method enabled one petroleum refinery to reduce its turnaround time by 10 to 20 percent below the average times required over the previous 25 years—with no increase in cost. The CPM enables the refinery manager to see the entire project as an entity. Critical jobs are pinpointed, impossible not to see immediately; these jobs then command management attention and can be expedited with overtime or any other technique management chooses. CPM enables a manager first to plan, then to schedule, and thereafter to control a project most efficiently. The manager is in a position to expedite the project judiciously and to apply additional resources only when and where they are needed. Thus, CPM can be one of his most effective tools for reducing project costs.

Management of a growing number of companies has turned to the Critical-Path Method during the past two years to help solve one of the most important problems confronting a firm—planning and controlling a project. The hundreds, sometimes thousands, of individual activities that go into a project, plus considerations of time, cost, manpower, and equipment, add up to a task far too great for any individual or any group to comprehend as a whole, let alone to manage as a whole. CPM enables the project manager to see the project in the round, to consider the relationship each activity has to the project, and to focus his attention on the relatively few jobs that are really critical.

Surprisingly, the number of critical jobs in any project—that is, those jobs in which a delay for any reason would result in a delay of the entire project—is small. In field application of CPM, the number of critical jobs has turned out to be as small as 2 percent of all activities, and it rarely goes above 10 percent. Once a manager knows on which jobs the project really depends for minimum-time, minimum-cost completion, he can apply any necessary overtime, expediting, greater supply of materials, or anything else he considers necessary, to those jobs only. This approach is vastly different from the common crash-everything approach that often results when a project is delayed by weather, strikes, materials shortages, or any other unpredictable cause. In other words, there is no panic button in the world of CPM.

The difficulty of choosing the most efficient plan for even a small project, using conventional techniques, has shown up [repeatedly] . . . .

Using the conventional bar-chart technique, the supervisor or manager decides when the job should be done. He considers alternatives and makes a decision. Then, he repeats this cut-and-try approach for each job until all jobs are set in their time slots. The resulting schedule represents his best efforts to take into account all the possibilities, but the number of possibilities can be enormous, even on a modest-size project.

Conventional methods are not methodical, even though they may seem to be so. CPM is methodical, and it will be worthwhile to look briefly at the way it works. The project is analyzed by being broken down into all the jobs of which it is composed. An arrow diagram is drawn showing the sequence in which the jobs are done, the way the start of one or more depends on the completion of a predecessor, and the extent to which they can be performed in parallel. When this closed network of arrows is completed, units of time are associated with each job, the time being based on the best available estimate. The critical path then is the longest path in time through the project. The jobs along this path allow no slack time; they cannot be reduced in time except by the application of more men, more time, or more equipment. The jobs not in the critical path—and the noncritical jobs, remember, are 90 to 98 percent of the total—do have slack time, or float; more time is available for their completion than is really necessary. Therefore, on a large-scale project, or even one of modest size, impressive savings can be realized by reserving manpower, overtime, or added equipment for the critical jobs, and taking advantage of slack time on all other jobs for emergency situations.

Example: Turnaround

An annual methanol unit turnaround was scheduled for the twenty-sixth time and this project was selected for trying out CPM. The turnaround by now apparently offered no new challenges to the plant management. The usual 120 men were scheduled to work the usual 8-hour shifts for the usual 12 days. The superintendent and foremen who were to lead the turnaround had taken charge many times in the past.

An arrow diagram was drawn representing the logical sequence of jobs necessary to complete the turnaround. . . . When the arrow network was complete, it underwent the all-important review, to be certain that no jobs were omitted and that the logic of the project was correct. At this point, only the planning had been done, not the scheduling. After the manager and his assistants reviewed the plan, deciding that the scope of work had been correctly interpreted and that the sequence of jobs was logical and technically accurate, the next step was to determine when the jobs should be done. The unit supervisor determined the time in manpower hours to perform each job, developing the time estimates with the help of some of the foremen who had worked on the methanol unit turnaround previously. Getting time estimates from the most experienced men on the job has a number of advantages in addition to that of reliability. The supervisors who contribute to the scheduling estimate feel that they are an integral part of the program; it is their estimate and they will try to meet it. (The arrow diagram shown [in figure 3-8] illustrates the recovery heater overhaul portion of the overall plan for the turnaround. The jobs along the path drawn with a heavy line are critical.)

Working with CPM, the planners came up with an 8¾-day project duration, using 120 men on 8-hour shifts. The plant manager decided to drop the schedule previously developed and to use the CPM schedule. All

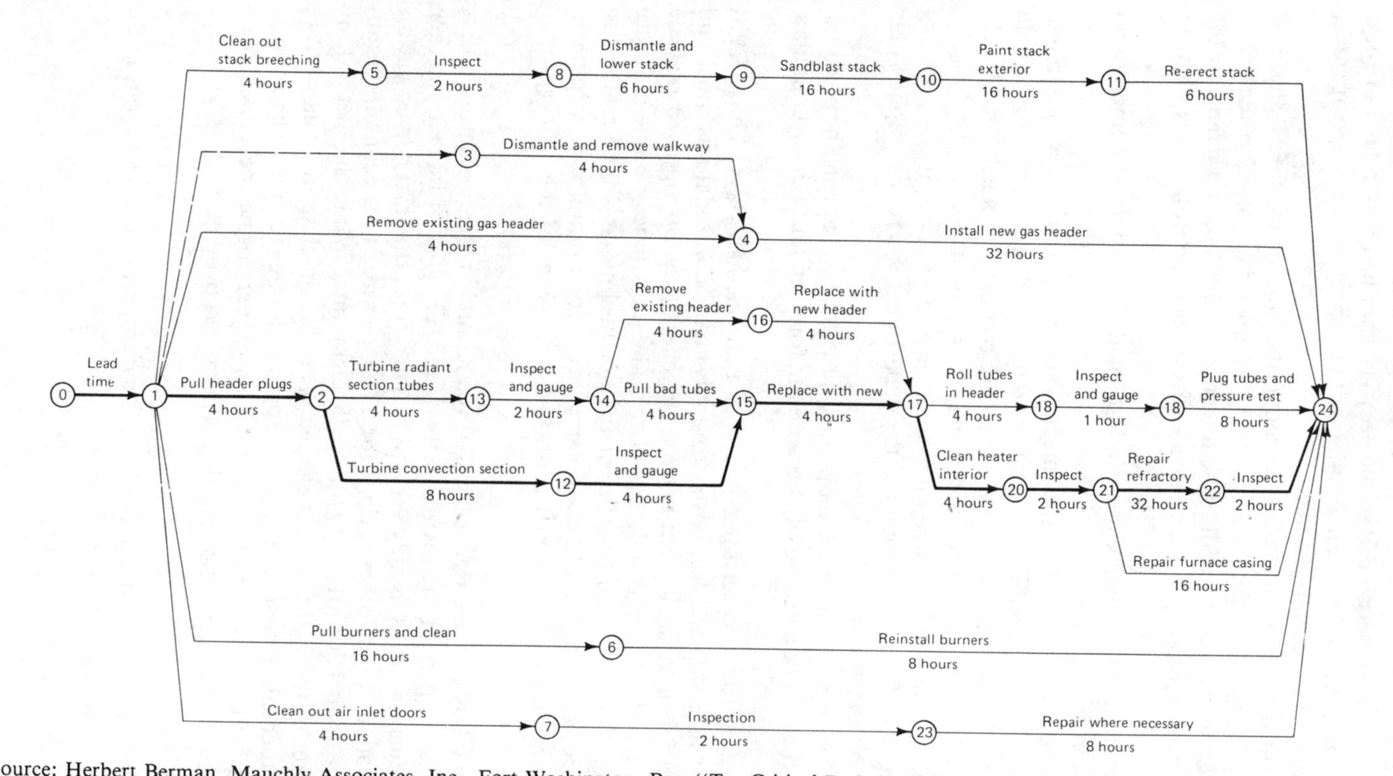

Source: Herbert Berman, Mauchly Associates, Inc., Fort Washington, Pa., "Try Critical Path Methods to Cut Turnaround Time 20 Percent," *Hydrocarbon Processing and Petroleum Refiner*, by permission.

Figure 3–8. Arrow Diagram of Methanol Unit Recovery-Heater Overhaul

new supervisors were assigned to the project. They had half a day to review the arrow network and become familiar with the project, the scope of work, and their specific responsibilities.

The turnaround then proceeded along the lines of the CPM plan and schedule. Completion time was nine days—three days or 25 percent less than the scheduled completion time under the old planning system. In the postturnaround report, the usefulness of this new approach was cited as follows:

To management: The CPM technique made it possible to pinpoint and justify overtime, actually to reduce it.

To area supervisors: The plan meant that the area supervisor no longer needed to carry so much detail in his head as he did previously. Also, if an area supervisor was taken off the job before completion, a new man would have the entire plan on paper and could easily pick it up.

To line supervisors: These men were relieved of some of the planning responsibility. The plan pinpointed the department where the emphasis lay at any moment in time.

To the shops: Individual shops, such as the Instrument Shop, could schedule their work more closely because the foremen knew the overall plan in detail.

Planning this project took just four and one-half days. Thus, CPM actually was easier and less costly to apply than traditional methods, and produced far more significant results. In general, the plan proved realistic and worth following in detail. The historical data will be of value in future turnarounds and the full benefits of the new technique will be realized in future projects. The benefits will carry over, for example, in helping to assess some standard work procedures in other areas of the refinery.

Downtime Losses Were Reduced. The reduction in actual turnaround time reflected only part of the saving to the plant. The loss of income for each lost production day is usually very high. Therefore, the saving in downtime must be added to the other savings for a true picture of the economy resulting from application of the critial-path technique. While the method applied here for the first time represented progress, much more remains to be done. Other refinements of the critical-path technique remain to be applied, for additional savings. Management of any refinery or petrochemical plant may want to consider following this generalized program:

First Requirement. Develop the arrow diagram, representing all jobs in the project, in sequence, with associated time estimates.

Apply Resources. Resources can be time, money, manpower, or equipment. Here again, basic CPM plus new extensions and techniques are applicable to project planning. A Resources Planning and Scheduling Method (RPSM) has been developed which is essentially a method of making the best use of resources on a project. Further, a computer program has been written which will handle resources of scheduling of equipment, space, dollars, and workers by craft. (In network-scheduling, this is the equivalent of getting the design off the drawing board and into production). RPSM techniques have been worked out that make it possible to develop a master plan for a complex of interdependent projects, possibly involving several subcontractors. All details of these projects can be displayed on the arrow diagram.

By taking advantage of the available slack time within the noncritical jobs, management can make the best use of manpower. Because CPM will show clearly the amount and location of free time, management can delay the start of noncritical jobs or move workers around, all for the better utilization of total manpower. This same technique can be applied to other resources.

Changing the Project. It is important to remember that CPM is a dynamic program. Once the turnaround (or any other project) begins, a project-reporting and -control system must begin to operate. If any part of the plan changes, the overall CPM program can be rearranged accordingly without difficulty—manually on relatively small projects, and by computer runs on larger undertakings. For example—assume that all preliminary information is on punched cards and the necessary first calculations have been made. When the towers, exchangers, drums, and other components are opened and inspected, the actual workload becomes known. It may be considerably greater or less than anticipated. It is necessary now only to change the information on the jobs actually affected and to reprocess the information. Within a few hours, the new plan will be available to management for decision making.

Actually, only by continually upgrading the plan and producing new information on which management can base its decisions, is it possible to realize substantial savings. The ability to continually produce current information is one of the most significant aspects of CPM.

Although the repeated development of numerical information in CPM makes the system naturally computer oriented, a computer may not be necessary for small, relatively stable projects. But on larger projects and in situations where management wants to sell all possible variations of all possible plans, computer runs are necessary.

Progress Reports

Even for day-to-day scheduling, the scheduler and the area or craft supervisor need to know the *status of each job in terms of work completed and work to be completed.*

In general, job status can be computed in two ways: (1) as the percentage of actual time against scheduled time or (2) as the percentage physically complete against wholly complete, in terms of man-hours. The first is usually of sufficient accuracy. The second is more costly, since either the planner or the area or shop supervisor must appraise the percentage completion of the job and convert it to man-hours.

If the intergral parts of a job are not too closely related, the work can be divided into several work orders so that each order can be completed in a shorter time. Thus, the need for personal observation of job status is greatly reduced. If the integral parts are closely related, however, and close coordination is required between parts, it would not be practical to separate the work into several jobs merely to facilitate the reporting of work accomplished.

Another means of obtaining work-progress reports is available as a by-product of scheduling by means of CPM arrow diagrams. Positive identification of nodes or jobs by number will accomplish the same purpose as writing additional orders and will permit simple reporting of activities finished or checkpoints reached.

Electronic data processing (EDP) of the information used in scheduling can greatly accelerate the reporting of work status. A single EDP report can show job-order number, job status, estimated hours, actual hours, estimated labor cost, actual labor cost, estimated material cost, actual material cost, estimated total cost, actual total cost, and variation total cost. This can be done in less time than a manual report would require.

Clerical work can also be greatly reduced by using EDP to compile work orders, manpower-availability reports, backlog reports, materials-status reports, production schedules, and tentative and final maintenance schedules.

Progress reports are especially important on construction, installation, and rearranging jobs. Progress reports are usually developed, at regular intervals, by the project coordinator or resident engineer on the job. The reports are often in the form of Gantt charts (see figure 3-7). When outside contractors are used, the progress report may be the basis for partial payments to the contractors.

Critical Path or Bar Chart?[3]

Critical-path scheduling has limitations, especially in industrial-maintenance control. Some of the most objectionable limitations are that it is difficult to interpret; that it schedules the job and not the man; that craft lines are not defined; that it requires considerable time to prepare; and that it is difficult to find estimated versus actual costs. The CPM is an excellent system for laying out the job in planning phase, however, and is useful in visualizing the status of a job and in eliminating bottlenecks.

To compare the critical-path method with one of the bar-type methods common to industrial-maintenance control, we will consider a fairly routine request—the removal, repair, and reinstallation of a vertical pump (figure 3-9).

Using either critical-path or bar-type methods of control, it is first necessary to determine

1. the required craft and number of men for each craft;
2. the time required, by craft, and the most appropriate time for craftsmen to be on the job;
3. the necessary tools and rigging;
4. the required parts; and
5. the necessary safety signs, tags, blocking, and the like.

Specific duties of each craft will vary according to labor policies. The crafts necessary to the overall function, however, will generally be the same in most cases. In the paper industry, for example, such a work request would require millwrights, pipe-fitters, electricians, and possibly machinists.

The sequence of the work would be the same for most industries and should proceed as follows:

1. Withdraw necessary parts from the storeroom.
2. Select the tools needed, including rigging, safety equipment, and conveyance for the pump.
3. Rope off the area, set up safety signs, and so forth.
4. Install rigging for handling the pump.
5. Tag the main power-supply switch.
6. Disconnect and mark the motor leads.
7. Disconnect the piping.
8. Remove the motor-foundation bolts.
9. Remove the motor.
10. Drain the oil from the drive mechanism.
11. Remove the pump-foundation bolts.
12. Remove the pump and deliver it to the shop.
13. Disassemble the pump and clean all parts.
14. Determine the extent of damage.
15. Secure or fabricate parts that had not been anticipated.
16. Reassemble the pump.

DEPARTMENT FILE

FORM F-150

MAINTENANCE JOB ORDER

WORK DONE

	NAME	DATE
ASSIGNED TO		
COMPLETED BY		
INSPECTED BY		
ACCEPTED BY		

JOB COST SUMMARY	EST.	ACT	VAR.
LABOR HRS.	24		
LABOR COST	96.00		
MAT'L COST	63.00		
TOTAL	$159.00		

MAINT RECORD: IBM : F-112

DATE WRITTEN	DATE WANTED	DATE SCHEDULED	COST CENTER	EQUIPMENT CODE	EQUIPMENT NO.	JOB ORDER NO.	CLASS	ACCT NO.	EMERGENCY YES	EMERGENCY NO
6/24/	6/30/		19	PUA	604	55342	12	1100		✓

EQUIPMENT NAME #1 PAPER MACHINE COUCH PIT PUMP

N.W. NO.

WORK TO BE DONE

REMOVE, REPAIR AND REINSTALL THE COUCH PIT PUMP, #1 KRAFT PAPER MACHINE

I.C. NO.

R. NO.

SHUT DOWN YES ☑ NO ☐

WHAT MUST BE DOWN MACH. ☑ DEPT. ☐ MILL ☐

DATE:

REQUESTED BY J. ROBINSON

APPROVED BY DEPT MGR. OR SUPT. W. D. Davis

OTHER APPROVALS WHEN REQUIRED JEM L.W.P.

EST	1 M W	2 WELD	3 ELEC	4 SHOP	5 PIPE	6 CARP	7 PAINT	8 P C	9 OIL	10 INST	11 YARD	12 P H.	13 A M	14
NO MEN	2		1	1	1									
HOURS	16		2	4	2									

$

Source: Bert Carson, "Critical Path or Bar Chart," Hudson Pulp and Paper Corp., Palanka, Fla., Part D, pp. 31–33.

Figure 3–9. Typical Repair-Request Order to Illustrate Operation of Critical-Path (figure 3–10) and Bar-Type (figure 3–11) Systems

17. Return the pump to the job site and reinstall it, bolting it to the foundation securely.
18. Put new lubricant in the drive mechanism.
19. Reinstall the motor.
20. Reconnect the wiring.
21. Reconnect the piping.
22. Remove the tag from the power switch.
23. Remove the rigging.
24. Make a test run.
25. Clean up, load the tools, remove the barricades and the signs.

Because of the number of details involved in this relatively simple job, an expert planner would require fifteen to thirty minutes to put the step-by-step outline on a critical-path form (figure 3–10). The personnel required for planning the daily work of a large maintenance crew in this manner would be economically prohibitive.

In the bar-chart method, it is not considered necessary to schedule job preparations, safety precautions, and cleanup for each job, since maintenance personnel are familiar with standard work practices. The bar chart depicts time, not method. This places the contractor at a decided disadvantage, because his workers come from different areas, have worked under different policies, and use different work methods.

Part of the regular weekly bar-chart schedule for the pump-repair job is shown in figure 3–11. From this schedule—which is made up and distributed every Thursday to all maintenance shops and covers the following week—the individual craftsman has no difficulty determining where he is to work, when to report, and what is expected of him. The schedule form also includes space for the estimated manpower by trades, the material requirements, and the anticipated and actual costs.

Upon completion of each job, maintenance control posts man-hours, labor dollars, and supply costs to the completed copy of the work order, which is then forwarded to the department supervisor for his records.

We have discussed two control systems, each specifically designed for a different purpose. Both systems are excellent when they are used for the purpose for which they were designed. Do not be misled by their apparent simplicity. Much time and effort are involved in the installation and operation of both systems.

The maintenance planner should not ignore the advantages of using the CPM for preplanning large maintenance of new work projects. It is recommended that a combination of systems be used for this kind of work—the CPM for planning and the conventional bar-chart method for scheduling.

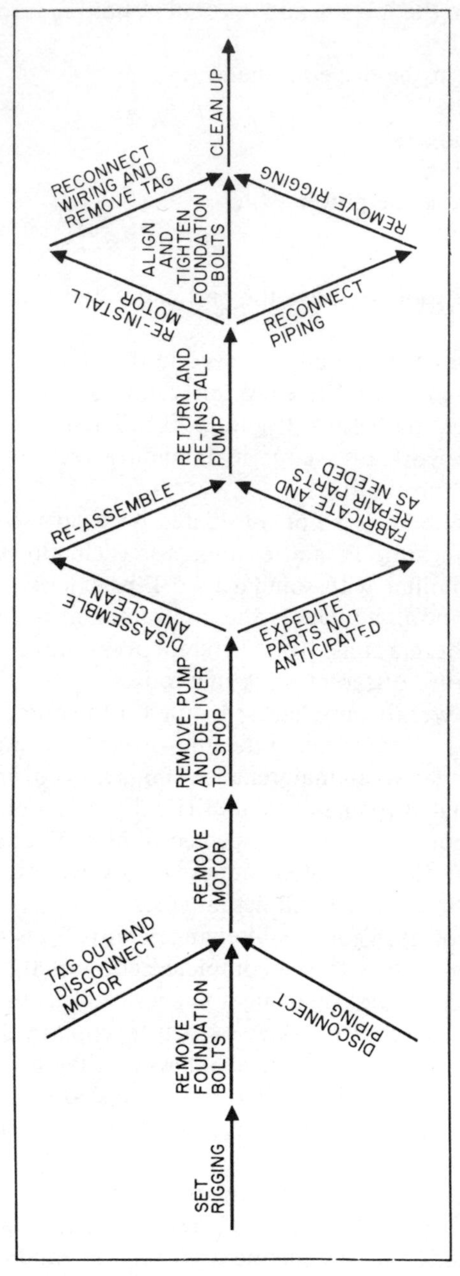

Source: Bert Carson, "Critical Path or Bar Chart," Hudson Pulp and Paper Corp., Palanka, Fla., Part D, pp. 31–33.

Figure 3–10. Step-by-Step Critical-Path Schedule for Overhauling a Pump

DATE JUNE 29-JULY 3, 19

HUDSON WORK SCHEDULE

AREA: KRAFT PAPER MILL SUPERVISOR: JAMES ROBINSON TYPE WORK: REGULAR MAINTENANCE

Cost Center	EQUIPMENT Code	Number	Job Order Number	Charge Account	DESCRIPTION OF WORK	Time Start	Time Stop	% Comp	Trade Code	TRADESMEN	MONDAY 6-29	TUESDAY 6-30	WEDNESDAY 7-1	THURSDAY 7-2	FRIDAY 7-3	Man-Hours Est.	Sch.
19	PUA	604	55342	12/1100	REMOVE, REPAIR AND RE-INSTALL COUCH PIT PUMP, #1 PAPER MACHINE	7^{00} 7^{00} 2^{30}	8^{00} 3^{30} 3^{30}		3, 5 1 3, 5	FORD & MORRIS SAPP & POPE FORD & MORRIS						24	24
19	UER	–	56874	02/1100	INSPECT BRUSHES IN MOTORS AND GENERATORS, #1 P.M. LIST BAD BRUSHES FOR WIRE CHANGE	8^{00}	2^{30}		3	FORD						6	6
19	UST	–	55346	04/1100	TAG STEAM LEAKS, #1 P.M. FOR WIRE CHANGE	8^{00}	12^{00}		5	MORRIS						4	4
21	UPB	–	63484	14/1100	CLEAN STRAINERS, #1 & 2 PAPER MACHINES	12^{30}	2^{30}		5	MORRIS						2	2

Source: Bert Carson, "Critical Path or Bar Chart," Hudson Pulp and Paper Corp., Palanka, Fla., Part D, pp. 31–33.

Figure 3–11. Part of Regular Weekly Schedule Showing Pump-Repair Job

Table 3–2
Shop-Planning Report

Period	*A* *Percentage of Scheduled Work (Aim: 80%)*	*B* *Percentage of Promised Work Delivered on Time (Aim: 95%)*	*C* *Percentage of Scheduled Work Promised and Delivered This Period (Aim: 76%)*	*D* *Percentage of of Excess Man-Hours (Aim: 0%)*
One year ago	67	49	16	65
Twenty-week rolling average	74	93	62	3
Two weeks, ending June 6	82	96	68	(−7)

Column A: Measures control over interruptions to scheduled flow of work through the shop (man-hours on scheduled work divided by total available man-hours).

Column B: Measures how well promises are being met (jobs delivered on time divided by total jobs).

Column C: Measures how well the man-hours available were both scheduled and utilized to obtain on-time deliveries (scheduled man-hours applied to on-time deliveries divided by total man-hours).

Column D: Measures excess man-hours used for shopwork, based on target of 80 percent performance (difference between actual shop man-hours and scheduled shop man-hours divided by scheduled man-hours based on 80 percent performance).

Copies to vice-president, operations; general manager, steel operations; district manager; plant engineer.

Report-Audit Findings

Since scheduling is a commitment to a projected result, comparisons must be made of actual versus projected plans, and explanations must be given for any major differences. Such comparisons may be continuous, and an exception report issued, or they may be made at intervals. Such audits of achievement enable the system to maintain its discipline and to provide maximum benefits.

It was stated earlier that improvements in scheduling come from changes in management thinking. One of the most effective ways to accomplish such a change, and to concentrate management emphasis on the areas where it is needed, is to issue a control report (see table 3–2). A control report should have the following important elements:

Simplicity: The report must be brief enough for a busy executive to read and yet complete enough to show what scheduling is—or is not—accomplishing.

Comparisons: Current results must be compared with the objectives set for the scheduling operation. In addition, recent results (such as a twenty-week moving average) and results in the past (for example, a year ago) should be shown, so that improvements can be evaluated.

Authority: Copies should be sent to the top-management personnel whose support is necessary for attaining optimum plant operation, especially the individual to whom the top production manager reports.

Notes

1. See Bernard T. Lewis, "Controlling Maintenance Backlog," *Plant Engineering,* February 22, 1973, p. 203.
2. This section is reprinted from Herbert Berman, Mauchly Associates, Inc., Fort Washington, Pa. "Try Critical Path Methods to Cut Turnaround Time 20 Percent," by permission of *Hydrocarbon Processing and Petroleum Refiner.*
3. This section is adapted from "Critical Path or Bar Chart," by Bert Carson, Hudson Pulp & Paper Corp., Palanka, Fla., part D. pp. 31–33.

4 Maintenance-Work Measurement and Standards

Historically, work measurement has been concerned with measuring direct labor. The primary reason for this is that direct-labor measurement is needed for product pricing and costing. The second major reason is that, in the past, direct labor accounted for most of the hourly labor in industry and was the most important cost to be measured and controlled. In addition, the measurement of direct labor is much easier and less costly than the measurement of indirect labor. The considerable changes taking place in our economy, however, require that more attention be paid to the measurement of indirect labor in general and the measurement of maintenance labor in particular.

Modern work measurement was introduced in 1881 by Frederick W. Taylor, chief engineer of the Midvale Steel Company. In 1903, he published "Shop Management" (Paper No. 1003, American Society of Mechanical Engineering), a paper that was concerned mainly with time study. The first modern motion study appears to have been done by Frank B. Gilbreth in the late 1800s. Gilbreth first reported this study in his *Brick Laying System,* published in 1909; in 1911, he published another book on the subject, *Motion Study*.

The basic techniques used to measure indirect labor are similar to those used to measure direct labor. The primary differences are in emphasis and in method of application. Other differences include the considerations of methods, standard-hour calculations, and crew application of standards. A closer look at some of these topics will reveal these differences.

Whether the concern is for direct or indirect labor, work measurement can be undertaken appropriately only after the process that involves the labor has been standardized and the job has been broken down into discrete activities. Production jobs, which are standardized and usually repetitive, have very short activity cycles. The activity cycles of maintenance jobs, however, are likely to be very long (perhaps from three to six hours), in part because the breakdown into job activities is not generally as detailed as it is for direct labor.

Thus, since maintenance-work measurement does not require the stopwatch precision of direct-labor measurement, which necessitates specifically trained personnel and extensive sampling to ensure statistical validity, it is generally less expensive.

The Why and How of Maintenance-Work Measurement[1]

Maintenance-management programs that include methods improvement and work measurement have reduced maintenance cost in many cases. (One company experienced a three-dollar reduction in maintenance cost for each dollar expended in work measurement.)

Such job-methods planning programs have also resulted in major reductions in equipment downtime. At one plant, for example, machine downtime for replacement of duct sections in a solvent-recovery system was reduced twelve hours, or about 40 percent. Another plant once required ninety-six hours to replace the installation on a reaction kettle; it now takes fifty-two hours—46 percent less time. Methods study helps engineering consider possible ways of doing a job and select the most effective way, which is then documented and becomes the standard for the job. Whenever that task is encountered in the future, the standard method can be pulled from the file and given to the maintenance planner.

Another area of improvement is the level of worker effectiveness. Before the installation of job-methods planning, levels of worker effectiveness usually range from 30 to 50 percent. After the institution of such a system, improvements of 65 to 75 percent are normally encountered. It is not generally considered economical to plan 100 percent of the maintenance jobs. Most plants consider the program complete when 75 to 80 percent of the maintenance man-hours are standardized.

A higher utilization of maintenance manpower is also realized. In this era of shortages of skilled mechanics, this is a very important consideration. Under work-measurement programs, many plants have found that their maintenance groups could assume additional duties without additional craftsmen or, in some cases, with fewer craftsmen.

Work measurement also results in a lower investment cost for new or improved facilities. The effect of lower unit-labor cost on project cost is self-evident. In one plant, the replacement of an installation was authorized before job-methods planning had reached full effectiveness. The work was completed under the new controls at two-thirds of the estimated cost. At another plant, the improvement in effectiveness reduced the cost of maintenance labor from $2.23 to $1.88 for every dollar of material.

Furthermore, work measurement improves both maintenance scheduling and the preventive-maintenance (PM) program, thus reducing time delays between jobs. It leads to closer project estimates and, therefore, to better decisions about their acceptability. Work measurement makes better control of crew sizes possible by providing a more accurate knowledge of the work backlog, which is normally categorized by craft or mechanic class, and work measurement allows more efficient manning of jobs by enabling the work planner to engineer exactly the crafts needed for specific activities.

A work-measurement program reveals areas in which additional training is needed. This can be seen when a worker or group of workers fails to meet the normal effectiveness level when following the suggested methods. Systemization, such as occurs with work-measurement programs, leads to more effective use of high-priced tools and equipment by minimizing standby time. In addition, delays caused by inaccurate supply or late arrival of materials are minimized. Work measurement helps justify new purchases of maintenance equipment by clearly showing the tangible savings in manhours that would result from the purchase of such equipment. Material specifications are improved when work measurement discloses operations that consume excessive time in application of material or in repair of equipment. Foremen, relieved of estimating duties by the existence of the work-measurement standards, are given more time for direct supervision. Finally, the results of work measurement provide a common unit of measurement, understood by both management and labor, for discussing cost, performance, and goals.

What Work Measurement Will Not Do

Work measurement is not a substitute for good supervision. If supervisors do not plan and schedule work properly, labor effectiveness will suffer. Work measurement does not train maintenance workers. Well-qualified workers are essential to maximum labor utilization. Work measurement does not guarantee work of high quality. Quality is still a supervisory responsibility, and good job planning, which gives proper consideration to methods, tools, and materials, encourages work of good quality. Work measurement does not guarantee good job engineering, but it does invite more careful review of specifications, drawings, and schedules.

The Goal of Maintenance-Work Measurement

The goal of work measurement is to reduce costs while improving both the adequacy and the quality of maintenance. Work measurement alone cannot achieve these objectives. A reliable work-order system, proper job planning and work scheduling, and qualified and skilled personnel with competent supervision are also essential to the effective utilization of labor. Furthermore, an adequate system for the utilization of tools, equipment, and materials is necessary. Assuming that these factors are under control, there is a need for a reliable means of measuring maintenance performance objectively, a means that can place maintenance management on a truly businesslike basis.

Without work measurement, management does not know where it stands in regard to worker performance. Plants without work-measurement programs typically find that their work force is only 35 percent effective. This does not necessarily mean that the work force is loafing; it does mean that it is working ineffectively. The real benefits of a work-measurement program can be realized when the reasons for this ineffectiveness are found and corrected. These reasons may include faulty supervision, poor tools, inefficient methods, poor organization, lack of training, inadequate planning, loose scheduling, and, least often, a poor worker.

The three elements necessary for the creation of a job-methods planning program appear to be a work-order system, a standard-time data system, and a proper organizational structure.

Where to Use Work Measurement

There does not seem to be any serious limitation to the use of work measurement, even when a highly refined method of measurement such as standard-time data is used. This method has been applied to the measurement of all types of mechanical work. It can be effective whether the plant has a traditional craft organization or the maintenance work force is built around the maintenance-mechanic concept. It is universally recognized that there are differences in job conditions for the same basic maintenance operation. Painting may be done at ground level or high on a scaffold. Lead burning may require working in an open space or in a cramped position.

It appears that very small plants can use work-measurement systems profitably. Many examples exist of plants with about fifty maintenance employees using work measurement successfully. A small plant must consider whether it can write off installation cost and pay for administrative cost out of direct and indirect savings. The larger the plant, the easier it is to amortize the installation cost of a work-measurement system.

Work measurement has been applied to maintenance jobs requiring as few as two man-hours. Jobs that require less than two man-hours are best handled by area-maintenance people and are impractical to measure. Work measurement for emergency work is also impractical, since time is not available for preplanning. The size of the job is not as important as a clear-cut definition of the scope of work and the availability of detailed plans.

Three Systems of Maintenance-Work Measurement

From Experience. The first of three primary ways of measuring work for maintenance is by experience or estimate. This is by far the most common approach used to establish labor control. Some companies estimate large jobs only. Others estimate all jobs over a fixed-minimum dollar value or number of man-hours. A few companies estimate all maintenance jobs.

This sort of work measurement is done by applying knowledge gained from experience to the description on the work order. Obviously, such estimating activity is present in practically every plant to varying degrees. In general, however, this type of work-measurement program is used in plants whose maintenance systems are rather rudimentary.

There are several advantages to the experience method of work measurement. An estimate of the total time required can be furnished before the start of any effort. That estimate can then be a basis for scheduling the activity and can be compared with actual man-hours charged to the job. The cost of such a program is not great. If estimators are used, it may require one estimator for every forty or fifty craftsmen. The estimating process forces the foremen to do some planning of a job before initiating work.

This form of work measurement also has a number of disadvantages. Estimates are often inconsistent and inadequate, for several reasons. The foreman may not know what a normal work pace is. He is expected to be judge and prosecutor at the same time. He may omit parts of a job because he has not analyzed it sufficiently. If the estimates are too loose, the workers lose confidence in the measuring program. If the estimates are too tight, the workers lose interest in trying to meet the goal. When methods change, estimates lag behind in their adjustments to new time requirements for the job. This is particularly true in areas of fast-moving technology or rapid acquisition of new materials and methods, where no experience exists for estimating the time required for new work.

Using Historical Data. This method calls for statistical analysis of the time data on completed jobs. The average time is considered a suitable point of departure for the realization of a standard. If the work is being performed satisfactorily and all parties are pleased with the effort, then the average time may serve as a standard. In the majority of cases, it is necessary to do a methods analysis of the activity so that everyone is assured that the job is being done as management anticipates. These methods studies need not be a major effort. In many cases, the correct method for performing a task can be established by informal discussion among a small number of experienced people. This can be done in as little as five to ten minutes.

I once participated in a series of meetings that resulted in a set of standards for cleaning storage tanks. A group of tank-cleaning foremen were called together, the sequence and methods used in tank cleaning were discussed, and a consensus was reached. When there was agreement on the specific steps, the foremen were asked to provide an estimate of the time and equipment necessary for each step. This information then became the standards.

In principle, work measurement can be performed by comparing actual man-hours spent on a task with the standard based on average historical time for the same type of work.

Typically, the historical method works in the following manner. A file of completed work orders is accumulated for a period of time—perhaps six to twelve months. The orders are then subdivided into about four categories. The first category might be routine work, such as inspection and lubrication; the second repetitive work that is irregular in frequency, such as packing pumps, replacing lamps, and testing safety valves in place; and the third, miscellaneous one-of-a-kind repair jobs that require less than 100 manhours of labor. The fourth category might include larger repair jobs that require more than 100 hours of labor and all new jobs. Standards are then derived for each of the four categories. For both routine and repetitive work, this is done by calculating their average time. Only jobs that are representative should be included. An analysis of the various work orders and other descriptions will indicate which jobs are unusual, and these should be excluded.

The miscellaneous jobs are subdivided into smaller groups according to the actual time charged—for example, 0–8 hours, 8–16 hours, 16–32 hours, 32–50 hours, and 50–100 hours. The average time is then determined for each subgroup. The average time for routine repetitive and miscellaneous activities is adjusted by leveling for the estimated or observed pace of the maintenance work force. This index of pace is closely related to productivity. If it is found that the local work force is 10 percent more productive than the national average, then the standards must be adjusted to reflect that.

The remaining category considers the handling of large jobs, those requiring more than 100 man-hours. Generally, these activities are estimated in the traditional manner. This estimate is then adjusted by a factor representing average total performance for a preceding period—perhaps two months—to obtain the standard time for the job. As an example, assume an estimate of 150 man-hours to install a new pump, with necessary piping, foundation, motor, conduit, and controls. Assume that the two months' average performance of the maintenance force was 80 percent. The standard for the job then becomes 150 × 0.80, or 120 hours. Note that the 120 hours is less than the estimated 150 hours when adjusted by the 80 percent performance factor. In this case, 80 percent indicated that the work force required 80 percent of the normal time to perform the task. If the job actually took 145 hours, the performance would be 120 divided by 145, or approximately 83 percent.

Normally, the application of standards is handled by a standards clerk, who calculates the standard time for each job order and compiles a weekly report of job performance by class of work and by total hours expended. In these reports, the total number of hours estimated is compared with the total hours expended to obtain an overall performance report. The tabulation can be broken down by craft to provide a performance report for subgroups as well as for the entire maintenance work force.

Perhaps the primary advantage of the historical method is its low cost of administration. After the standards have been developed, one standards clerk is required for about every fifty maintenance workers. The standards program is most adaptable, of course, to routine and repetitive activities. It is easily installed in a few months in any plant that currently uses a work-order system. The installation of such a program will begin to improve labor effectiveness almost immediately, and the advantage will accelerate as the program progresses.

Some disadvantages exist, and a knowledge of them will contribute to effective use of the program and will help minimize the inaccuracies of the method. The standards used are based on past performance, with all its inefficiencies, and may therefore be inaccurate, especially in relationship to one another. The standards for miscellaneous activities are based on time charged, not on job content. In addition, measurement on large jobs for which little history exists must be done by the conventional estimating procedures.

A work-measurement system based on historical data appears to be better than no system at all, and it is easy to install and administer. The limitations of the historical-data approach are great, however, and care should be exercised in using it for a serious effort of maintenance-work measurement.

Using Standard Data. The standard-data approach to maintenance-work measurement relies on the use of time study much as does measurement of the direct labor of production operations, with some important differences. Stopwatch studies do not appear to be applicable to most maintenance operations. A regular clock or wristwatch appears to be accurate enough to time cycles for maintenance operations. Similarly, the breakdown of the different activities need not be as detailed as that usually found in production operations.

The recommended method for time study of maintenance work is to choose operations that are performed under normal plant conditions and to record the time used for each element of the job. As in production studies, it is necessary to adjust the recorded times in accordance with the observer's estimate of the worker's skill and effort. The results are compiled into a table of standard-time data.

Some routine jobs require additional considerations. There is a tendency to make analysis of maintenance jobs too detailed, which is expensive. There may also be a tendency to develop standards that are too loose and therefore unreliable. A number of factors must be resolved. The initial cost for developing these systems increases in proportion to the desired accuracy. In addition, the more detailed the breakdown of the data, the greater the cost of the system. The degree of acceptance of work measurement by management, supervisors, and workers is related to the accuracy of the time allowed for jobs. This, in turn, depends on the accuracy of the standard data. If standards are built up from detailed studies, the initial cost will be greater, but it will be easier to keep the standards current. Workup of studies and transposition to analysis forms will take about seven hours for each hour of time study. Application of standard data will require about one hour for about every ten standard hours established. Upkeep of standards will be easy, however.

At the other extreme, one can develop standards data from studies taken on such broad elements as working time, personal time, and delays. The workup of these studies will take about one hour for each time-study hour. Since few variables are considered because of lack of job breakdown, one can apply the standard-data tables rapidly—about one hour of application to establish fifty standard hours of work. Upkeep of standards would be difficult, however. One would have to restudy the entire process to set up a new standard for any change in method.

The middle-of-the-road approach appears to be best. The following procedures, with minor modifications needed for specific installations, will provide good standards data at reasonable cost:

1. Determine in advance the long or gross time-study elements that are common to each type of maintenance operation.
2. Make time studies of these long elements and of personal delays and foreign elements.
3. Work up, analyze, and chart the results.

Workup and analysis take about two hours for each time-study hour. Figure 4–1 shows one of a family of curves developed for making up standard-screw pipe joints. The values are approximate and are given for illustrative purposes only. This curve shows only the elemental time (less allowance) for making up a standard-screw pipe joint. Each curve in the family was individually identified and timed. The study also identified all foreign elements, delays, and personal time. In figure 4–2, these elements are combined to develop the standard time for making up a two-inch standard-screw pipe joint. All elements have been resolved into a common unit of measurement.

Figure 4–3 illustrates the development of a standard-data table for a variety of pipe sizes (¼ inch to 2 inch). Costs of applying this table were carefully weighed against the accuracy gained by identifying more variables. Preparation of such a chart requires about one hour to develop twenty-five standard hours of work. Standard data developed in this way are easy to maintain and should never become obsolete.

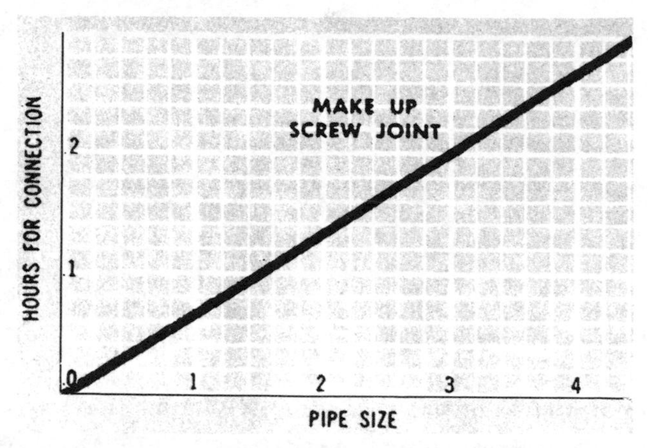

Source: Based on "Work Measurement in Maintenance," by J.O. Heritage, Kenneth Digney, R.E. Deem in *Factory* 113 (April 1955), copyright Morgan-Grampian, Inc., by permission.

Figure 4–1. Elemental Time Standards

All data developed in the manner described here should be statistically tested two ways before being used. The data should also be checked against time studies. Inevitably, errors accumulate in values as the result of averaging time-study measurements. In order to correct for these, the standard data for a specific operation must be applied to each of the jobs composing the time-study material for that operation. Following this, the standard-data time is then compared with leveled allowed time from the studies. Figure 4–4 is an example of how to adjust the standard-data tables.

The other way of statistically testing the data is to check for acceptability. Recompute the comparison indexes in figure 4–4 on the basis of adjusted data. If variations above and below 100 percent still exist, then these variations must be checked to ensure statistical reliability. One authority suggests that, in production work, application of standard data must yield standard times that come within ± 5 percent of the true time.[2] Such accuracy is not needed for maintenance activities.

Figure 4–5 shows an empirical curve for use in testing the reliability of data. The reasoning behind this curve is as follows. A worker or foreman accepts a standard according to the job size. For a job that normally requires one hour, he generally will not question an allowance from almost nothing up to two hours, or 100 percent deviation. For a job that normally requires ten hours, however, he will accept eight to twelve hours as allowed time, or 20 percent deviation. For a 100-man-hour job, the allowance should be between 90 and 110 hours, or 10 percent deviation. It should be emphasized that the variations in figure 4–5 are measured in terms of two standard deviations, since, by statistical proof, 95 percent of the jobs should then fall within the range indicated. Referring to the curve in figure

MAKE UP 2-IN, STANDARD SCREW PIPE JOINT

Material delivered to job site; working from ground, permanent platform, scaffold or ladder; standard pipefitters' tool kit, post vise, stock and die.

	Elemental Time Standard	Frequency per 1000 Pieces of Pipe	Standard Man-Hours
Handle pipe horizontally (50' avg.)	.012/2 pcs	2/1000	6.0
Handle pipe vertically (7' avg.)	.009/pc.	70	.6
Handle fittings horizontally (50' avg.)	.012/5pcs.	5/1073	12.9
Handle fittings vertically (7' avg.)	.005/pc.	75	.4
Measure and plan	.052/pc. pipe	1000	52.0
Cut pipe (hack saw)	.100/cut	21	2.1
Cut pipe (pipe cutter)	.031/cut	635	19.7
Thread pipe	.068/thread	821	55.9
Make up joint (screw)	.134/joint	1053	141.1
			290.7
		6% allowance	17.4
		Total standard hours per 1000 pieces	308.1
		Standard hours per piece of pipe	.31

Source: Based on "Work Measurement in Maintenance," by J.O. Heritage, Kenneth Digney, R.E. Deem in *Factory* 113 (April 1955), copyright Morgan-Grampian, Inc., by permission.

Figure 4–2. Development of Summary Standard-Time Value

4–5 at 6.4 hours (average true time), one will find twice the acceptable limit to be 34.2 percent. Since 16.5 percent is less than 17.1 percent (one-half of 34.2 percent), these standard data are satisfactory.

Standard Allowances. In figure 4–2, note that an allowance was introduced into the standard. This was developed by determining the percentage of direct working time taken for foreign work elements. Foreign work elements are miscellaneous duties that are related to the operation. They do not necessarily occur in every cycle of operation. Foreign-element allowances usually amount to approximately 10 percent or less of the standard. If an allowance greater than 10 percent is encountered, the job should be restudied to see if the work elements are properly identified. Do not worry about unusual accuracy for time studies when an allowance of 30 to 40 per-

PIPEFITTING (standard-screw pipe joint)

PIPE SIZE	STANDARD MAN-HOURS PER PIECE OF PIPE	
	MAKE	BREAK
1/4 - 1/2	.12	.08
3/4	.17	.11
1-	.20	.13
1 1/4	.23	.15
1 1/2	26	.18
2-	.31	.23

Source: Based on "Work Measurement in Maintenance," by J.O. Heritage, Kenneth Digney, R.E. Deem in *Factory* 113 (April 1955), copyright Morgan-Grampian, Inc., by permission.

Figure 4–3. Standard-Time Data

cent might be added to the task. Idle time should not be built into the standard. The standard should not be idealized, however; it should truly represent what the average worker can do. Frequently, the standard cannot be met because, in the creation of standards, the allowances necessary for the job were not considered. A methods study is also essential for setting a proper standard. It is ineffective to create a standard for the way the job is being done at present when the job is being done inefficiently at present.

All allowances for delays attributable either to the work or to management must be included. Only in this way can it be realized that many delays are not worker caused. When the delays are engineered out of the task, a more efficient method results. It is important to remember that the objective is to measure the work, not the worker.

Typical allowances include travel time for material; waiting for hot-work permits; consulting with the operator or area supervisor; waiting for mechanical equipment; and, of course, the personal allowances required on every job.

The Standards Program

The introduction of a standards program is crucial to the success of any maintenance-management system. (Figure 4–6 shows a typical standards format.) To make the most effective use of developed standards, it is necessary to be able to quantify maintenance efficiency; and some productivity index must be established before a standards program is used. This can be done in a number of ways. Work sampling seems to be the most popular

STANDARD HOURS FOR WORK STUDIED

TIME STUDY NO.	FROM STANDARD DATA	FROM TIME STUDY	INDEX OF COMPARISON OF STANDARD TIMES %
1	3.3	4.2	78.5
2	7.6	6.6	115.1
3	3.8	4.6	82.6
4	11.3	10.1	111.8
5	1.8	2.9	62.0
6	7.3	6.6	110.6
7	5.8	5.5	105.4
8	11.2	11.6	96.5
9	4.3	4.1	104.9
10	7.3	8.2	86.5
TOTALS	63.5	64.4	953.9

Average of all indexes of comparison = 95.39%

Index of comparison of totals (63.5 ÷ 64.4) = 98.6%

To adjust standard data for errors caused by averaging out data from time studies, adjust all values in standard data table by the factor 100/98.6

Source: Based on "Work Measurement in Maintenance," by J.O. Heritage, Kenneth Digney, R.E. Deem in *Factory* 113 (April 1955), copyright Morgan-Grampian, Inc., by permission.

Figure 4–4. How to Adjust a Standard-Data Table for Errors Caused by Averaging Time Studies

method at present; it will be covered in detail in chapter 6. Another method compares the number of hours actually required to carry out a work order with the estimated number of hours assigned to that task. Application of a productivity index to the individual standards is an important part of any standards program; for instance, if the productivity index is 0.94, then 6 percent must be added to each standard (assuming that the standards are based on a productivity index of 1.00).

Plants that start operations with a standards program appear to have a much greater degree of success with it than plants that introduce such a program after operations have started. The primary reason for the failure of standards programs has been failure to win acceptance of the program by

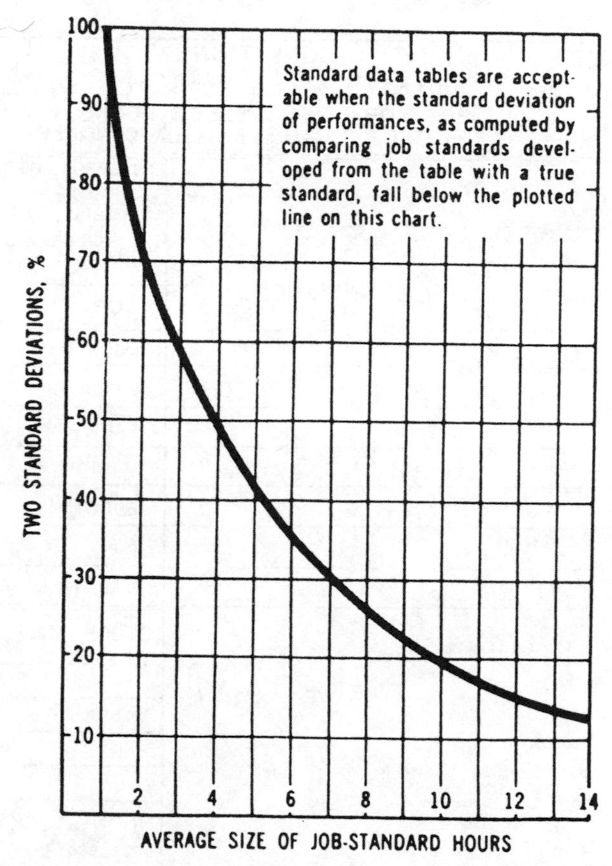

Source: Based on "Work Measurement in Maintenance," by J.O. Heritage, Kenneth Digney, R.E. Deem in *Factory* 113 (April 1955), copyright Morgan-Grampian, Inc., by permission.

Figure 4–5. Standard-Data Acceptability Limits

the individuals who must work with it. Perhaps the best method of avoiding this problem is to conduct a number of meetings with maintenance planners, schedulers, supervisors, and craftsmen. Such meetings should include an initial discussion of what standards are, with emphasis on the fact that standards of some sort have always been part of the supervisors' thinking and have always been used in the estimating process, and that, therefore, the program involves formalizing or documenting that which is known and already used in the plant.

The next topic for discussion should be the manner in which the standards are to be documented and used. It should be pointed out that standards are the basis for an estimate and that appropriate additions or subtractions can be made if there are nonstandard elements to a job. Further-

MAINTENANCE STANDARD	CONDITIONS		NO.	
COORDINATOR	☐ SHUTDOWN ☐ HOT WORK ☐ PROTECT. CLOTH.	☐ RESPIRATOR ☐ ELECT. LOCKOUT ☐ ENVIRON. CONSIDER.	AREA	DATE

DESCRIPTION OF LOCATION				EST. LABOR COST	EQUIP. NO.
				EST. MAT'L. COST	ADVANCED NOTICE
				TOTAL EST. COST	JOB DURATION
FOREMAN	SUPT.	PROCESS SUPT.	MAINT. MGR.	CRAFT	EST. MH
SEQUENCE:				MECH.	
				WELD.	
				CARP.	
				M & I	
				ELECT.	
				P. F.	
				UTILITY	

SKETCH	STOCK MAT'L	NO.
	NON-STOCK MAT'L	P.O. NO.

Figure 4–6. Typical Maintenance-Standard Form

more, it should be emphasized that standards are an ever-evolving file and that, as methods, equipment, and materials change, standards must be altered to reflect what actually occurs in the field. The meetings should terminate with a discussion of the improvement in control of scheduling and expenditures that will result from a successful standards program.

Standards programs usually are introduced into the plant by an internal group, such as industrial engineers; or they may be introduced by an outside consultant who comes into the plant and designs and sets up the program. Whether the program originates with consultants or with plant personnel, it is essential that the planners and schedulers who are to work with the standards program be involved from the very beginning. Far too many programs have failed because the plant personnel who will work with the day-to-day operation of a standards program are not involved with its development. The number of planners and schedulers required to work with the system varies considerably from plant to plant. It must be realized that those planners and schedulers have duties other than those associated with the standards system. When work is planned in considerable detail, the average planner can handle between fifteen and twenty craftsmen. When the work is planned on a more overall basis rather than a detail basis, one planner can handle between thirty and forty craftsmen.

The job of the planner is greatly facilitated by a standards program. It has been pointed out that the planner who estimates a job is, in effect, engineering that job. When he takes an appropriate standard from the file and uses it to estimate his work order, the planner is dispatching a work order that has been preengineered. Using the standards augments the planner's training in that he sees how representative work orders should be engineered.

It should be obvious at this point that clerical personnel who deal with standard data can do a much more efficient job if they are entirely familiar with the work that is done by the maintenance forces in the field. Even clerks who are newly hired should be rotated into the field periodically so that they become familiar with the tasks and jobs and the manner in which they are performed by the various craftsmen.

Much has been said about the portion of maintenance activity that should be reduced to standard data and planned. Obviously, it would be uneconomical to plan every small effort that is done by maintenance. Many practitioners indicate that a maximum of approximately 85 percent of the work orders performed should be planned and the remaining 15 percent, which normally account for very short increments of time, should not be reduced to standards. The best coverage is in central-shop work. The lowest coverage is probably in instrument maintenance because of the thousands of possible combinations of operations on a large variety of instruments and also because much of this work consists of inspection and troubleshooting.

For a program to be successful, it is necessary to audit at least some planned work orders after the work is completed to compare the actual hours spent with the planned hours. Seldom are all of the work orders audited continually; this would be too time-consuming. Normally, the auditing is done periodically, and then only on certain representative activities. If done on a regular basis, these audits will reveal obsolete or inaccurate standards. If job methods or materials have been changed, or if new equipment has been obtained, the standard must be revised. If a standard is inaccurate because the original concept of the job was in error, that standard must be revised.

The standards program may be refined by including a methods-improvement phase that continually reviews maintenance jobs and determines better ways to perform these jobs. Often, the methods group works with safety standards, which have an effect on the man-hour standards. The Occupational Safety and Health Act of 1970 necessitated review of each maintenance effort from the safety standpoint. In many cases, it was necessary to alter the number of men (therefore the number of man-hours) specified for the job and therefore to revise the standards.

To minimize maintenance costs, a methods-engineering group should continually study activities that account for a significant number of man-hours. This group should be interested in tools, materials, and equipment as well as methods; its functions should include evaluation of alternative materials, feasibility studies, and justification of additional equipment.

Establishing and Justifying Maintenance Standards[3]

Plant engineers should seriously consider the use of maintenance standards for measuring indirect work. Direct-labor and material costs make up about 52 percent of the total cost of all manufacturing industries, and indirect costs account for the remaining 48 percent. It is expected that indirect costs will increase more rapidly than direct-labor costs and that they will soon account for more than 50 percent of total cost. An appreciable part of an industrial plant's indirect costs is related to maintenance labor.

It is becoming increasingly important that a plant have a maintenance-work standards program that is applicable to this large portion of total cost. If standards are to be used for scheduling and estimating in maintenance, reliable standards that can easily be maintained at a minimum cost must be employed.

The objective of any indirect-labor standards program is to provide management with a tool that can premeasure the work done by its service work force. Using reliable standards, management can evaluate the effectiveness and improve the efficiency of indirect labor. Specifically, a well-conceived standards program increases productivity, decreases cost, and improves the maintenance effort through better methods, scheduling, and control.

Figure 4–7 shows all the information needed for establishing a maintenance standard, including the sequence of events, a sketch (when applicable), manpower and tools, materials list, and safety considerations.

A planning program to reduce indirect costs should give full attention to inspections, turnarounds, normal maintenance, and other off-line functions, including promotion of safety and coordination with other operations.

Determining the Need for Standards

Three types of input into the maintenance system help determine if a standard should be written (see figure 4–8). When a work order is initiated, the maintenance manager should question whether a standard should be created for the job. To arrive at an answer, he must consider the number of times that job is requested, the degree of criticality of the equipment, the cost of labor and materials, and the priority of the component in relation to overall production operations.

The first requirement in the preparation of a maintenance standard is to list all actions that must take place for the work order to be completed. It may be beneficial to include a sketch in the standard.

The second task is to estimate the number of crafts man-hours and the number of hours of tool use necessary to complete the sequences listed in the first section of the standard. Estimates may be obtained from the area-maintenance supervisor, from previously recorded data on similar work orders, or from other information that may have been collected by the maintenance planner.

Third, all the materials needed to take care of the work must be identified, including items that will come from the storehouse and items that must be purchased. A number in the purchase-order-number column would indicate that the material is not in the storehouse and must be purchased.

Finally, safety requirements must be covered, including such needs as gas-blanketing a pipeline in order to burn into it, gas-testing an area before welding, and providing special clothing for acid-area work.

Developing Standards

If standards are developed from such detailed studies, the initial cost will be greater, but it will be easier to keep the standards current. If some standard

MAINTENANCE STANDARD SCHEDULES

INSTALLING 8 IN. GATE VALVE IN 150 LB INSULATED LINE

Sequence

1. BUILD 25 FT SCAFFOLD
2. STRIP INSULATION
3. CUT AND BEVEL PIPE
4. INSPECT
5. FIT AND WELD 2 FLANGES
6. INSTALL VALVE
7. REINSULATE
8. INSTALL SPROCKET AND CHAIN
9. REMOVE SCAFFOLD
10.
11.
12.

Sketch

EXISTING 8" STD. WT. C.S. PIPE

8"-GA. VA.

8"-R.F., W.N. FLG.S

Manpower & Tools

Craft/Tool	Man-hours	Hours
CARPENTERS	8	
PIPEFITTERS	2	
INSULATORS	4	
WELDERS	2	
WELDING MACHINES		2

Materials List

Quantity	Item	Stock Number	P.O. Number
2	FLANGES	12-1078-5	
1	VALVE	12-8153-3	

Safety and Special Considerations

NITROGEN BLANKET LINE BEFORE CUTTING PIPE
GAS TEST BEFORE WELDING

Source: Lawrence Mann, Jr., "Establishing and Justifying Maintenance Standards," *Plant Engineering,* April 30, 1981, p. 43.

Figure 4-7. Maintenance-Standard Schedules

data are available from historical data, creating standards will be faster and less expensive, but the standards should be reviewed and modified to account for differences between previous and current work orders.

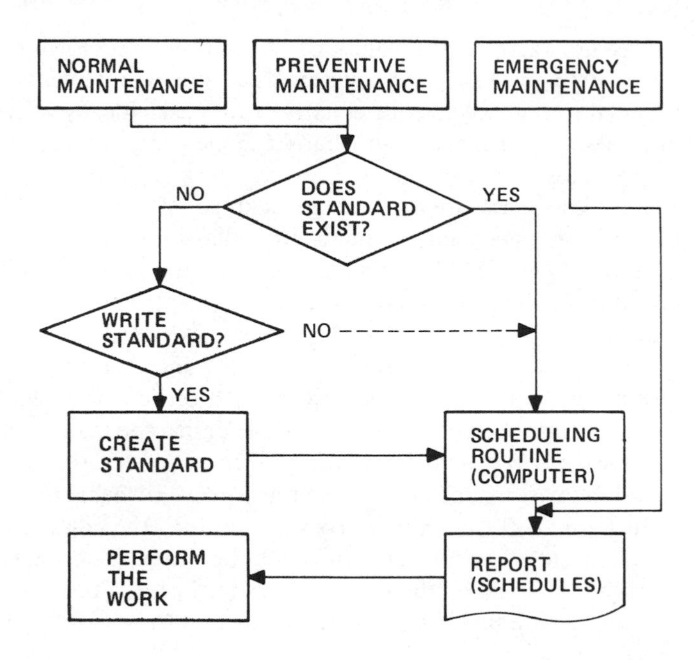

Source: Lawrence Mann, Jr., "Establishing and Justifying Maintenance Standards," *Plant Engineering,* April 30, 1981, p. 44.

Figure 4–8. Portion of a Standards Flow Chart

Once a time estimate is established, the results are transformed into a standards form (similar to some work-order forms). To convert the raw data into a standard, the worker's performance level must be determined and the raw data must be adjusted to standard time, using the formula $P = R/E$, where P is performance in minutes, R is raw data in minutes and E is efficiency percentage, expressed as a decimal.

To determine standard time, the formula $S = (P)(1 + A)$ is used, where S is standard time in minutes and A is allowance percentage, expressed as a decimal. If the efficiency of the system is 80 percent, for example, allowance is 10 percent, and the time required for a four-inch weld on a schedule 160 pipe is 25 minutes, we obtain $P = 25/0.8 = 31.25$, and $S = 31.25(1 + 0.10) \cong 34.4$ minutes.

The efficiency factor is the ability of the worker to perform relative to the average craftsman. Allowances are external time requirements—that is, time not related to the direct requirements of the task. Ideally, efficiency is significantly above 100 percent—perhaps 125 percent; 7 to 18 percent is acceptable for allowances.

Standard allowances are for the percentage of direct working time taken for foreign work elements—that is, miscellaneous duties related to the direct operation. Typical allowances include travel time for material, 5 to 15 percent; waiting for hot-work permits, 2 to 6 percent; and consulting with the operator or area supervisor, 4 to 12 percent. Personal allowances, also required on every job, average 8 percent. Allowance factors for maintenance work are generally higher than those used on repetitive work.

The unit cost and total cost can be plotted against the man-hours scheduled (see figure 4–9). When data are being collected for creating a new standard, the indirect cost will be larger than the cost of job scheduling with no standard. In figure 4–9, the curves cross at point *X*, the equal-cost point. The area to the right of *X* is the area in which the standards program begins to earn dividends; it is called the justified standard area. The cost of scheduling jobs with no planning increases, while the cost of scheduling jobs with standards decreases.

Justifying Standards

The variables used for creating standards and the functions that save money in using standards are shown in figure 4–10. To justify a maintenance standard, the maintenance manager should total all expenses required to create a standards program, the total expected cost, and compare that total with the savings expected from using the standards, the total expected savings. Current industrial practice indicates that at least a 20 percent return on investment (saving) is necessary for the activity to be financially attractive.

Example: Piping Activity Standard[4]

The following example of pipe-fitting work includes cutting, threading, and joint installation. These operations should require approximately the same standard time regardless of location. Figure 4–11 shows a typical page from a set of direct-work standard data for pipe fitting, covering the installation of flange joints. The normal minutes per joint are identified by pipe size, working height, and length of pipe. An outline of the job steps included in the data is also shown.

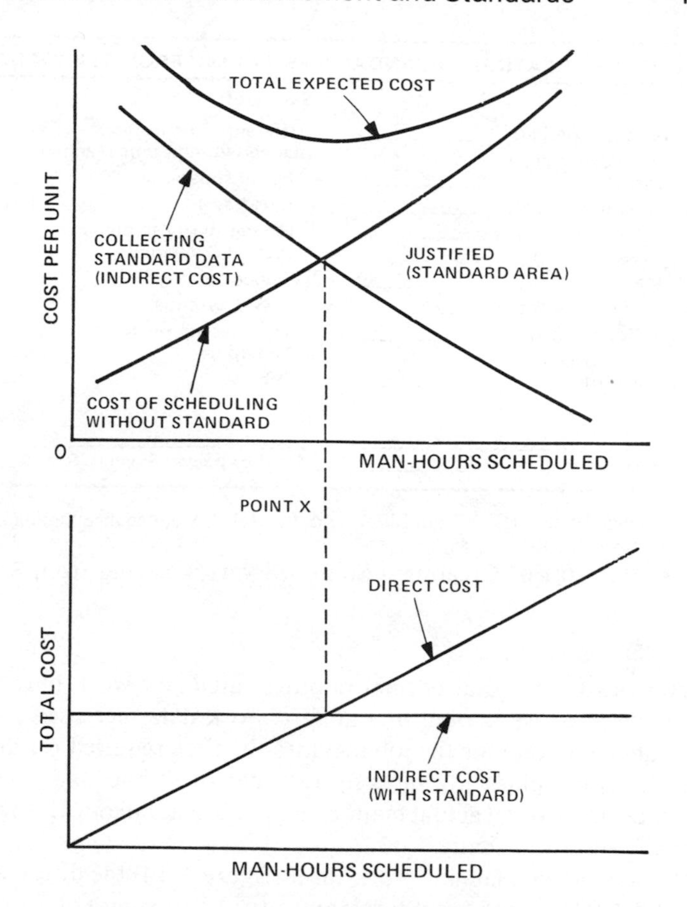

Source: Lawrence Mann, Jr., "Establishing and Justifying Maintenance Standards," *Plant Engineering,* April 30, 1981, p. 44.

Figure 4–9. Costs versus Man-Hours Scheduled

Indirect or auxiliary functions are those that are part of the job but are not apparent after the work is completed. The amount of auxiliary work is affected by the physical facilities—the plant layout, the method of material delivery, and so on.

Figure 4–12 shows typical relationships between auxiliary- and direct-work functions for pipe-fitting work performed under field conditions. It should be noted that the percentage relationship obtained by comparing

COST OF CREATING A STANDARD VS. SAVING FROM STANDARD			
Cost of:	**$**	**Saving of:**	**$**
Time to plan a job, design	______	Differentiation of time planned vs. non-time planned:	
Time to break down the job into its elements	______	Fewer supervisors	______
Applying times to elements	______	Less travel	______
Looking up past work orders	______	Having ability to predict costs and schedules	______
Applying predetermined standard data	______	Fewer backlogs	______
Applying performance rating	______	Less downtime	______
Developing allowances	______	Tools, equipment tied up less	______
Maintaining standards and keeping them current	______	Others	______
Clerical work	______		
Others	______		
Total Expected Cost	$______	**Total Expected Saving**	$______

Source: Lawrence Mann, Jr., "Establishing and Justifying Maintenance Standards," *Plant Engineering,* April 30, 1981, p. 45.

Figure 4–10. Cost of Creating a Standard versus Saving from Standard

auxiliary work to the total normal minutes (auxiliary work plus standard data) decreases as the amount of standard-work data increases.

The labor estimate for the job includes the time required for direct and auxiliary functions plus allowances for rest, personal time, and unavoidable delay. An example of an actual maintenance job evaluated with engineering standards is shown in figure 4–13.

Note that, in the planning work sheet (figure 4–14) the direct-work elements total 1,083.0 minutes. By reference to the auxiliary chart shown in figure 4–12, it was found that another 673.2 minutes were required for auxiliary work. An additional allowance of 15 percent was added to cover rest, personal time, and other unavoidable delays. Thus, the standard for the job is 33.7 hours.

The data developed with the labor standards provide performance guidelines for maintenance management. Figure 4–15 shows a typical work-order performance report. Each plant determines its own best index, but all have some common yardstick to judge results. Some of the advantages to using engineering standards for maintenance control are as follows.

Consistent Estimating. Since all the maintenance planners use the same standard data for estimating job content, the man-hours required for various maintenance jobs will be consistent. Proposed expenditures can be evaluated with some assurance of reliability. Performance by individuals or by crafts can be measured accurately and reliably.

Department Pipe Fitting

Operation Install standard flange joint Operation No. 7-2

Pipe Size	No. of Bolts	Normal Minutes Per Joint					
		Standard Pipe			XH Pipe		
		A	B	C	A	B	C
1"	4-1/2"	5.3	8.4	14.6	6.6	10.5	18.3
1-1/4	4-1/2	5.5	8.6	14.8	6.8	10.7	18.5
1-1/2	4-1/2	5.6	8.7	15.0	7.0	10.9	18.7
2	4-5/8	5.9	9.0	15.3	7.4	11.3	19.1
2-1/2	4-5/8	6.3	9.3	15.6	7.8	11.7	19.5
3	4-5/8	6.6	9.6	15.9	8.2	12.0	19.9
3-1/2	8-5/8	12.3	15.5	21.6	15.4	19.4	27.1
4	8-5/8	12.5	15.7	21.8	15.7	19.6	27.3
5	8-3/4	12.9	16.0	22.2	16.1	20.0	27.7
6	8-3/4	13.2	16.3	22.5	16.5	20.4	28.1
8	8-3/4	13.9	17.0	23.2	17.4	21.3	29.0
10	12-7/8	23.3	26.4	32.6	29.1	33.0	40.8
12	12-7/8	24.7	27.8	34.0	30.9	34.8	42.5
14	12-1	26.1	29.2	35.4	32.6	36.5	44.2
16	16-1	34.8	37.9	44.1	43.5	47.4	55.1
18	16-1-1/8	39.0	42.1	48.3	48.7	52.6	60.3
20	20-1-1/8	48.7	51.8	58.0	60.9	64.8	72.5

Note: Class A - Both working height and length of pipe under 6'.
Class B - One under 6' and the other 6' or over.
Class C - Both 6' or over.

Includes:

1. Clean flange face and install new gasket.
2. Dope bolts.
3. Install bolts and bolt up joint with wrenches.
4. Handle pipe and fittings from floor to working height by hand, rope or chain hoist, or in and out of vise.
5. Support pipe and fittings for installation with wires, blocks, etc.
6. Line up pipe and fittings.
7. Climb up and down ladders, scaffolds and equipment or crawl under machines and equipment.
8. Install pipe in hangers.
9. Move ladders, scaffolds and chain hoists, handle tools and equipment.

Source: Ralph E. Renken, "Performance Evaluation and Improvement Using Basic Maintenance Management Tools," *Techniques of Plant Engineering and Maintenance* 17, p. 148. Copyright 1966, Clapp and Poliak, Inc.

Figure 4-11. Piping Standard

Method Analysis. A job can be evaluated before it is started. Alternate methods of installation can be properly compared. Decisions can be made regarding whether jobs should be contracted, prefabricated in the shops, or performed by maintenance personnel. The need for improved facilities is easily recognized, since handicaps that exist from a physical facility standpoint must be allowed for in planning the labor content of an activity.

Operation General Auxiliary Credit Operation No. 1-2

Normal Minutes Per Job		Normal Minutes Per Job		Normal Minutes Per Job	
Standard Data	Auxiliary Work	Standard Data	Auxiliary Work	Standard Data	Auxiliary Work
1- 3	15	91- 94	160	208-212	305
4- 6	20	95- 97	165	213-217	310
7- 9	25	98-101	170	218-222	315
10-12	30	102-105	175	223-228	320
13-15	35	106-108	180	229-234	325
16-18	40	109-112	185	235-241	330
19-21	45	113-115	190	242-248	335
22-24	50	116-119	195	249-256	340
25-27	55	120-123	200	257-265	345
28-30	60	124-127	205	266-275	350
31-33	65	128-131	210	276-286	355
34-36	70	132-135	215	287-300	360
37-39	75	136-139	220	301-313	365
40-42	80	140-143	225	314-325	370
43-45	85	144-147	230	326-338	375
46-48	90	148-151	235	339-350	380
49-51	95	152-155	240	351-363	385
52-54	100	156-159	245	364-375	390
55-57	105	160-163	250	376-388	395
58-60	110	164-167	255	389-400	400
61-63	115	168-171	260	401-413	405
64-66	120	172-175	265	414-425	410
67-69	125	176-179	270	426-438	415
70-73	130	180-183	275	439-450	420
74-76	135	184-188	280	451-463	425
77-80	140	189-192	285	464-475	430
81-83	145	193-197	290	476-488	435
84-87	150	198-202	295	489-500	440
88-90	155	203-207	300		

Note: When Standatd Data normal minutes exceed 500, Auxiliary normal minutes are equal to 440 plus 40% of Standard Data normal minutes in excess of 500.

Applies to:

1. All jobs which do not require a Special Auxiliary Credit

Source: Ralph E. Renken, "Performance Evaluation and Improvement Using Basic Maintenance Management Tools," *Techniques of Plant Engineering and Maintenance* 17, p. 149. Copyright 1966, Clapp and Poliak, Inc.

Figure 4–12. Auxiliary Credit

Elimination of Delays. Since the duration of an activity can be predicted, the maintenance planner and the maintenance supervisor are aware of the starting and finishing point for each job. This improves scheduling by eliminating delays brought about by inaccurate estimating and results in reduced labor costs through improved effectiveness.

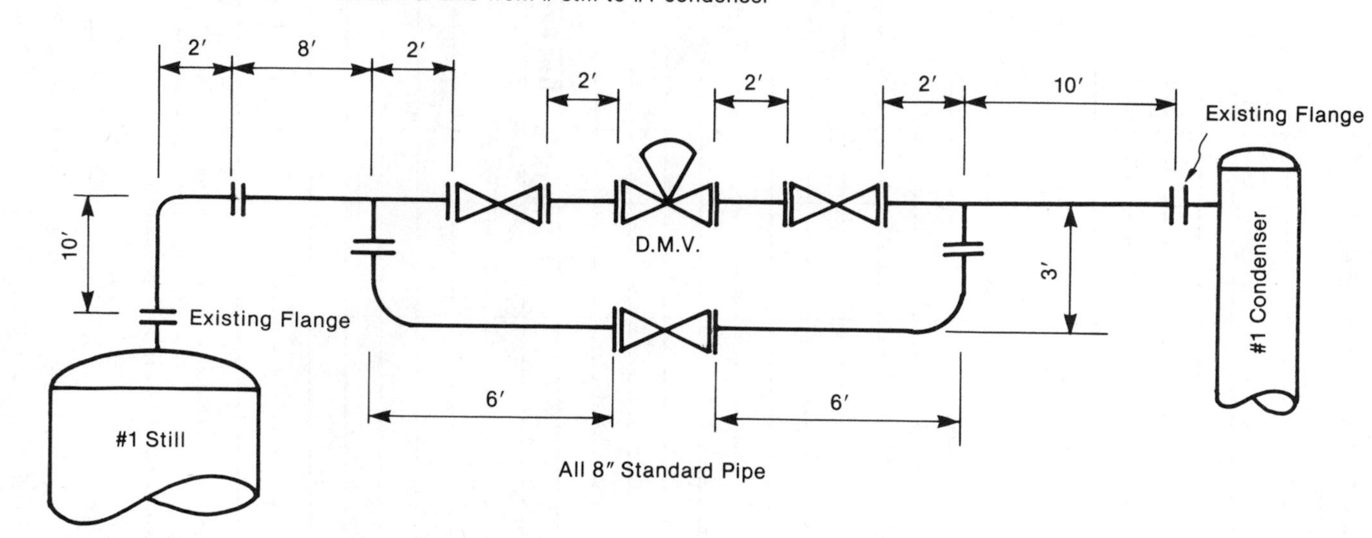

Source: Ralph E. Renken, "Performance Evaluation Using Basic Maintenance Management Tools," *Techniques of Plant Engineering and Maintenance* 17, p. 150. Copyright 1966, Clapp and Poliak, Inc.

Notes: All pipe to be machine cut and threaded in shop. Prefabricated ells to be used on 90° turns. Electric arc welding. All pipe to be installed on existing pipe racks with temporary hangers-permanent hangers to be installed when all pipe runs are completed. All work over 6′ off the working surface.

Figure 4–13. Piping Sketch

JOB: INSTALL D.M.V. STATION & LINE FROM #1 STILL TO #1 CONDENSER

PLANNING WORK SHEET

Job Description	Section and Page Number	No. of Occur.	Time Per Occur.	Total Time
Install 8" Std. Flanged Joint - B	PF 7-2	7	17.0	119.0
(See Exhibit I) - C		6	23.2	139.2
Machine Cut 8" Std. Pipe	PF 2-1	12	6.9	82.8
Machine Thread 8" Std. Pipe	PF 3-1	16	8.8	140.8
Install Flange on 8" Std. Pipe with Pipe Machine	PF 5-1	16	14.2	227.2
Cut Hole in 8" Std. Pipe for Saddle Joint	W 2-11	2	14.0	28.0
Cut Saddle on 8" Std. Pipe	W 2-10	2	9.7	19.4
Arc Weld Saddle Joint in 8" Std. Pipe	W 5-11	2	40.6	81.2
Chamfer 8" Std. Pipe to Weld Prefab Ells	W 2-9	6	6.6	39.6
Arc Weld 8" - 90° Prefab Ells to 8" Std. Pipe	W 5-11	6	34.3	205.8
Total Direct Work Minutes				1083.0
Total Auxiliary Minutes				673.2
Sub-Total				1756.2
15% Allowance for Rest, Personal and				
Miscellaneous Delay				263.4
Total Std. Minutes				2019.6
Total Std. Hours				33.7

Source: Ralph E. Renken, "Performance Evaluation Using Basic Maintenance Management Tools," *Techniques of Plant Engineering and Maintenance* 17, p. 150. Copyright 1966, Clapp and Poliak, Inc.

Figure 4–14. Planning Work Sheet

Coordination of Crafts. Since the craft content of each job is established, it is possible to coordinate the start and end times for each craft so that craftsmen do not arrive too early or too late at the job site.

FOR WEEK BEGINNING ____________________

Priority	W.O. Number	Estimated Manhours	Actual Manhours	Performance Index
1	16234	58	56	1.04
1	16248	12	14	.86
2	16198	4	4	1.00
2	16310	16	16	1.00
2	16280	12	14	.86
1	17272	50	52	.96
2	16430	48	46	1.04
1	16600	36	30	1.20
1	16602	24	26	.92
2	16593	20	18	1.11
2	16482	26	30	.87
1	16486	18	16	1.13
3	16399	12	10	1.20
1	16220	8	6	1.33
1	16319	8	8	1.00
1	16311	6	6	1.00
2	16720	12	12	1.00
2	16718	4	5	.80
1	16809	24	20	1.20
3	16306	4	4	1.00
3	16605	6	8	.75
1	16478	40	36	1.11
3	16510	42	40	1.05

Source: Ralph E. Renken, "Performance Evaluation and Improvement Using Basic Maintenance Management Tools," *Techniques of Plant Engineering and Maintenance* 17, p. 148. Copyright 1966, Clapp and Poliak, Inc.

Figure 4–15. Work-Order Performance Report

Improved Supervisory Controls. Any improvement program should permit more direct supervision on each job through elimination of nonsupervisory duties. In addition, if each maintenance job is consistently estimated, supervisors have an additional control tool to use in directing personnel. Accountability for improper performance can be established if realistic labor estimates are provided.

Training. The planners receive invaluable training in the techniques of supervision and management as preparation for future promotions. Also, inadequate training of craftsmen is brought to light through the application of proper job-content estimates.

Materials Control. Material costs are assembled in addition to labor costs on each job. Since a more detailed material list is possible when the job is planned in detail, the maintenance planner, in addition to controlling labor content, contributes to the control of the material.

Work-Backlog Control. Since maintenance work that can be planned in advance is estimated with regard to labor content, the work backlog for each craft can be controlled. In many cases, the work backlog is a major determining factor in deciding whether to contract or to add to or subtract from the work force.

Controlled Manning. With proper estimating data available, supervisors and management can more readily handle fluctuations in maintenance requirements. Personnel transfers or replacements and the volume of work done by outside contractors are more easily managed.

Cost Control. The result of the program is better maintenance-cost control. When informed of actual maintenance costs and standard maintenance cost, management has a basis for accurately evaluating, forecasting, and controlling maintenance expenditures.

Universal Maintenance Standards[5]

Universal maintenance standards (UMS) was developed specifically to solve the work-measurement problems of nonrepetitive jobs in maintenance operations.

Setting accurate maintenance standards with time study is too costly to be practical, requiring almost a one-to-one ratio of time-study men to maintenance men. Standard data or time formulas, based on time study, offer a better solution. This approach has been known for years but has not been used extensively because of the high cost involved in development and application. More recently, the high cost of maintenance operations has resulted in rendering maintenance standards economically feasible.

The techniques of time study, standard data, and historical records have been used in an attempt to develop accurate standards to represent more exactly the time required to do a given job. However, it is not practical to expect every maintenance worker to do a given job with exactly the same motion pattern and in exactly the same time. Therefore, it is difficult to set a standard *in advance* that will be exact for each maintenance job. It is more feasible to set a standard based upon a range of time. Whereas most maintenance men will not accept, as a fact, that a standard for replacing a valve in a pipeline, for example, is 28½ minutes, they will agree that such a job can be performed in, say, between 20 and 40 minutes. Thus one can follow this principle to set a reliable standard *in advance* based upon sound engineering data.

Once this concept is understood, the basis for a practical and workable maintenance work-measurement program using UMS exists. Its introduction into a plant is not difficult if the following approach is used.

1. Develop accurate time formulas for every type of work performed by the maintenance department.
2. Establish standard work groupings.
3. Establish bench-mark jobs.

Although it is impractical to strive for exactness in the standards applied to individual jobs, it is important to have accurate basic data from which to work. Time formulas are, therefore, developed for most types of work the maintenance department is likely to do.

Fortunately, developments in the field of predetermined-motion-time systems, such as methods-time measurements, have made it possible to establish standards accurately and rapidly, not only for direct labor, but also for indirect labor.

For example, it takes several man-days of time study to establish data for changing the mechanical seals in a process pump. A competent, predetermined-time-system analyst can get the same results in about four hours.

In addition to the technique of predetermined times in the development of maintenance time formulas, the technique of work sampling may be used to determine allowances. The time-study technique is used to time machine or process times. All these techniques are essential in the management and control of maintenance costs.

If used for the direct establishment of each standard, time formulas, however, could entail too high an application cost to be used directly for establishing time standards for each maintenance job performed in a plant. In some cases, there are just too many standards to set.

Standard Work Groupings

Studies of maintenance work almost invariably show that about 80 percent of the maintenance jobs require less than eight hours to perform. These numerous short jobs cause the real standards-setting problem. One answer to this problem is to develop what may be called standard work groupings. Using these groupings, an analyst can establish a time standard in about two minutes, on the average.

The concept of standard work groupings may be seen in figure 4–16.

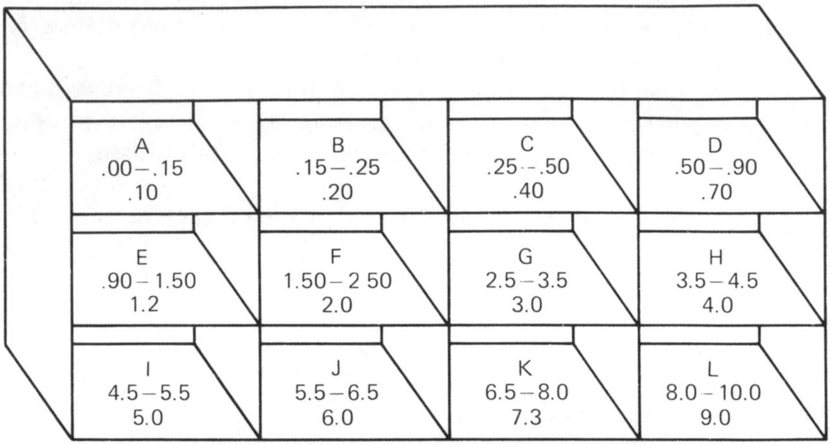

A
.00–.15
.10

B
.15–.25
.20

C
.25–.50
.40

D
.50–.90
.70

E
.90–1.50
1.2

F
1.50–2 50
2.0

G
2.5–3.5
3.0

H
3.5–4.5
4.0

I
4.5–5.5
5.0

J
5.5–6.5
6.0

K
6.5–8.0
7.3

L
8.0–10.0
9.0

Source: From H.B. Maynard, *Industrial Engineering Handbook,* 2nd Ed. Copyright 1963 by the McGraw Hill Book Co.; used with permission of McGraw-Hill Book Company.

Figure 4–16. Standard Work Groupings

The sketch shows a number of pigeonholes, A to L, inclusive. Each pigeonhole represents a range of time. The total range covered by all the pigeonholes in this illustration is from 0.00 to 10.00 hours. Pigeonhole A is for the range from 0.00 to 0.15 hour. Pigeonhole D is for the range from 0.50 to 0.900 hour, and so on.

The job of the analyst, in essence, is to take work orders and place them in the proper pigeonhole. He does not attempt to say exactly how long it will take to do a given job. He merely says that it will take, for example, from 0.50 to 0.90 hour to do a particular job, then places the order in Pigeonhole D. He does not actually do this, of course, but he accomplishes the same thing when he decides that the job is a class D job.

It is impractical to assign a range of time as a standard, so he uses the average of the range. The time values he uses are shown below the pigeonholes in the future. For a class D job, the time value assigned would be 0.70 hour.

The pigeonholes illustrated in figure 4–16 are those which were established for a specific installation. The classes and time ranges can be different for other installations because of different conditions. The principle followed in establishing standard work groupings is that the range should be sufficiently long so that it will not be difficult for the analyst to place a given job in its proper classification, and yet short enough so that this method of measuring performance is accurate over a week's period. That is, the total time allowed for all jobs performed by a craft group over a week's period should be less than ±5 percent of the total time allowed if each job were accurately measured.

Bench-Mark Jobs

With standard work groupings established, the next step is to establish a number of bench-mark jobs to guide the analyst in slotting each job in the correct pigeonhole. A typical example of bench-mark jobs for the millwright craft is shown in figure 4–17. The sheet covers some typical pump-repair jobs performed by the millwrights in standard work groupings C, D, E, F, and G. Figure 4–17 is one page from a complete application manual. Additional jobs in these and other groups, H, I, and so on, are on other pages of the manual and cover a time range up to 32 hours. In most instances, manuals are prepared for, and limited to, a certain type of equipment, for example, repairs on a given machine, or spray painting, or instrument repair, or pump repair.

Obviously we cannot include in a manual of this sort every object that millwrights, electricians, pipefitters, and so forth, may be called upon to repair. We can, however, indicate typical jobs, as we have done in this case. By using these typical jobs as points of reference, or bench marks, a good analyst can do a remarkably fine job of classifying all other jobs into their proper work groups. In the checks that we have made, we seldom find that an analyst misses the true work group, and when he does, it is by one group, one side or the other. The number of bench-mark jobs vary up to 300 or more for each craft.

Here is the way the spread sheet is used. Assume that an analyst is using the sheet to help him set standards on repair work. He receives a work order calling for replacing an impeller in a medium-sized horizontal split pump. He looks at the sheet and sees that this particular job is listed as a bench-mark job. It is bench-mark job 0290-113 in group G. So the job belongs in the G pigeonhole and will receive a time standard of 3.0 hours.

Now assume that the next work order calls for making an installation or a repair to a piece of equipment for which there is no bench-mark job on the sheet. The analyst then considers the key features of the job to be done and looks for a bench-mark job with similar features. First, he considers the type of repairs and the size of the repair job to be done. These affect not only repairing time but also preparation time as well. He then looks for a bench-mark job where the type and size of repairs are about the same as the one to be done, and finally, he classifies the repair job accordingly.

Actually, it takes more time to describe the mental process that the analyst goes through in establishing a standard than it takes the analyst to come up with the answer. In actual practice, he will usually be able to classify similar items from memory after he has gained some experience in working with the data, and then he will not have to refer to the sheets of bench-mark jobs in many cases.

Now, what about the bench-mark jobs themselves? How are the standards for them established? This is where the time formulas make their con-

CODE 0295
CRAFT Millwright

Task Area: Pumps				
GROUP C (0.25) 0.4 (0.50)	*GROUP D* (0.50) 0.7 (0.90)	*GROUP E* (0.90) 1.2 (1.5)	*GROUP F* (1.5) 2.0 (2.5)	*GROUP G* (2.5) 3.0 (3.5)
0290-70 Impeller, medium diagonal split pump, remove and replace from shaft (2 men)				0290-113 Impeller, medium horizontal split pump, remove and replace from shaft (2 men)
0290-86 Impeller, small horizontal split, remove and replace from shaft		0290-195 Water well, remove and replace 1 stage (2 men)		
0290-72 Impeller, medium vertical split, remove and replace from shaft (2 men)		0290-194 Water well, pull additional 10-foot section of pipe and shaft (3 men)	0290-191 Pump, water well disassemble and assemble 10-foot section of discharge and oil pipe and shaft (3 men)	0290-193 Water well, pull 1st section of pipe and shaft (3 men)
0290-61 Motor and pump, medium, realign 0290-65 Motor and pump, large, realign	0290-91 Rotating assembly small vertical split, disassemble and assemble		0290-81 Rotating assembly medium diagonal split, disassemble and assemble (2 men)	0290-87 Rotating assembly medium horizontal split, dis assemble and assemble (2 men)
	0290-82 Rotating assembly small diagonal split, disassemble and assemble	0290-47 Rotating assembly small diagonal split, remove and replace	0290-94 Rotating assembly medium vertical split, disassemble and assemble	
0290-86 Rotating assembly small horizontal split, disassemble and assemble	0290-59 Rotating assembly medium vertical split, belt drive, remove and replace (2 men)	0290-49 Rotating assembly small horizontal split, remove and replace	0290-60 Rotating assembly medium diagonal split, remove and replace (2 men)	
		0290-57 Rotating assembly medium vertical split, coupling drive, remove and replace (2 men)	0290-62 Rotating assembly medium horizontal split, remove and replace (2 men)	0290-122 Rotating assembly medium horizontal split, obstructed location, remove and replace (2 men)
0290-50 Pump, small vertical split, remove and replace		0290-120 Rotating assembly small horizontal split, obstructed location, remove and replace	0290-119 Rotating assembly small diagonal split, obstructed location, remove and replace	0290-121 Rotating assembly medium diagonal split, obstructed location, remove and replace (2 men)

Source: H.B. Maynard, *Industrial Engineering Handbook*, 2nd Ed. Copyright 1963 by the McGraw Hill Book Co.; used with permission of McGraw-Hill Book Company.

Figure 4-17. Typical Spread Sheet

tribution. Time formulas and standard data are developed in the conventional way and are then used for establishing the times for the bench-mark jobs.

Three other points to be considered in compiling UMS standards are job preparation time, travel time, and allowances.

Job-preparation times are normally covered by one or two time formulas allowing time for such items as receiving instructions, gathering tools and equipment, preparation at job site, collecting and cleaning tools, and cleaning the job site.

The time for traveling from the maintenance shops to the job site is also covered by time formulas. Generally, one travel-time formula is good for all crafts. In most cases, the travel time will be condensed to one table. It provides the time required for a round trip to the various locations. On larger installations, the area is usually divided into zones, and round-trip times are computed for each zone.

The third item, allowances, is comparable to the allowance percentage normally used in establishing conventional time standards, but it does have some features peculiar to maintenance work. The allowance factor covers many types of delays and activities. They include personal delays, including fatigue, unavoidable delays, balancing delays, and planning time. It is also necessary to consider additional variables such as differences in crafts and differences in complexity of work within crafts.

Personal delays include the usual ones, such as getting a drink of water. Fatigue can be included in this same category or listed as a separate item.

Unavoidable delays include such things as talking to a supervisor or waiting for a hot-work permit.

Balancing delays are seldom encountered on repetitive work but are quite common to maintenance work. A balancing delay occurs whenever the sequence of operations is such that one man of a crew is delayed from performing useful work until other members of the crew have completed their phase of the operation. A pipefitter, for example, experiences a balancing delay while he waits for his helper to move a piece of pipe into position.

Planning is normally an incidental part of most work, but on maintenance work it can account for a good portion of the total working time.

Because of the difficulties experienced in trying to measure planning time on each individual job as it is required, it is measured by craft and the general complexity of the job within a craft, and it is included in the allowance percentage.

Since the allowance factor for maintenance work includes the two items, balancing delay and planning, not normally encountered in production work, the allowance factors for maintenance work are generally somewhat higher than those used on repetitive work.

Experienced Personnel Necessary

In order to apply the UMS procedure, skilled, practical craftsmen should be selected as applicators. They should be men who have performed many of the maintenance jobs themselves.

Consequently, when a work order is received, the analyst will be familiar with the job or will have sufficient knowledge about it so that he will experience little difficulty in determining what work is required. With this knowledge, and a complete understanding of the bench-mark jobs, the analyst will be able to perform the slotting procedure with little difficulty. Where suitable bench-mark jobs do not exist, his familiarity with the work called for on the work order and his knowledge of the data will permit him to develop a new bench-mark job. In fact, after a comparatively short period of time each analyst will have sufficient jobs set up to cover practically any new job that might come along.

In one installation, for example, an electric-shop analyst received a major work order. He broke this order down into smaller tasks and bench-marked each of these tasks. The total time allowed was 1780 hours. The total time taken was 1720 hours. This same analyst now processes about 2500 work orders each month.

Now, what about the problems that may arise, and the steps that can be taken to minimize them? Let us use a specific installation to describe briefly a procedure that produced excellent results.

Installing the UMS Procedure

The first major and positive step taken by Company X after the decision to proceed with a UMS installation was made in the area of communications. All management people in manufacturing, engineering, industrial relations, and the like were thoroughly informed about this program. Their active cooperation and support was obtained.

The line of communication was then extended to all maintenance foremen, supervisors, and craftsmen. A series of group meetings was held at various times over a period of several days to permit everyone to participate. The president, executive vice-president, and several other members of their executive committee participated in every session. This permitted explanation of the complete program to everyone who would be affected by it, and it also allowed opportunity to answer all questions and to demonstrate beyond a doubt that the program for better maintenance had the full support of management. Further details were provided to the craftsmen by the team members as they began their work in each shop.

A series of training sessions was held for the maintenance foremen and supervisors over a period of several months. These sessions were in the nature of both appreciation training and practical application. All aspects relative to the program were explored with them.

The first problem that arose had to do with the work-order system. Company X agreed that no maintenance work should be performed without a written work order. They had used both verbal and written work orders in the past but did not have an established procedure. Consequently, when the analyst in the electrical shop, where the program originated, began his application of standards to the work orders, he found considerable information missing.

For example, the analyst received one work order with the description "repair switch." Past practice had been for the foreman or one or two craftsmen (the "buddy system") to follow up on such an order to determine precisely the nature of the work, and only then to go about making the repairs. The solution of this was to designate certain people in the company as having the authority to prepare work orders. In cooperation with manufacturing, these men were then trained in how to prepare work orders. The result was that, when these orders came through to the analyst, the information required to do the work and set the standard was well spelled out.

This procedure eliminated time that was previously unutilized and at the same time paved the way for the solution of the next problem—that of planning and scheduling maintenance work. Because standards were applied to the work orders coming through as well as to the existing backlog orders, all the work was specified and could be, and was, properly scheduled. As new orders came through, they were placed into the schedule according to their priority.

At this time, however, emergency work orders were coming into the analyst over the phone at a fairly high rate. The analyst wrote up the appropriate order, passed it on to the foreman, and the work was accomplished. These emergency orders disrupted the schedule and/or required that too much open time be left in the schedule.

Company X solved this problem by making improvements and refinements in their preventive-maintenance program. The nature of the emergency work was analyzed carefully and preventive-maintenance practices were instituted to reduce these emergencies. Gradually, the concept of breakdown maintenance was changed to preventive maintenance. The setting of standards for known preventive-maintenance operations was simpler. Slowly the percent of the fixed portion of the maintenance schedule increased, and the variable portion decreased.

The installation of the standards on a measured-daywork basis at Company X proceeded well in each of the craft shops. Toward the latter stages of the development and installation, management decided to introduce a wage-incentive plan for the shops. The standard-hour plan was selected, with payment on a group basis and performance of the group measured over a week's period. [Figure 4–18] is part of the performance chart of the electrical shop in Company X. An interesting fact here is one that has occurred in other installations; namely, an increase in performance materialized during the data-development period, after the work sampling and before the installation of standards. That is attributable to the awareness of both the supervisor and the craftsmen that their work was being measured.

Figure 4–18 shows the situation in each of the major shops, before the start of the program. In the pipe shop, for example, approximately 2.3 hours were required to perform one standard hour of work. Three months after the start of the data development by one of the teams, the dry run of the standards in the pipe shop was initiated. Initial coverage of pipe-shop hours was about 5 percent. This percentage increased steadily to the point where Company X had 98 percent of its total pipe-shop manhours covered by engineered standards.

In like manner, the preinstallation ratio of 2.3 actual hours for each standard hour in the pipe shop decreased gradually over a period of four to six months as controls were applied. At the end of this period, effectiveness

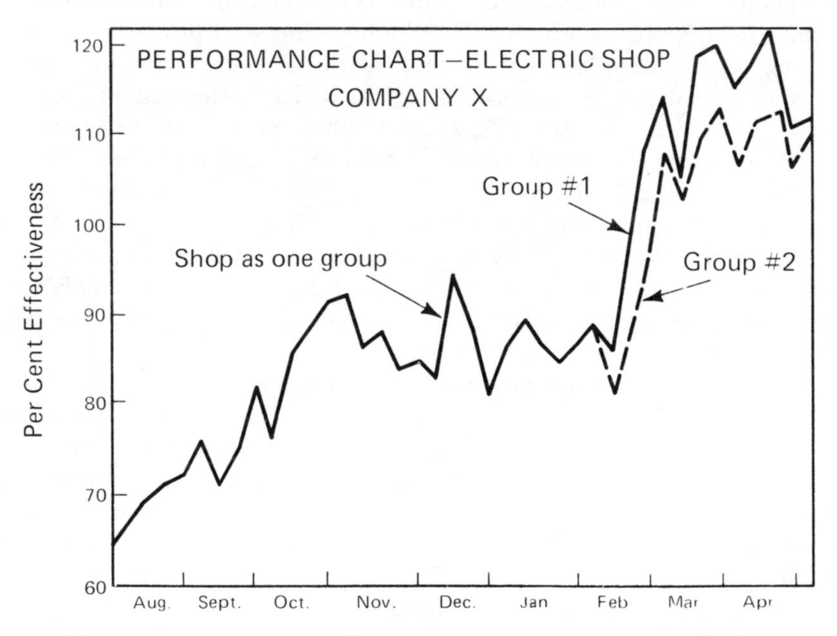

Source: H.B. Maynard, *Industrial Engineering Handbook,* 2nd Ed. Copyright 1963 by the McGraw Hill Book Co.; used with permission of McGraw-Hill Book Company.

Figure 4–18. Performance Chart—Company X Electrical Shop

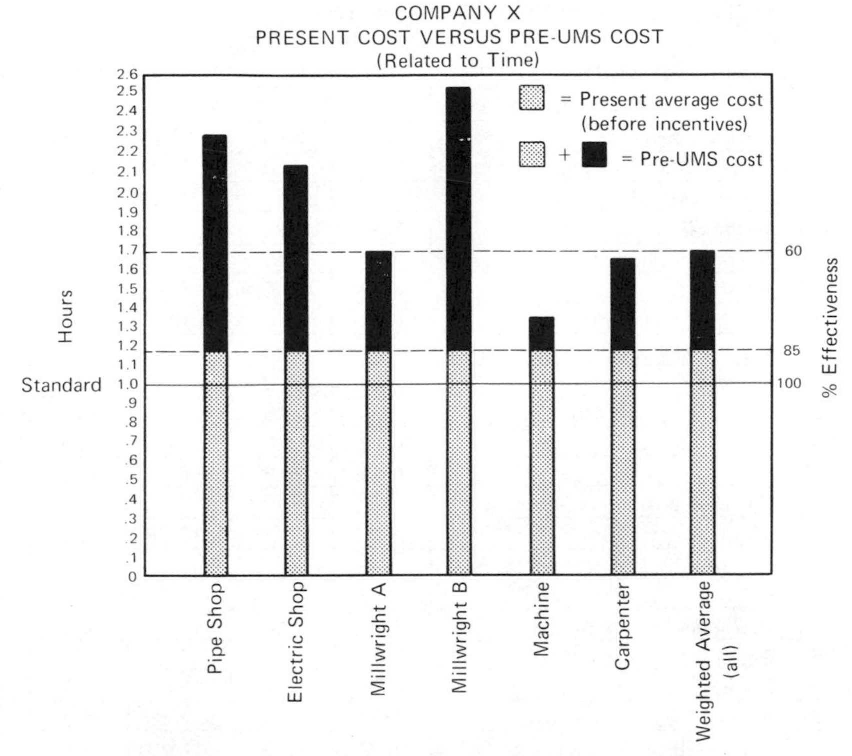

Source: H.B. Maynard, *Industrial Engineering Handbook,* 2nd Ed. Copyright 1963 by the McGraw Hill Book Co.; used with permission of McGraw-Hill Book Company.

Figure 4–19. Company X—Present Cost versus pre-UMS Cost

was ranging from 80 to 90 percent; or put another way, one standard hour of pipefitting work was now being accomplished in somewhat less than 1.2 hours. This is a savings of approximately 50 percent.

The average increase in production was 40 percent and savings 30 percent. As can be seen in [figure 4–19], some of the shops realized savings considerably in excess of this amount.

[Figure 4–20] shows a simple comparison of like jobs in a millwright shop. These tangible results indicate what can be attained with work measurement applied to maintenance activities with its resulting control.

Other benefits realized by Company X include improved employee morale, foremen who have advanced to maintenance managers, a minimum of delay in handling emergency work, less downtime of production equipment, an effective preventive-maintenance system, a practical work-order system, and effective planning and scheduling.

MILLWRIGHT SHOP PREPARED BY John Doe

#	JOB DESCRIPTION	Date	Hours	Date	Hours
1	Cover cylinder X machine two				
	wires and sew ends	7/14	53.0	10/13	19.0
2	Cover extractor cylinder	2/16	64.0	9/10	11.9
3	Cover spare cylinder for Y				
	machine and sew ends	7/12	16.4	11/12	15.1
6	Pull Jordan, fill with new plug				
	and shell	9/22	39.6	10/15	32.5
8	Seam felt low mark	6/28	6.6	10/14	6.0
11	Top Primary roll Journal				
	replaced	7/26	31.3	8/29	12.1
12	Fill 1/4 in. X 66 Jordan plug	3/30	28.5	7/15	18.8
13	Grind in Jordan	8/22	6.8	11/19	6.4
17	Felt roll journals	6/11	26.5	9/8	21.3
22	Change inboard bearing, stock				
	pump	8/7	3.8	12/20	3.6
23	Change outboard bearing, stock				
	pump	12/2	1.2	12/20	.8
30	Repair agiflow pump	10/16	14.6	10/23	16.4
35	Replace handle in sledge				
	hammer	8/22	1.1	10/9	.6
37	Replace 2 vanes on pulper	9/29	1.8	9/11	1.2
39	Replace slitter roll hournals	7/17	45.5	11/29	43.5

Source: H.B. Maynard, *Industrial Engineering Handbook,* 2nd Ed. Copyright 1963 by the McGraw Hill Book Co.; used with permission of McGraw-Hill Book Company.

Figure 4–20. UMS Job-Comparison Sheet

Example: Using Universal Maintenance Standards[6]

Universal Maintenance Standards can be purchased from the Office of Technical Services, U.S. Department of Commerce, Washington D.C. These standards, developed by the U.S. Navy Bureau of Yards and Docks, cover over 85 percent of the maintenance jobs in most plants. The standard data cover craftwork from electronics to carpentry. These universal standards are possible because pipefitting, electrical work, and so on, are basically the same in a Navy facility as in any other plant.

Checking Out the Job

The planner must check a job out before he can plan it accurately. He will be guessing unless he looks at the job in the field. Before the planner can check a job, a work order must be written. In figure 4–21, the pipe fitters are the major craft, and the pipe welders, pipe coverers, and laborers support the effort. While the planner is checking the job out, he should fill in any information that is missing from the order. If necessary, he must expand the definition of the work and note changes on the work order.

Locating Material

Locating materials and tools is one of the planner's most important responsibilities. A planner should never take it for granted that a critical item is on hand. An extensive search is sometimes necessary to locate a specific item, and it is more efficient to have the planner, rather than the craftsmen, do the searching. When an item is located, it should be tagged with work-order numbers or otherwise designated for the specific job.

If materials or tools have to be purchased, this should be done before the planner releases the work order to the shop; if necessary, the planner may hold the work order until the material arrives. The shop does not want the work order until the work can be done.

Preparing Sketches

Often a job cannot be described with words, but a simple sketch will clarify what is to be done. It is recommended that isometric sketches be used where possible. This type of drawing gives a clearer picture of the job than do the conventional one-, two-, or three-view projections. If isometric sketches are not made with care, however, they can be misleading.

A sketch for the job of fabricating and installing six-inch stainless-steel pipe for a wastewater still appears in figure 4–22. The sketch clears up questions left by the wording of the work order. The following can be seen from the sketch:

1. The laborers must dig and backfill the ditch.
2. The pipe coverers must remove the covering and reinstall it.
3. The pipe welders must fabricate the new pipe and modify the existing pipe.
4. The pipe fitters must remove the existing pipe and install the new segments.

It would have been impossible for a planner to plan this job if he had not checked it out first. By the same token, it would be impossible for the craftsmen to do the job without the sketch unless one of them checked it out. Obviously, having the craftsmen check out the job is not desirable. The checking must be done by the planner before he releases the work order.

HERCULES POWDER COMPANY
INCORPORATED
WORK ORDER

REF.	PRIORITY	CROSS ORDER NO.	WORK ORDER NO.
54679	3		15039

REQUESTED BY: CARTWRIGHT — PHONE: 317

BLDG. NAME: EXTRACTOR — BLDG. NO.: 2006

CODE: 1050/70

DATE ISSUED: 10-28- — DATE REQ.: 11-14-

EQUIP. NAME: WASTE WATER — EQUIP. NO.: STILL LINE

DESCRIPTION AND LOCATION OF WORK: PLEASE EXTEND WASTE WATER STILL LINE TO THE MIDDLE OF THE MAIN DITCH

TYPE WORK

- ☒ MAINT.
- ☐ IMPROVE
- ☐ TANK CAR
- ☐ SAFETY
- ☐ PROJECT

MAINTENANCE USE ONLY

CRAFT	STD.	CRAFT	STD.
BOILER		✓ PIPEFIT	
CARP		(PIPE WELD)	
ELEC		REFRIG	
INSTR		RIGGER	
(LABOR)		OILER	
MACH		RAILROAD LABOR	
MILLWR			
PAINT			
(PIPE COVER)			

JUSTIFICATION

ACTUAL COST
HOURS ______
MAT'L ______
TOTAL ______

DATE COMPLETED

FOREMAN

OPERATING APPROVAL

MAINTENANCE APPROVAL

Source: Gordon W. Smith, "Using Universal Maintenance Standards," *Techniques of Plant Engineering and Maintenance* 20:159–166. Copyright 1969, Clapp and Poliak, Inc.

Figure 4–21. Work Order 126

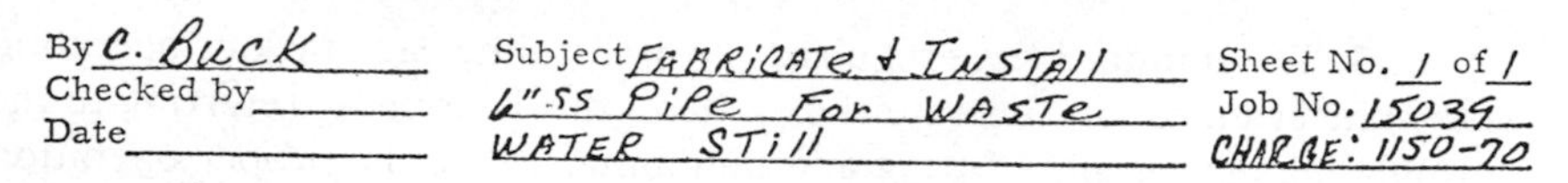

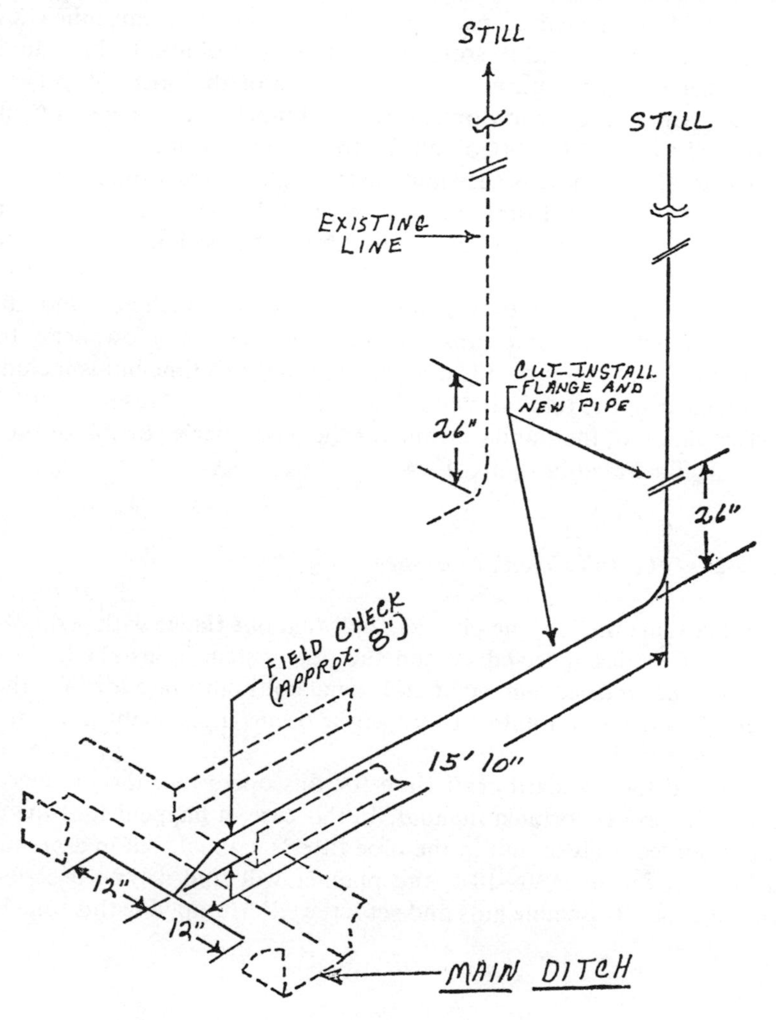

Source: Gordon W. Smith, "Using Universal Maintenance Standards," *Techniques of Plant Engineering and Maintenance* 20:159–166. Copyright 1969, Clapp and Poliak, Inc.

Figure 4–22. Piping Sketch

Assigning Standard Time

On the job-calculation sheet, the planner first lists brief descriptions of the operations that make up the entire job. One such task is shown in A in figure 4–23. The standard time required for performing each task or operation is then determined. In the case shown here, this is done by referring to the formula manual on pipe fitting and plumbing, where the proper operation is PWQ-3-II (illustrated on page 11 of the manual). The unit time shown is entered in Column C and the reference number in Column B. In Column D, the planner enters the number of occurrences of the operation. Then the unit time (Column C) is multiplied by the number of occurrences (Column D) to yield the craft time (Column E) for the operation.

The total craft time is the summation of all the craft times for the job. To obtain the standard time, allowances, travel time, and job-preparation time must be added to the total craft time. This is done by using factor tables.

Column F is used for time estimates of tasks for which no standard data exist. This estimated time must include any time for allowances, travel, preparation, and cleanup. It is not added to the craft time but is included as an increment in the total allowed time.

Derivation of the standard time for the pipe fitter's portion of the sample job is illustrated in figure 4–24.

Analysis of the Job Calculation Sheet

After analyzing the job, the planner finds that one flange with eight ¾-inch bolts must be disconnected. When the new system is installed, the eight bolts must be replaced, and eight additional bolts must be added for the new flange. This makes a total of twenty-four occurrences: eight bolts out and in, and eight bolts in.

To find the standard craft time for this operation, the planner must consult the proper formula manual. In this case, it happens that the information needed is given, not in the pipe fitter's manual, but in a manual on machine repair. In PWN-10-I, the planner will find a time described as "assemble or disassemble nuts and set screws." Multiplying this time by the

JOB CALCULATION SHEET

Reference	Task Description	Unit Hours	Occurrence	Craft Time	Estimated Time
PWQ-3-II	Measure and mark pipe with exact measurements	0.0495	3	0.148	
(B)	(A)	(C)	(D)	(E)	(F)

Source: Gordon W. Smith,[a] Using Universal Maintenance Standards," *Techniques of Plant Engineering and Maintenance,* 20:159–166. Copyright 1969, Clapp and Poliak, Inc.

Figure 4–23. Partial Job-Calculation Sheet

Hercules Powder Company | Industrial Engineering Department

JOB CALCULATION SHEET

Job Reference No: 15039

Work Center: PIPEFITTER | Date: 10-29-

Job description: EXTEND WASTE WATER STILL LINE TO MIDDLE OF MAIN DITCH AT EXTRACTOR HOUSE

Reference	Task Description	Unit Hours	Occur-rence	Craft time	Estimated time
PWN-10-I	REMOVE AND REINSTALL 3/4" BOLTS	.016	24	.384	①
PWN-10-II	REMOVE AND REINSTALL GASKETS	.012	3	.036	②
PWN-10-II	REMOVE AND REINSTALL PLATE OVER DITCH	.030	8	.240	③
PWN-10-II	POSITION PIPE	.148	2	.296	④
PWQ-4-E-F-G	DOPE BOLTS	.003	16	.048	⑤
PWN-8-P-E-Z-C1-B1	CLEAN FACE OF FLANGES	.015	4	.060	⑥
PWQ-3-M-Q	CLIMB UP AND DOWN AND MOVE LADDER	.013	2	.026	⑦
PWA-5	MATERIAL HANDLING PER CWT	.110	3	.330	⑧
PWA-5	MATERIAL HANDLING PER ARMLOAD	.035	3	.105	⑨
HPC-TR	EXTRA TRAVEL TO BRING PIPE TO SHOP (TWO MEN ONE WAY)			⑩	.250
	EXTRA TRAVEL				.250
	ADD FACTOR				1.040

Multiplying factor: 1.56
Add Factor per man: .52
Travel Zone: 2
No. of craftsmen: 2

	Craft time	Estimated time
Totals	1.525	2.379
Standard Hours Calculated		3.669
Standard Hours Allowed		3.75

Source: Gordon W. Smith, "Using Universal Maintenance Standards," *Techniques of Plant Engineering and Maintenance,* 20:159–166, copyright 1969, Clapp and Poliak, Inc.

Figure 4–24. Job-Calculation Sheet

number of occurrences (twenty-four) gives the craft time for removing and installing bolts.

One old gasket must be removed from the pipe system, and when the new pipe system goes back in, two new gaskets (one for each flange) must be installed. Here again PWN-10 must be consulted. This standard time will be found under "remove or install machine parts or components" (in the small, close-fit category).

Before the new pipe can be installed, the cover plates over the ditch must be removed. Because there are four plates to be removed and replaced, there are eight occurrences. The standard time is obtained from PWN-10-II.

Large pieces of pipe must be positioned before they can be assembled. PWN-10-II gives the standard time for positioning large objects, such as 6-inch pipe segments. In this case, two pieces of pipe must be positioned.

Sometimes, the operation to be performed is not specified in any of the manuals, and it is necessary to arrive at a standard time in some other way. Then the time can be estimated on the basis of experience with similar jobs, or determined more exactly by one of three methods: (1) using time study, (2) using Methods-Time-Measurement (MTM), or (3) finding analogous elements in the manuals.

The first approach to try, and the one used in this example, is the search for an analogy. In this case, a check of the manuals reveals that no elements are listed for doping bolts, but there is a listing in PWQ-4-E-F-G for doping pipe threads. The close resemblance between pipe threads and bolt threads is obvious, and there is no need to look further. The operation time is developed in the usual way from elements E, F, and G in PWQ-4, and it must be documented and filed for future use. Note, however, that when this approach is used the planner should make sure that the analogy is actually as close as it appears.

Before new gasket faces can be mated, they must be cleaned. PWN-8 gives the time for cleaning metal surfaces. However, PWN-8 times include certain elements that are not required here and exclude some that are. Naturally, the elements not required are omitted and those needed are added in calculating the craft time. The calculated time will permit more extensive cleaning than is sometimes required; but flanges that are old and covered with chemicals are often used and must be thoroughly cleaned.

Ladder time is a frequent item. In this case, the ladder must be used and moved twice.

When standard time is being allowed, each operation should be studied so that the amount of material- or tool-handling time, if any, is known. Only then should material-handling time be allowed. After the old piece of pipe is removed, it must be taken to the shop; when the two new pieces are fabricated, they must be taken to the job site. Thus, a total of three material handlings are required for the six-inch pipe.

This six-inch pipe is more than an armload, but it can be handled by two men. The material-handling time allowed for the pipe is 0.110 hour per piece, and a total of three pieces must be moved. Therefore, the craft time is 0.330 hour.

There are usually movements of material that are not covered in the operation formula. In this case, the ladder must be removed from the job and returned to the shop rack, accounting for two material handlings. The third material handling is an armload consisting of bolts, gaskets, and so on.

The old segment of pipe must be transported by truck to the shop for modification. This means that the craftsmen working on this job are forced to make an extra trip to the shop. No allowances are given for travel time, so it is recorded in the estimate column (see figure 4–24).

After all the operations required to perform the job have been recorded, they are totaled, and the total is called the total craft time. The total craft time is then multiplied by the multiplying factor (1.525 × 1.56) to obtain the figure 2.379 hours; and adding the 0.250 hour extra travel and the 1.040 hours add factor gives an allowed time of 3.669 hours.

Finally, the standard work classification chart (figure 4–25) is scanned and a job-phase time of 3.75 standard hours is allowed.

It is interesting to note that, in this case, the time allowed by the add factor is actually required. When the craftsmen bring the old section of pipe to the shop for welding, they will go on another job until the new sections

STANDARD WORK CLASSIFICATIONS

Hrs.	0	1	2	3	4.5	6.5	9	12	15.5	19.5	24
	A	B	C	D	E	F	G	H	I	J	K
	0.5	1.5	2.5	3.75	5.5	7.75	10.5	13.75	17.5	21.75	25

Hrs.	26	31.75	36	41.75	48	54.5	62	70	78	87	96
	L	M	N	O	P	Q	R	S	T	U	V
	28.75	33.5	38.75	44.75	51.25	58.75	66	74	82.5	91.5	100

Source: Gordon W. Smith, "Using Universal Maintenance Standards," *Techniques of Plant Engineering and Maintenance* 20:159–166. Copyright 1969, Clapp and Poliak, Inc.

Figure 4–25. Standard Work Classifications

are ready. This means that additional travel, job preparation, and cleanup time will be required to complete the job. This time is allowed for in the add factor. In actuality, this job was not completed at one time or at the end of the eight-hour day.

When the standard time has been derived, it is placed on the work order. A complete package includes: (1) the craftsman's copy of the work order, (2) a material list, and (3) a sketch.

Cross Orders

Cross orders are written for the various crafts and are planned exactly like the work order last described. The sketches, material lists, and white copy are treated like the original work order.

The only difference between a cross order and an original order is that the cross order bears the original work-order number in the cross-order blank and has a circle in which the craft that will do the work is shown.

Management Reports

The reports submitted to management by maintenance planning and scheduling may cover a wide range, but they all depend on one basic figure: productivity. Productivity is the level of efficiency of the group being measured.

Productivity is calculated by comparing the actual time expended on planned jobs to the standard time calculated for those jobs. The actual hours used for this calculation are derived from the back of the white copy of the work order or from other sources that show the actual time per job.

As an example, let us assume that the pipe fitters worked a total of 3½ man-hours on the wastewater still-line job (1¾ hours for each man), compared with the 3¾ man-hours allowed. The productivity is calculated as follows:

$$\text{Productivity \%} = \frac{\text{standard hours}}{\text{actual hours} \times 100} = \frac{3.75 \text{ hours}}{3.50 \text{ hours}} \times 100 = 107\%$$

Very seldom is the productivity calculated for a single job. Generally, it is calculated for a craft for a period of time, such as a week; actual times from completed work orders for a defined period of time are totaled and divided into the total standard time for the same jobs.

Adjustments

Occasionally, the standard time for a job turns out to be wrong, usually because the scope of the job changed after the job was planned. The planner must review each completed work order to determine whether any adjustment or additional investigation is required. Any standard time that varies appreciably from the actual time must be investigated.

The back of the white copy of the work order provides space for comments by the craftsmen and the foreman and a description of any additional work done. From this standpoint alone, it can be seen how important it is that the planner define what the standard time includes. The definition must be in the form of a request for work, outlined on the face of the work order, and the planner must review any deviation from the work described.

When an adjustment must be made, the planner either adds to this original job-calculation sheet or deletes items from it. Under no circumstance is an adjustment made unless it is calculated using times from the data manuals.

When additional time is allowed, the amount of the adjustment is credited to the original standard. When the original standard is to be reduced, it is debited by the amount of the adjustment. In either case, the adjustment is added to or subtracted from the standard hours calculated, not the standard hours allowed.

The new job-phase time and the total adjustment are recorded on the work order; no adjustment will be made unless it is fully described on the back of the work order. The new standard time allowed is used when calculating the productivity of any job that has been adjusted.

Standard Time on Work Order

The standard hours allowed are the average time required by a qualified craftsman to do a job. In figure 4–25, the top rows of numerals are the time limits for each classification; the numerals in the boxes are the standard times for the classifications.

Notes

1. This section is based on "Work Measurement in Maintenance," by J.O. Heritage, Kenneth Digney, R.E. Deem, in *Factory,* Jan. 1955, pp. 86–101, copyright April 1955, by permission of Morgan-Grampian, Inc.

2. Richard Deem, supervisor of maintenance standard, Lukens Steel

Co., Coatesville, Pa., "Work Measurement by Standard Time Data," *Factory* (January 1965).

3. See Lawrence Mann, Jr., "Establishing and Justifying Maintenance Standards," *Plant Engineering,* April 30, 1981, pp. 43–45.

4. This section is based on Ralph E. Renken, "Performance Evaluation and Improvement Using Basic Maintenance Management Tools," *Techniques of Plant Engineering and Maintenance* 17:147–159. Copyright 1966, Clapp and Poliak, Inc.

5. This section is reprinted from H.B. Maynard, *Industrial Engineering Handbook,* 2nd ed. Copyright 1963 by the McGraw-Hill Book Co. Used with permission of McGraw-Hill Book Company.

6. This section is based on Gordon W. Smith, "Using Universal Maintenance Standards," *Techniques of Plant Engineering and Maintenance* 20: 159–166. Copyright 1969, Clapp and Poliak, Inc.

5 Preventive Maintenance

Definition and Philosophy

Ideally, all maintenance should be preventive; that is, maintenance should be performed before any equipment failure. In these terms, *failure* means the point at which there is deterioration in the quality or quantity of the product. At the outset of any preventive maintenance (PM) program, it is quickly realized having all maintenance preventive would not be economically feasible; equipment would be grossly over maintained.

The concept PM is dynamic; and the decisions regarding what constitutes adequate PM are constantly changing. If a production unit is necessary in that all of its products can be sold, for example, then PM usually is performed at a high level. When the output of that equipment is no longer needed, or when a sufficient amount of its output is in storage, its level of PM becomes considerably lower. Some aspects of PM are referred to as repair maintenance, predictive maintenance, or overhaul maintenance. For purposes of this discussion, PM will be considered to be any maintenance that occurs before the quality or quantity of the equipment's product deteriorates. Maintenance can then be divided into PM and emergency maintenance.

Plant management views preventive maintenance in the context of days (or hours) out of service. Therefore, if plant management desires equipment to be operative a larger percentage of the time, the amount of PM applied to that equipment might be increased, assuming, of course, that maintenance personnel are capable of diagnosing the faults that cause the equipment to fail. This type of knowledge is not easy to acquire in the normal, multifaceted process plant.

Some managements view PM programs merely as the lubrication, painting, and cleaning operations normally conducted on a scheduled basis. Others expect PM programs not only to prevent downtime but also to minimize maintenance costs, to improve output, and to influence the quality of the product.

Preventive maintenance actually starts before the equipment is built or purchased. Plant maintenance should review proposed specifications and plans in order to assure that the plant is not engineering, constructing, or purchasing equipment that will have abnormally large requirements for maintenance. This review will assure that such factors as equipment spacing, access to components, and location of facilities such as furnaces have been considered from the maintenance standpoint.

Preventive maintenance programs are not restricted to large-scale, multicraft activities. Although PM might include such items as periodically cleaning the strainer in a pump, testing the high-level alarms in tanks and spheres, changing filters in the intake of an air compressor, or changing the relief valves on pressure vessels, it might also include activities as extensive as rebuilding compressors, repacking a fractionating tower, or the complete turnaround of a process unit. Some plants find, however, that their equipment that brings itself down does not cost maintenance any more to repair than if the work is done as PM. A plant might be so configured that the random removal of equipment from service does not place production at any disadvantage. If either of these two situations exist, it might be reasonable to question whether a PM program should be instituted.

The term *planned maintenance* is often used as a synonym for PM. The term is a poor one in that it does not necessarily denote preventive maintenance. Some plants maintain so-called emergency manuals that are, in effect, specifications for maintenance jobs that *might* occur and that are of an emergency nature. Thus, the turmoil that usually accompanies emergency operations is reduced because the job has already been planned.

All plants, whether or not they have a specific PM program, carry on some PM activities, including refilling ink containers in automatic recorders, lubricating, changing valves, painting, and similar activities.

The Benefits of a PM Program

Preventive maintenance programs yield numerous benefits, although few plants realize all of the following advantages:

1. *Minimum maintenance cost:* Maintenance can be planned, standards can be used, and materials can be obtained prior to the start of the work order.
2. *Maintenance performed when convenient:* Decisions can be made about when in the production cycle it is most advantageous for equipment to be removed from service. In addition, the work can be done when it is most convenient to maintenance in terms of availability of materials, equipment, and personnel.
3. *Ability to contract maintenance:* When a number of PM jobs can be packaged, it becomes feasible for maintenance management to have the alternative of contracting the work.

4. *Less downtime:* If the job can be engineered before removal of the equipment from service, the time that the equipment is out of service can be minimized.
5. *Minimum spare parts inventory:* If it is possible to maximize PM, the work order can be anticipated and spares can be obtained from the supplier. This minimizes the number of spares that must be purchased and stored in anticipation of emergency maintenance.
6. *Less disruption through emergency maintenance:* When the work order is anticipated, the sequence of operations can be documented better so that future work orders of a similar nature can be written with more knowledge than if the job had been an emergency that was not documented.
7. *Less standby equipment needed:* When it is possible to anticipate maintenance, equipment can be taken out of service at a time convenient to operations, and standby equipment need not be used.
8. *Less overtime needed:* Preventive maintenance makes it possible to plan and schedule maintenance jobs with a greater degree of accuracy, and knowing how long the job should require reduces overtime.
9. *Increased safety:* Rules of safety can be applied better to anticipated work than to emergency work.
10. *Less pollution:* Many of the emergency problems that result in flaring of smoke-producing chemicals are eliminated by PM.

The process industry is particularly adapted to reap the benefits of PM programs. In general, the more capital-intensive a plant is, the more it can benefit from a PM program. This is because the capital-intensive plant requires more maintenance than does the labor-intensive plant and is therefore more vulnerable to unforeseen shutdowns. In addition, since the capital-intensive plant generally does not have the redundant systems that usually exist in labor-intensive industries, it is more vulnerable to unforeseen shutdowns of facilities for which there is no alternative.

It is difficult, of course, to prove these benefits before the fact. They can only be predicted. Production uses the same process of prediction, however, to justify investment in production equipment. Many of the benefits are demonstrated by data on PM programs in other plants. Indeed, the impetus for initiation of a PM program in one plant often comes from the published report of a particularly successful program in another.

Plant management must see a return on any additional investment for maintenance. For the plant that has an ongoing maintenance program and does not give much attention to PM, this creates the problem of whether to begin by supplying additional resources for a PM program, which will provide a higher level of maintenance and thereby justify the cost, or by depriving the conventional maintenance program of resources to be used in a PM program in order to justify further PM investment. If the first approach is used, management should realize that, at first, maintenance cost will rise, as more craftsmen, material, and spare parts—and possibly more power tools, material-handling equipment, and so forth, are acquired. Returns from a PM program do not occur immediately. In large plants, as much as twenty-four months might elapse before substantial returns are realized. Many question whether this lag is necessary and whether a consultant might help. There is no doubt that consultants who have had experience with other PM programs will help shorten this period; unless unusual precautions are taken, however, the more involved consultants become in the program, the less likely the program will be successfully transferred from the consultants to the customer. The plant must involve its own people in the program from the very beginning, and not just on a token basis. Management should be willing to assign the necessary personnel, on a full-time basis, to work with the consultants from the first day. If this is done, the program has a greater likelihood of success and the time lag can be reduced.

Before Starting a PM Program

To begin or to enlarge a PM program, it is necessary that much groundwork be done before starting. Many a PM program has failed because management did not give sufficient thought to its appropriate introduction in the plant.

Selling the Program

The first step must be to sell the PM program to all levels in the plant. Operations should be persuaded that the PM program will make the equipment more responsive to their needs. The maintenance craftsmen should be persuaded that the program will be made easier, less hectic, and safer.

It has been said that the maintenance craftsman does not know whether the work he is doing is PM or emergency maintenance. Unfortunately, this is true in many cases, and the incentive for a craftsman to become involved in his job is thus removed somewhat. The craftsman should understand when he is doing a PM job and should be motivated by management's efforts to upgrade the maintenance program. Inevitably, procedures will be designed into the program that will be counterproductive, and a number of discussion sessions having to do with the PM program will give supervisors and craftsmen the opportunity to provide input to the system. This input further motivates the craftsman, since he is aware that his ideas are considered and that he has had an opportunity to assist in the design of the program in which he is working. Finally, it is necessary to sell the program to the maintenance supervisors. The maintenance supervisor not only controls the most important element of the PM program, he also trains the craftsmen who must learn to live with the program. His cooperation and enthusiasm are essential.

Next, management must decide whether the program will be designed and implemented from within or whether a consultant will be used. The third preparatory step is a series of orientation programs designed to inform every level of management and the craftsmen of the purpose of the program and the degree of involvement expected at each level.

Indexes

The final step before initiation of a PM program is to develop some means of evaluating the increased effectiveness of maintenance in general and the progress of the PM program itself.

Obviously, one of the purposes of introducing a PM program is to minimize maintenance costs. As the program begins to become effective, the important consideration for maintenance management will be the downward trend of maintenance costs rather than the isolation of specific savings. Thus, instead of attempting to cost out the program, it is better to use one or more of the several indexes worked out by maintenance departments to evaluate the effect of the PM program on the overall maintenance operation. (We will discuss indexes further in chapter 6.)

A primary index is the cost of maintenance related to some variable, such as product throughput, dollars invested, or percentage of sales. Another index is percentage of equipment downtime. This is a realistic indicator as long as the plant is running at full capacity; when reduced demand results in production equipment being even partly idle, this index becomes irrelevant; that is, the index should reflect the percentage of time that equipment is not ready for production rather than the percentage of time that the equipment is not operating.

A third index is the ratio of planned work orders to emergency work orders—or, approached in another way, the ratio of man-hours for planned work to man-hours for emergency work. Most plants find that about 80 percent of the total maintenance man-hours are devoted to work that would be classified as PM—inspections, lubrication, work usually done during overhauls or turnarounds, and the like. The other 20 percent is usually in the category of emergency or rush orders and maintenance that does not find its way onto a work order. Still another index might be the number of material stock-outs recorded. The philosophy behind this index is that a planned work order has, as a portion of its planning, the obtaining of necessary material, whereas, if the material for emergency work is not in the plant, the work order was not planned, and the work is thus not PM. This index is not necessarily accurate, however, since a plant that is overinvested in the storehouse is not as likely to run into stock-outs.

Indeed, none of these indexes is entirely satisfactory alone. A steady increase in the percentage of planned man-hours and PM work orders along with a corresponding decrease in emergency work provides the best indication that the PM program is progressing satisfactorily. It should be emphasized from the beginning that the aim is not to have 100 percent of plant maintenance performed on a PM basis. If a plant is overmaintained, too much of the firm's resources are being devoted to maintenance.

Usually, the more established the technology, the greater the degree of PM. As the plant acquires new and unusual operating conditions, the percentage of emergency or unforeseen maintenance increases. A recent study indicates that approximately 53 percent of the maintenance man-hours in the chemical and allied industries are planned. This figure drops to approximately 44 percent for paper industries and 40 percent for food-products industries. Other research indicates that 80 percent of the rotating equipment, 65 percent of the automatic controls, 60 percent of the material-handling equipment, and 55 percent of the process equipment receive PM. If the plant is operating to the satisfaction of production, the amount of PM can be considered satisfactory.

A last prerequisite to embarking on any PM program is an inventory of the condition of maintenance. This would include, among other items, some quantitative consideration of the level of maintenance that exists in the plant at the time the PM program is initiated.

Initiating a PM Program

Although the steps taken to create a PM program are fairly similar throughout the process industry, there is no standard procedure. The PM program must complement the existing operation, which is unique for each plant.

The history of PM programs reveals that most were started without being identified as preventive maintenance. When a planner and scheduler became aware of the repetitive nature of some work orders, we would assign resoùrces to the facilities concerned before initiating repeat work orders. Sometimes PM was born when operating personnel who did not want their production units to become inoperative at inconvenient times began to initiate work orders when, in their opinion, it was time to perform necessary maintenance.

It is almost a universal axiom that the PM program should start small;

once progress has been shown, it can then be made more comprehensive. Whether the program is based initially on functions—that is, lubrication, valve changing, inspection, and the like—or on a certain geographical area of the plant (one with a poor production or safety record, for example) will be dictated by the needs of each plant. No matter what the reason or where the program is started, an important part of it is the ability to predict when maintenance should be done. In plants with an extensive inspection program, this capability is obtained fairly easily. In plants that do not have such a program and do not have an extensive equipment-history system (discussed elsewhere), initiation of the PM program becomes a more difficult task.

PM Coverage

A new PM program may cover process equipment only, whereas coverage by a developed program may be so extensive as to include office buildings, fencing, yard lighting, and the like. Typically, in a process plant, PM covers a number of categories, including process equipment and its instrumentation; conservation and safety, the steam and power plant or, if the plant purchases one or both, the equipment for transmitting steam and power from the source; equipment concerned with process and cooling water; storage equipment; and fire-fighting facilities, which, of necessity, receive considerable PM attention in process plants.

Plants in colder climates have a considerable seasonal variation in maintenance work load, most of which is preventive—for example, inspecting insulation on water piping; draining equipment needed mainly in the summer, such as cooling towers; placing refrigeration equipment on standby; and using antifreeze in water-cooled, internal-combustion engines. Plants in the Gulf Coast area of the United States, in the Caribbean area, and on the East Coast of the United States are subject to hurricanes, which usually occur between July and September; those plants immediately adjacent to the shore in these areas will find that a little PM goes a long way. During a recent hurricane in the Gulf of Mexico, for example, a chemical plant was off stream for fifteen days. At the end of the first three days, the plant was ready to go back in service, but the remaining twelve days were required to clean and check electric-switch gear, transformers, and breakers. Had these items been placed on the second tier rather than the ground level, the plant would have been back on stream twelve days sooner.

The most popular approach to the question of what to include in a PM program seems to be to begin with the equipment that accounts for the most maintenance. A review of past work orders would be helpful here.

Economic rather than engineering considerations should control the decision-making process. Maintenance management must be apprised of operation's long-range plans for production units so that process units with a life of ten to fifteen years are not maintained as though they were going to be in production indefinitely. Also replacement materials should not be selected for durability if the total cost of the material is greater than the total cost of more-frequent replacements.

Standby items or redundant equipment receive the least amount of PM. If, for example, there is a standby motor for a certain pump so that, when one motor becomes inoperative, the other one takes over, PM on the original motor could be eliminated, provided that the damage done to the motor in bringing itself down is not excessive.

The following questions form the basis for determining the degree to which PM should become involved:

1. Will the faulty item put the plant or a portion of the plant out of service?
2. Are spares or standby equipment available?
3. Can the necessary maintenance work be subcontracted locally?
4. Can equipment be purchased or rented locally?
5. Is it cheaper to maintain the equipment or to replace it?
6. Can the equipment wait for the next normal turnaround?
7. Can the equipment be redesigned to extend its run life?
8. Will the equipment last longer than the system in which it is placed? If so, why not let the equipment run to destruction?

When equipment that operates intermittently or equipment for which there is no current demand requires PM, it is necessary that maintenance management meet with the operations people to forecast the future needs for that equipment. (In most cases it would be logical to perform PM immediately, before the equipment is scheduled to return on stream).

PM Inspection

Having established what equipment is to receive PM, it is then necessary to establish what specific items are to be inspected. One should begin with the maintenance procedures recommended by the manufacturer, but these recommendations must be revised in view of how the plant will use the equipment. Equipment that operates intermittently needs more maintenance than equipment that operates continuously. Specific gravity, amount of erosion and corrosion caused by the material being handled, phase differences (such as sometimes liquid, sometimes gas), methods of pumping and handling sludges; and attempts to handle more than that for which the equipment is designed are some factors that would require more extensive PM than that normally suggested by the manufacturer. Unique equipment requires more-extensive PM than equipment with which the maintenance forces are more familiar, at least until there is an accumulated history on it.

Perhaps the best source of information regarding what to inspect should come from the individual craftsmen who will maintain the equipment and who have previously maintained similar equipment. These craftsmen should be given the opportunity to comment on the manufacturer's recommendations and should be consulted about the list of spares to be stocked.

Equipment inspection is not the responsibility of the PM group alone. Everyone in the plant has the responsibility to report any faults they notice in equipment during the course of their business. The prudent operator periodically walks around his unit looking for possible problems. The maintenance craftsman performing his regularly assigned work always comes into contact with other items of equipment, and it is his responsibility to report any situation he believes is not normal. Many plants have energy-conservation programs that include reporting steam leaks and faulty steam traps, water drips or water leaks, and product leakage. Although the purpose of these programs is conservation, they often spot faults before they become major, thus allowing PM to be applied before a true problem develops.

By their nature, PM inspections are repetitive, and good methods and procedures should yield dividends. Some approaches to making inspection more effective include the following:

1. Analysis of the methods and procedures used in specific work orders might reveal better and more effective ways of inspection.
2. The inspection procedure should be designed so that travel time is at a minimum and so that, when other work orders are performed on the equipment, the equipment-inspection work orders are automatically considered if not performed.
3. When equipment is out of service for maintenance, it should be inspected for purposes of recording some evaluation of the work that should be done when the equipment next comes out of service.
4. Consideration should be given to determining the best inspection methods from the standpoint of tools and equipment. The inspection department should be knowledgable about all new developments in the testing-equipment field, particularly nondestructive testing, vibrations analysis, and thermography.
5. Equipment should be designed to facilitate inspection. For this reason, the equipment-inspection group should have an opportunity to review plans and drawings of new equipment.

The Checklist

It is the responsibility of the inspection group to supply maintenance with a checklist of the items to be monitored in the PM program. One of the advantages of this checklist is that it can be used as a training aid, in that it informs the maintenance mechanic who is not familiar with the equipment what items should be looked at. There is a close relationship between the checklist and the maintenance standards discussed in chapter 4. The standard is the sequence of events that make up the work order, and the checklist is the sequence of events that make up the PM work order. Both the checklist and the standards should include the list of spares and material required for the job as well as some estimate of the time and crafts needed to perform the work. The checklist is also a permanent record of what is done at each inspection interval and, in some cases, who does the work. This latter is important in execution of the work and in pinpointing individuals for further training when improper execution of the work results in future problems. An entry should be made on the equipment history from the checklist after the PM work has been accomplished. The checklist has a disadvantage, in that maintenance craftsmen may restrict their attention only to items on the list and may ignore service needs that were unforeseen or that result from aging of the equipment. The checklist should be reviewed periodically in light of the work orders and equipment history to ensure that it is current and that it includes all items necessary for an adequate PM program.

Frequency

It is also the responsibility of the inspection group to specify the frequency of inspection for the equipment receiving PM. Optimum frequency is established on the basis of the following considerations:

1. *Criticality of equipment:* The more critical the equipment, the more PM attention it requires. In most process plants, the series aspect of equipment means that failure of any individual item would cause problems both upstream and downstream from that equipment. Equipment is also considered critical if commitments have been made or if the product of the equipment is to go directly to the customer rather than into storage.
2. *Experience on similar equipment:* This is an important factor in determining the frequency of PM operations on new items of equipment. Two considerations are important here. Experience not only indicates the amount of experience that the plant itself has in dealing with similar

items of equipment, it also indicates how well the craftsmen are trained to perform the PM program.

3. *Operating characteristics:* Identical items of equipment will be used for different types of service in different plants. Operating characteristics include the severity of the service in view of the type of material being handled, its specific gravity and phase, the pressure and temperature, and whether the equipment is operating continuously or intermittently.
4. *Age:* As the equipment becomes older, the work force knows better how to maintain it and, in many cases, has built out some of its weak spots so that the frequency of PM can be extended rather than decreased. Of course, as equipment begins to wear out and as its useful life is approaching an end, a drastic redefinition of the frequency of PM should be made.
5. *Safety and pollution requirements:* Maintenance of equipment related to the safety and pollution requirements of the plant should also be considered in light of those programs.

Service Manuals

It does not seem necessary to stress the need for service manuals, which should accompany the acquisition of all new equipment. What should be stressed, however, is the systematic filing of and use of manuals. These manuals are often mislaid or misfiled so that they become difficult to locate. Whether service manuals are located in the operating-unit files or in the maintenance-planning-office files, or are distributed to various locations in the plant, the system should be well thought out and consistent, so that all who need the information know where to find it and, more important, where to file the manuals when the work has been completed.

Scheduling PM

By its nature, PM is usually scheduled with a time range in which the work should be done. There is a great tendency to let PM work slide and to give all attention to emergency and higher-priority work orders. Inevitably, when the PM work slides long enough, it becomes emergency work, and all the disadvantages of emergency-work orders then accrue.

A specialized category of PM scheduling is the major overhaul or turnaround. This is a normal operation in the process plant. In the usual case, PM work orders are accumulated until the backlog is sufficiently large to warrant removal of the equipment from service for a package of work orders. Normally, any construction work that has been anticipated for the unit is performed at this time. To minimize equipment downtime, work orders that will be performed during the turnaround are given special handling to assure that scheduling is appropriate and that material and equipment are on hand before the unit is removed from service. This process requires patience and understanding on the part of the scheduler. He must realize that production has many pressures on it; if production must change the schedule immediately before the turnaround, the scheduler must be tolerant of that situation. Similarly, the production people must be aware of the time spent by maintenance in planning the turnaround.

Scheduling of PM work orders should be made with the following principles in mind:

1. *Minimized downtime:* If the downtime brought on by emergency situations is approximately equal to the downtime for PM, little advantage accrues from the PM program. The location in the production cycle is significant, however, since downtime becomes less important if the demand for the equipment is decreasing.
2. *Relation of PM to the total maintenance work load:* The maintenance scheduler must continually evaluate the ratio of PM to emergency maintenance. He must consider turnarounds and other foreseeable demands for maintenance, but, if he continually lets the PM program slide, emergency maintenance will increase. This scheduling problem is difficult to quantify. The scheduler must have a detailed knowledge of the system to make optimum use of maintenance resources.
3. *Day-shift operations:* Since supervision and productivity are greatest during the normal shift, the scheduler should make every effort to see that preventive maintenance is done during the normal day shift and that a minimum is done on overtime or on off shifts. This does not apply to the major overhaul or turnaround.

Normally, the PM work order is scheduled along with emergency and other work orders to be executed during the immediate and short-term future. Nevertheless, it would seem wise to have a schedule of PM jobs only, if for no other reason than as a basis against which to evaluate the actual PM work that gets done during a particular time period. This schedule can be broken down according to area, individual items of equipment, or class of equipment. An overall chart of all PM work that should be scheduled in the immediate future appears to be the best control mechanism for these purposes. Blanket orders should be included in this schedule. These include daily or periodic maintenance such as lubrication, instrument calibration, or testing of safety and pollution-preventing equipment.

When PM scheduling is dictated by the number of equipment operating hours, there should be a systematic way of informing the maintenance planner and scheduler when the total operating hours are approaching the point for a PM work order. This is particularly true of equipment that operates intermittently.

Should every item of PM be listed on a work order? Most plants use an open or blanket work order, which covers a number of minor or miscellaneous items. Since the total man-hours applied to PM should be recorded, a real problem arises in deciding how detailed the data-accumulation system should be. Admittedly, there might be opportunities for improvement in the system if all items were identified. Nevertheless, there is some balance between identifying every item of maintenance and the cost of maintaining records for small efforts. Whenever the total of the blanket or open work orders amounts to an excess of perhaps 5 percent of the PM effort, the system probably should be reviewed to determine if the open-work-order system is being abused.

By its very nature, PM is more easily scheduled than emergency maintenance. For this reason the PM work orders have a greater probability of being completed during the time allocated for them. Nevertheless, there inevitably will be times when the order is not completed and must be continued either on an overtime basis or the next day. When the PM program is extensive, the likelihood that the orders will overrun is great, and the scheduler must adjust the schedule so that these work orders may be completed.

Cooperation with the production people is even more essential for the PM program than for the emergency program. Emergency work orders are usually initiated by production. The PM work order is usually initiated by the maintenance department itself, and production should cooperate by releasing the equipment at the scheduled time. Obviously, production should be consulted for appropriate scheduling. If production decides that it is not feasible to release the equipment on schedule, the maintenance department must be informed immediately and alternative plans must be made. Each group should appreciate the scheduling problems of the other.

Where there are area foremen of supervisors, this is not a problem, because the area foreman and the operating personnel will have developed a working relationship that enables the area foreman to see problems from both points of view. Most plants advocate a weekly meeting between production and maintenance to establish the following week's PM schedule. The schedule can be published as a tentative one and given the production department in advance for review. Accompanying this schedule should be the reasons for requesting release of the equipment and the possible results of not meeting the tentative schedule. Maintenance must be realistic about these possible results so that production will not see them simply as a method of forcing adherence to a schedule.

Establishing a PM Inspection Schedule[1]

The inspection of plant equipment is an important phase of a comprehensive preventive maintenance program. Inspection can be defined as examining equipment to:

- Ensure that it is performing as designed
- Evaluate the mechanical, pneumatic, hydraulic, and electrical mechanisms in terms of potential problems
- Estimate when a breakdown could occur
- Identify the component or function that may precipitate a breakdown
- Schedule repairs at a convenient time to prevent a breakdown at an undesirable time.

Inspecting every piece of equipment in the plant is as unrealistic as not inspecting any equipment. Although these two extremes are practiced, a middle-of-the-road approach is recommended. An objective engineering approach [figure 5–1] should be used to determine which equipment to inspect.

An inspection program can include several approaches. In the "first breakdown" approach, equipment is operated until it breaks down. Then, while repairs are being made, the machine is inspected for potential maintenance problems.

This approach assumes that running some types of equipment—most fractional-horsepower motors, most lighting systems, and equipment and systems with backup units—until it breaks down is good business.

The "no alternative" approach includes equipment that must be operative when its use is required. Examples are plant fire pumps, major electrical substations, emergency lighting systems, and some sump pumps.

One approach—when the law says "do it"—must be a part of every plant's inspection program: all equipment covered by OSHA and insurance company regulations must be inspected.

Figure 5–2 illustrates how one company classified its critical equipment as to maintenance priority.

Although these three approaches categorize some plant equipment according to whether or not it should be inspected, a lot of equipment remains undesignated. The breakdown-dollar-versus-inspection-dollar approach can be used in deciding if such equipment should be inspected.

The formula used in this approach is:

$$PM = \frac{D(A+B+C)}{(E \times F)}$$

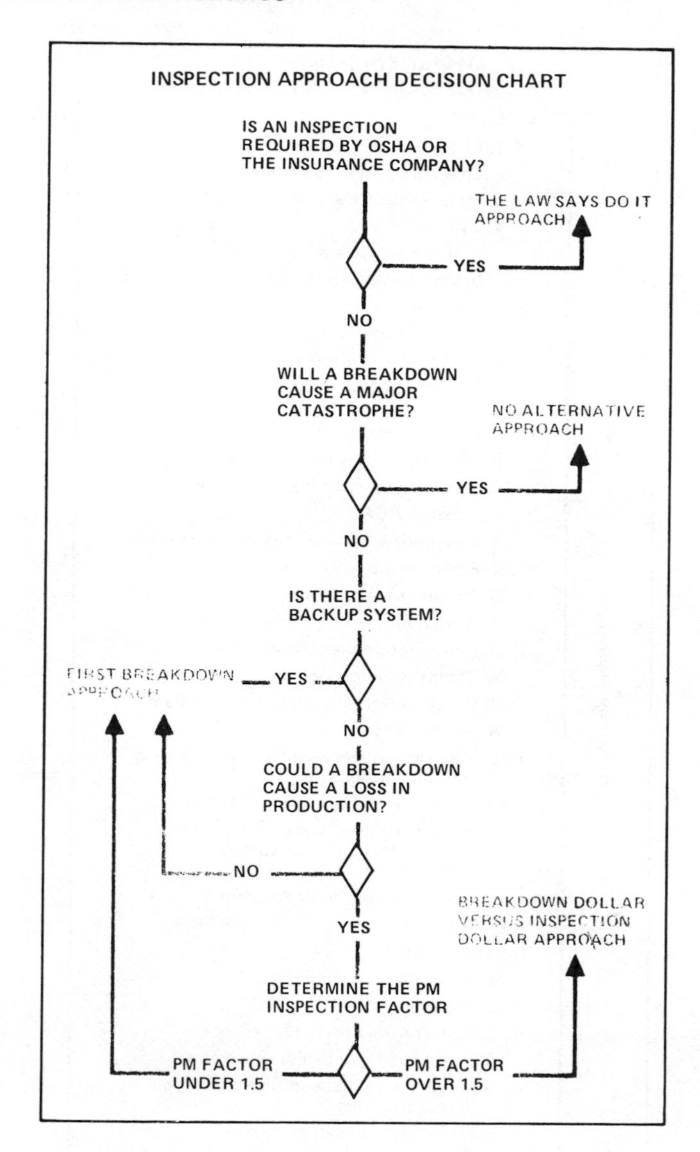

Source: J.D. Andrica, "Establishing a PM Inspection Program," *Plant Engineering*, August 9, 1979, p. 109.

Note: The engineering approaches that should be considered before setting up a PM inspection program are shown on the chart. A "law" or "no alternative" approach should take precedence over all others.

Figure 5-1. Inspection-Approach Decision Chart

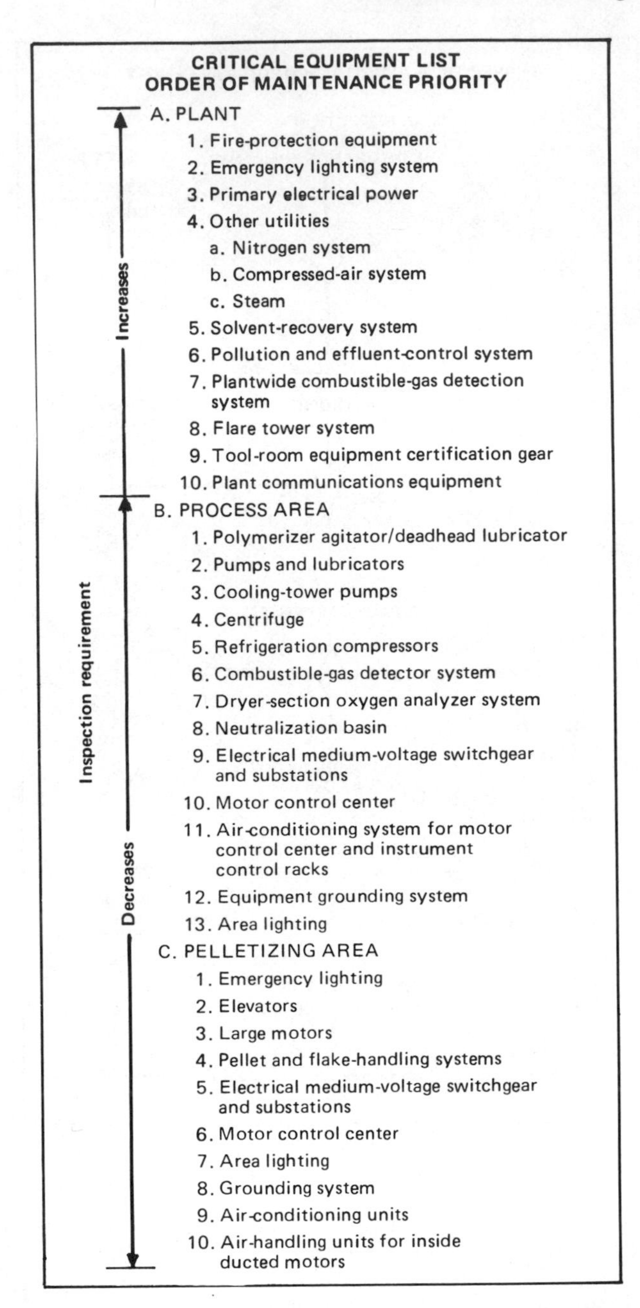

Source: J.D. Andrica, "Establishing a PM Inspection Program," *Plant Engineering,* August 9, 1979, p. 110.

Note: This method, used by one plant to rank its equipment on a maintenance priority basis, can be used in determining the PM inspection sequence.

Figure 5–2. Critical-Equipment List: Order of Maintenance Priority

where PM = inspection factor

D = number of breakdowns per year (based on past experience or estimate)

A = cost of repairs caused by the breakdown, dollars

B = production loss resulting from breakdown, dollars (use standard cost dollars)

Determining the production dollar loss is a sensitive area and should be done in an objective manner. Points that should be considered are:

1. Is there an in-process inventory after the equipment breakdown? If so, will the breakdown deplete this inventory to an unsafe level? If yes, determine production dollar loss by multiplying the number of units lost (that is, not built for sale) by manufacturing standard cost. If no, determine production dollar loss by multiplying the number of machine hours (over and above normal production) required to replenish the inventory by the machine hour cost.

2. While determining production dollar loss, beware of the manager who says he needs this equipment 24 hours a day and 7 days a week, but is running only one shift.

C = cost of repairs to other equipment damaged by the breakdown, dollars

E = average cost of PM activity (include time and material), dollars

F = number of planned PM activities per year

Applying the formula to a particular machine will give the inspection factor for that unit. For example, the factor for a plastic-forming machine is found to be:

$$PM = \frac{14(150 + 100 + 0)}{(20 \times 50)} = 3.5$$

where D = 14

A = \$150

B = \$100 per hour

C = zero

E = 1 hour per week × \$15 per hour + \$5 for material = \$20

F = 50

When calculations have been completed for each piece of equipment in question, machines can be ranked by order of importance: ranking begins with units having the highest inspection factors and ends with those having the lowest. One question must still be answered: How much equipment can be inspected with the funds available?

It can be answered in two ways:

1. Start at the top of the ranking list and accumulate inspection costs until they equal the amount budgeted for PM. Equipment included to that point will be inspected.

2. Inspect only equipment having an inspection factor of 1.5 or higher. Use the first-breakdown approach for equipment having an inspection factor lower than 1.5.

In any PM inspection program, the next question is: Who will do the inspections? Making this decision requires consideration of:

- Inspection frequencies
- Importance of the equipment and inspection
- Complexity of the equipment and the inspection
- Reliability of the equipment
- Cost of repairs normally associated with the equipment
- Follow-up procedures.

These considerations will determine whether one or all of the following people should be used to perform inspections: maintenance foreman, production worker/foreman, qualified maintenance craftsman, maintenance engineer, industrial engineer, lubricator, and contractor. Most PM inspections can be assigned to a maintenance craftsman. However, assigning one person to do all inspections may be a mistake—the need must dictate the inspector.

Proper inspection frequencies cannot be established until several sources have been checked for recommendations:

- Equipment manuals for specific PM tasks
- Mechanics for their opinions on frequency of breakdowns and repairs
- Suppliers (they may have a computer program for determining frequencies)
- Equipment builders.

Once all the available data are gathered, establish schedules based on the findings. One key point to remember in establishing frequencies: Never make them absolute. Experiment with them. Increase or decrease a frequency when it seems appropriate.

The soundness of the scheduled frequencies can be determined by tracking the progress of the PM inspection system. Tracking families of equipment will simplify data collection.

The best way to ensure the success of the PM inspection program is to limit the amount of handwriting involved. Each form should be identified by name or color. The forms used should:

- Minimize writing and maximize checking or coding
- List each item
- Leave space for action taken
- Remind the inspector of important items
- Provide enough space for entries
- Ask questions.

Set down specific do's and don'ts for filling out the forms. Generalizations should not be allowed. For example, an entry such as "Checked all components for loose bolts, wires, and other problems" is insufficient. Instead, a form such as that in figure 5-3 should be used.

And instructions such as "Insert the proper term where appropriate" should be replaced with "Use the following code:

✓ = Okay
O = Immediate attention required
X = Schedule required repairs
+ = Does not apply
• = We were too late."

Another important part of the communication network is a record of action taken as a result of the inspection. Such records will be of interest to the OSHA inspector and insurance agent; the records can add a high degree of credibility to the PM inspection program. The records should contain such items as:

- Was a work order required? If so, what was its number?
- Was the foreman informed? If so, at what time?
- If a serious safety hazard was involved, was the safety expert informed? If so, list his name and when (date and time) he was informed.

The example in figure 5-4 illustrates the classic system breakdown. The inspector took the time to inspect the crane cable, determined that it required some form of repair, but never told anyone. And he failed to indicate what repair was required (A). Instead, the inspector should have filled out the report as shown in example B.

A mechanic sent to make a major repair on a key piece of process equipment should be given:

	OK	Repair by Inspector	Repair Needed by: Next Week	Next Inspection	Next Month	Other Action Required
Crane cable	✓					
Upper limit stop		ADJUST UP ½ IN.				
Driver north truck wheel			X			W/O #12468 WRITTEN
Driver south truck wheel				CHECK BEARINGS		

Source: J.D. Andrica, "Establishing a PM Inspection Program," *Plant Engineering,* August 9, 1979, p. 111.

Note: An inspection report need not be too detailed, but it should be specific as to what was done and what should be repaired.

Figure 5–3. Example of Inspection Report

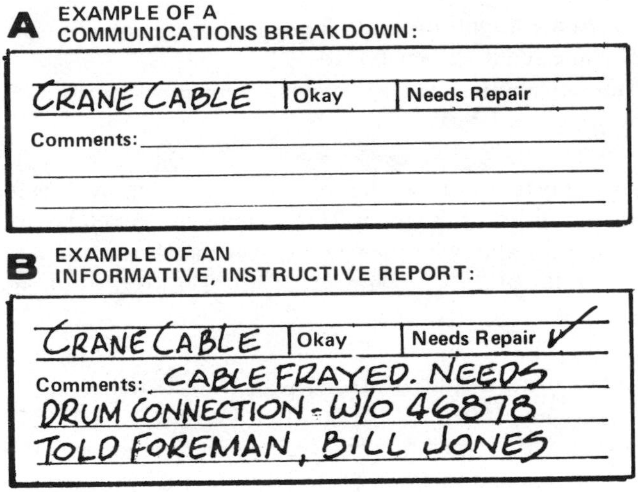

A EXAMPLE OF A COMMUNICATIONS BREAKDOWN:

CRANE CABLE	Okay	Needs Repair

Comments: __________

B EXAMPLE OF AN INFORMATIVE, INSTRUCTIVE REPORT:

CRANE CABLE	Okay	Needs Repair ✓

Comments: CABLE FRAYED. NEEDS DRUM CONNECTION - W/O 46878 TOLD FOREMAN, BILL JONES

Source: J.D. Andrica, "Establishing a PM Inspection Program," *Plant Engineering,* August 9, 1979, p. 111.

Note: If an inspector fills out a form as A, he has failed and the system will fail. A good PM inspection report is shown at B.

Figure 5–4. Inspection-Communication Reports

- A description of the repair so that he can prepare for the job
- The equipment's location
- A description of the condition of the area where the repair is to be performed
- A standard time for accomplishing the repair.

The instructions required for an inspection should be even more detailed than those required for a maintenance repair. In addition to the normal maintenance job instructions, specifics about the equipment that must be checked should be included. The mechanic should be instructed to indicate:

- What is tight
- What is loose
- How hot is hot
- How much of a leak is normal
- What is the right gap.

Other items that should be considered when preparing inspection instructions are:

- Normal operating temperature (high and low)
- Equipment location
- Item name
- How to remove inspection covers
- Locations of inspection covers
- Sounds to listen for
- Locations of sounds
- Normal operating speeds and feeds
- Description of normal wear on parts
- Lubrication points and type of lubricant used
- Where to obtain lubricant samples for analysis
- Most frequent breakdowns and types of repairs normally performed
- Sequence of inspection steps
- Estimate of time required to complete inspection
- Frequency of inspection.

Figure 5-5 illustrates some of the pertinent steps that should be included in a PM inspection.

The next question is: Is the program making money? Before the first inspection is performed, the engineer can prepare to answer this question by keeping track of:

Steps	Frequency	Inspection Points Page 1
6	2Y	AIR-COOLED CONDENSERS: Inspect for leaks; dust, dirt, or other accumulations; excessive noise and vibrations; loose, missing, or damaged parts. Tighten loose parts.
8	Y	WIRING AND ELECTRICAL CONTROLS: Inspect for loose connections; short circuits; loose or weak contact springs; improper adjustment of control and parts; worn or damaged contacts; defective operation; correct fuses. Tighten connections; make adjustments and minor repairs.
9	Y	STEAM AND HOT-WATER HEATING UNITS: Inspect for clogging, dirty heat-transfer surfaces; leaking, loose connections and parts; bent fins; misalignment; water hammer; air binding; nonuniform heat spread; below normal temperature readings. Tighten loose connections and parts. (Before heating season.)
11	Y	PIPING: Inspect for leaks; loose or damaged fittings, flanges, nuts and bolts, defective gaskets, cracked or faulty welds, undue strain on lines, lack of proper support; missing or damaged insulation. Tighten loose parts.
13	Y	SELF-PROPORTIONING AND MODULATING STEAM VALVES (ELECTRIC AND PNEUMATIC): Inspect for leaks, improper valve and linkage adjustments, valve control settings and pointer indicator readings; improper operation of valves resulting from defective diaphragms, bellows, discs, seats, electric valve motors, and wiring.

Source: J.D. Andrica, "Establishing a PM Inspection Program," *Plant Engineering,* August 9, 1979, p. 112.

Note: A sample inspection instruction sheet lists some steps in an inspection. List the safe operating temperature, the type of lubricant to be applied, the amount of torque needed, etc., in preparing instructions.

Figure 5–5. Inspection-Instruction Sheet

1. The number of breakdowns in each area of the plant by major unit, by department, and by equipment group
2. The costs of idle labor, of scrap, of lost sales, of extra maintenance per breakdown.

By the time the first inspection is performed, a base period will be available for comparison with the results of the program.

Starting a PM inspection program is more than walking out and inspecting equipment; it requires a well-thought-out, professional approach.

The Organizational Structure of PM

Although not absolutely necessary, a special equipment-inspection group is of great assistance in any PM program. The responsibilities of this group are to operate and interpret vibrations analysis and nondestructive testing equipment and, on the basis of the resulting data, to keep maintenance and production personnel apprised of what is necessary to maintain the plant adequately. This group is also responsible for locating safety and pollution hazards. Some plants require that the equipment-inspection group also be a methods-analysis group, in which case they might more appropriately be called the mechanical-technical service department.

Occasionally, one finds a plant in which the organizational structure of PM and that of general maintenance are entirely different. This creates very real problems. Although emergency and preventive maintenance may be going on side-by-side, there is little or no communication between the two groups because they perceive their goals to be different. The most that can be said for this dual type of maintenance organization is that PM usually gets accomplished, since the tendency to ignore PM when general maintenance problems arise is minimized.

Some plants budget PM separately from general maintenance. This appears to be a needless separation of accounting charges, although it does provide a convenient way of isolating PM costs and man-hours for purposes of evaluating the PM program.

PM Personnel

Administrators

The question often arises regarding at what point the PM program needs at least one full-time administrator. It would appear wise at the very outset to assign the PM function to one individual and to have that be his only responsibility. When one considers the necessary data analysis, scheduling, planning, and evaluation that should go with the PM program, it does not appear unreasonable to assign at least one individual on a full-time basis, even if very few craftsmen are assigned to the PM system.

Inspectors

The equipment inspector should be an individual who likes to persevere until he obtains an answer to a problem. He must be able to imagine all possible situations that might cause an item of equipment to need maintenance, and he must systematically track down an appropriate sequence of events to prevent its unforeseen breakdown. He must possess patience; he should be entirely familiar with the type of equipment he will be required to inspect; and he must be trained in the operation and interpretation of an increasing number of sophisticated inspection instruments. Most plants naturally seek a highly skilled craftsman with plenty of experience for the job of inspector.

Some inspection functions are furnished by local laboratories—such work, for example, as soil inspection or monitoring concrete quality. Often, the equipment-inspection group must work with outside inspectors representing insurance companies or state or local governments. These inspectors usually look at such equipment as boilers, pressure vessels, or fire-fighting equipment, and equipment whose operation would cause pollution problems.

It is generally agreed that the equipment inspector should have the rank and pay of a maintenance foreman or supervisor. His responsibility certainly is more than that of the craftsmen, since, in many cases, the inspector is evaluating the work of the craftsmen. (This is not to say that the inspectors are the only individuals responsible for quality control. Maintenance management would like every participant in the maintenance program to feel responsible for the quality of the program). In addition, equipment-inspection personnel are normally required to operate expensive and complicated equipment in which the plant has a great investment. This places the inspector in a technician's role, and he should be compensated commensurately.

The PM Paperwork System

It must be admitted that the introduction of a PM program will require considerably more paperwork than do the emergency and normal work-order systems. By its very nature, the PM program is guided by accumulated data on the equipment it is to monitor. Since management tends to view this aspect of the PM program as a disadvantage, the manager of the PM program must emphasize the necessity for it. Management must be made to realize that planning, scheduling, and forecasting are as necessary to the operation of maintenance as they are to the operation of production.

It is difficult to forecast the number of clerical and filing personnel needed for the adequate functioning of a PM program. The correct approach is to design an appropriate paperwork system and then obtain sufficient personnel to ensure that the work remains current. The degree of involvement with electronic data-processing equipment will, to some extent, dictate the number of clerical positions needed to maintain a PM paperwork system. The clerical and filing time required to maintain an established PM program is less than that required to get the program started.

Seldom is the PM clerical staff restricted to handling only PM work orders. Most frequently, these individuals also work with the normal maintenance system. It has been said that two clerks are needed for every one hundred maintenance men to administer the PM system, excluding standards and estimating.

The rules to follow in establishing the PM paperwork system are little different from the rules for establishing any paperwork system.

1. Minimize the amount of data collected. Too many PM systems have adopted forms from other operations. Too often, this results in collection of data that are not needed in a form that will not be useful. Forms that ask for useless information will not be completed conscientiously. A good axiom to follow is to design the system and then design the paperwork, rather than letting the forms dictate the system.

2. Do not needlessly separate the PM and normal maintenance systems. Since work-order routes, reports, material requisitions, and cost are all handled in the same manner, such separation can only increase overall maintenance cost needlessly.

3. Keep the cost of the equipment-inspection function separate. The equipment inspection group not only is a necessary part of the PM program, it also has responsibilities to maintenance and operations. It is a point of contention whether the cost of an equipment-inspection group should be charged, entirely or in part, to the PM program. It is true that many PM work orders result from the work of the equipment-inspection group. Nevertheless, its relationship with safety, pollution, and operations dictate that this function would exist even if the PM program did not. There appears to be no uniformity in the ways process plants charge the cost of the equipment-inspection function.

4. Use performance indexes. No system is complete without some way to evaluate its performance. Periodic reports concerned with such matters as work orders closed out on schedule, overtime, or percentage of maintenance man-hours worked on PM will reflect PM performance.

Specifically, what might management expect to be added to the system in the way of forms for the PM programs? First, an inspection report is needed. From the inspection report, a checklist or standard is written. The work order is then introduced into the overall maintenance schedule, and a performance report follows execution of the work order.

Flow is an important aspect of the paperwork system. A flow chart should be constructed (see any industrial-engineering handbook). The horizontal flow chart usually used by maintenance provides easy access to such information as what functions receive copies of forms and reports, what information each function derives from the forms and what each adds to the forms in their travels, and the final destination and disposition of the forms.

Miscellaneous Considerations

It should be emphasized that the PM system is not one that can be established and then assumed to remain current. The ever-changing nature of the typical process plant requires that the PM program be monitored continually to ensure that it is always responsive to the current needs of the plant and that the system continues to fulfill the needs of management. It is necessary, however, to guard against overmaintenance. Undermaintenance will be brought forcibly to the attention of maintenance management, but overmaintenance will not immediately make itself known. The important index here is the amount of maintenance performed under the PM system relative to the entire maintenance system, rather than the degree of coverage on equipment. Accurate reporting, not only of man-hours but also of cost, is necessary for successful monitoring of a PM program.

I was once a maintenance estimator in a plant that had a procedure whereby the area supervisors could overrun or underrun any work order by plus or minus 10 percent and not have to explain the variation. Upon checking with the accounting department, it was found that most of the work orders approached 109 percent and then were closed out. A field check was made with the area supervisors. It was found that they performed any maintenance task they wanted to and charged the job to the oldest work-order number issued to them until it reached 109 percent. Then they chose the next-oldest work order and continued performing the work they deemed advisable. Naturally, any report received from these supervisors was meaningless. Constant vigilance must be exercised by maintenance management to assure that cases such as this do not occur.

Inspection frequencies must be continually reviewed. In any maintenance operation, maintenance craftsmen and supervisors are always looking for better methods and better materials. This results in a longer run life for equipment. If an inspection frequency is not continually monitored, overmaintenance will occur, since the inspection interval will remain the same and maintenance personnel will always by taking steps to extend the run life. The introduction of nondestructive testing and vibration analysis techniques have greatly facilitated the diagnosing of possible faults and the resulting alteration of inspection intervals.

And finally, the preventive maintenance program will make the best possible use of the data it collects when it uses industrial engineering techniques such as method analysis, time study, statistical analysis, operations research and critical-path methods for manning and scheduling maintenance, control chart monitoring of equipment, and the emerging developments of reliability engineering.

Statistical Methods in PM Analysis and Scheduling[2]

Nomenclature

λ = mean arrival rate of units into system
L = length of waiting line for maintenance
μ = mean service rate
σ = the standard deviation
W = idle time of unit until it is maintained
x = any data point
$\bar{x}$ = the arithmetic mean of all data
z = standardized variable of normal distribution

Single-Component Systems

In preventive maintenance scheduling, the primary problem is: "When should the equipment be taken off stream for maintenance?" In most plants, a large portion of the equipment can be grouped into more or less homogeneous units, for instance, all high and low-level alarms in hydrocarbon service, all continuously operating centrifugal pumps within size ranges, all compressors, and so forth.

No statistical analysis can be undertaken without first having collected data. Not just data, but a planned program to collect the needed information in the form and frequency needed. This includes such items as length of run, duration of downtime, details of what maintenance work was necessary, and causes of downtime.

Therefore, to initiate the program, design the maintenance reporting forms to collect the data in a form that can be useful in statistical analysis.

Component Parts

Assume a bank of compressors with the data given in table 5–1 accumulated for a certain shaft bearing. The first step to planning a maintenance schedule is that of plotting the frequence of run-length occurrences. To do this, the data must be grouped into classes. Examination of table 5–1 indicates a range of from 870 to 1140 hr, or 1140 − 879 = 270 hr. Divide this into, say, 9 classes: 270/9 = 30 hr per class. Table 5–2 arranges the data from table 5–1 into a frequency distribution. Figure 5–6 shows the data from table 5–2 plotted to give a graph of the frequency distribution. Examination of figure 5–6 indicates that the distribution is approximately normal. This will be the case with many component parts or items with a small number of parts. The average or mean of the data shown in table 5–1 is 1000 hr. Both table 5–2 and figure 5–6 confirm this. The mean is, by definition, that point above and below which 50 percent of the cases occur. Therefore, if it was decided to tolerate half of the bearing failures at unforseen or unplanned times, then the decision would be to let all bearing run for 1000 hr, then take the compressors down for replacement.

A more realistic decision rule would be to arrange a preventive maintenance program so that only 10 percent of the failures would be of the emergency or unforseen type. The percentge of the area under a normal curve between the mean and any point to the left or right is approximately defined by the z static, which is

$$z = \frac{x - \bar{x}}{\sigma}$$

when x = any data point

$\bar{x}$ = the mean of all data—1000 hr in this case

σ = the standard deviation

The standard deviation is defined as

$$\sigma = \left[\frac{\Sigma(x - \bar{x})^2}{n - 1}\right]^{1/2}$$

where n = the number of data points.

The standard deviation for the data shown in table 5–1, without showing the computation, is 57.2 hr. Nothing only the abscissa of figure 5–6, it can be seen that essentially no bearing will fail before 874 hr. If zero percent will fail before 874 hr and 50 percent will fail at about 1000 hr, then 10 percent will fail somewhere in between, at say x hr.

A greatly abbreviated table of z-values is shown in table 5–3. These values are the area under one half (the right half) of the normal curve. Since the curve is symmetrical about the mean (z = 0 at the mean), a table of one half the curve suffices. From table 5–3, the z-value corresponding to x in

Table 5-1
Compressor-Run Duration to Bearing Failure
(hours)

Compress. No.	*Length of Run*	*Compress. No.*	*Length of Run*	*Compress. No.*	*Length of Run*	*Compress. No.*	*Length of Run*
8	967	7	1059	6	1022	7	984
4	909	5	1035	2	987	4	1019
6	978	3	926	8	998	2	1041
3	1018	8	1002	5	1038	5	917
5	938	4	964	7	955	1	1024
2	1010	3	874	4	1015	8	959
1	1008	2	1014	3	982	1	1047
6	1054	6	940	8	992	7	882
2	1104	1	996	1	1077	3	1089
7	1040	5	1138	4	1028	6	970

Source: Lawrence Mann, Jr., "Statistical Methods Facilitate Planning and Scheduling," Amerian Society of Mechanical Engineers Report 66-PET-12, 1966.

Table 5-2
Frequency Distribution of Data in Table 5-1

Class Interval (Hours)	*Frequency of Occurrence*
870-900	2
900-930	3
930-960	4
960-990	7
990-1020	10
1020-1050	8
1050-1080	3
1080-1110	2
1110-1140	1

Source: Lawrence Mann, Jr., "Statistical Methods Facilitate Planning and Scheduling," American Society of Mechanical Engineers Report 66-PET-12, 1966.

figure 5-7 is 1.28. Substituting in (z is negative because area of interest is to left of the mean).

$$-z = \frac{x - \bar{x}}{\sigma}$$

$$-1.28 = \frac{x - 1000}{57.2}$$

$$x = 926.8$$

Therefore, by removing each unit from service at approximately 927 hr, on the average, 90 percent of the unscheduled failures will be averted.

Multicomponent Systems

Consider the foregoing problem from another point of view. Interest is now centered around the failure of a compressor which is, in effect, a multicomponent system. The phenomenon present here is that, even though the life of each individual component follows the normal distribution, the increments of time between system failures usually follow a Poisson distribution. For the behavior analysis, an additional group of data is necessary. The second distribution is that of the length of time required to repair each compressor. Care should be exercised so that only repair time is recorded, not idle time plus repair time. Figures 5-8 and 5-9 illustrate, respectively, these two distributions. Assume that the average in figure 5-8 is 84 hr (3.5 days)

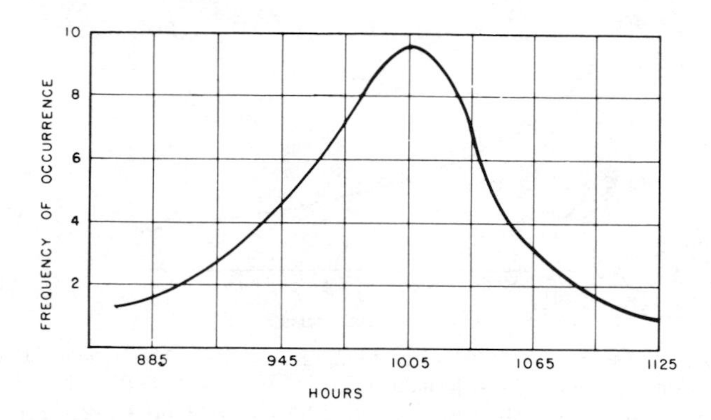

Source: Lawrence Mann, Jr., "Statistical Methods Facilitate Planning and Scheduling," American Society of Mechanical Engineers Report 66-PET-12, 1966.

Figure 5-6. Frequency Distribution of Bearing Failures

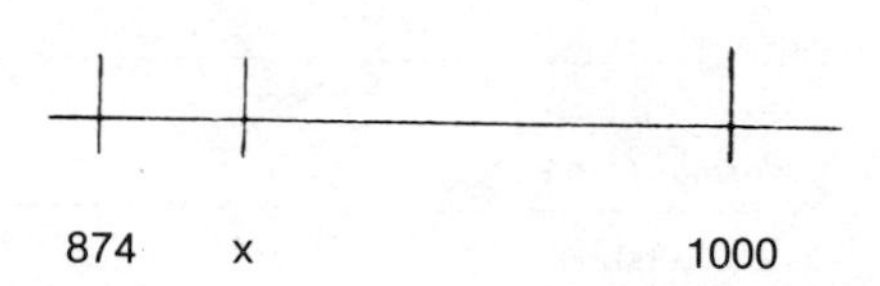

Source: Lawrence Mann, Jr., "Statistical Methods Facilitate Planning and Scheduling," American Society of Mechanical Engineers Report 66-PET-12, 1966.

Figure 5-7. Hours in Service

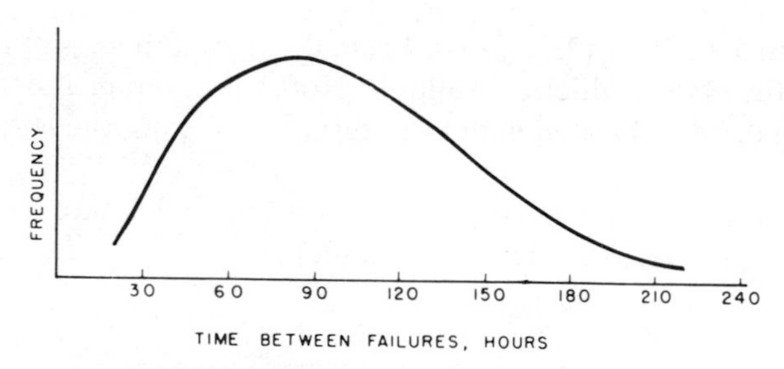

Source: Lawrence Mann, Jr., "Statistical Methods Facilitate Planning and Scheduling," American Society of Mechanical Engineers Report 66-PET-12, 1966.

Figure 5-8. Frequency Distribution of Time Intervals between System Failures

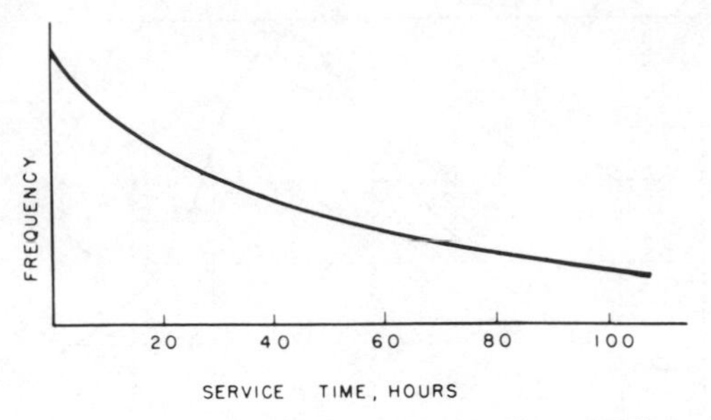

Source: Lawrence Mann, Jr., "Statistical Methods Facilitate Maintenance Planning and Scheduling," American Society of Mechanical Engineers Report 66-PET-12, 1966.

Figure 5-9. Frequency Distribution of Time Required to Repair Compressors

Table 5-3
***Z* Values**

z	*Area under Normal Curve*		*Area under*
0	.0000	1.6	.4452
.2	.0793	1.8	.4641
.4	.1554	2.0	.4772
.8	.2881	2.4	.4918
1.0	.3413	2.6	.4953
1.2	.3849	2.8	.4974
1.4	.4192	3.0	.4987

Source: Lawrence Mann, Jr., "Statistical Methods Facilitate Planning and Scheduling," American Society of Mechanical Engineers Report 66-PET-12, 1966.

and in figure 5-9, 30 hr (2.25 days). From these data there is much information about the system which is available. Note that average is defined as that point which divides the area under the curve in half, not the midpoint of the *x*-axis.

The average number of compressors waiting to be maintained (assuming only one maintenance crew is available) is

$$L = \frac{\lambda^2}{\mu(\mu - \lambda)}$$

where λ = mean arrival rate (units are in arrivals per time unit), in this case 1/3.5 or 2/7 days, and

μ = mean service rate (units are in services per time unit), in this case 1/2.25 or 4/9 days.

Therefore

$$L = \frac{(2/7)^2}{4/9(4/9 - 2/7)} = 1.16 \text{ days}$$

It is of interest to know the average length of time an idle compressor must wait until the maintenance crew starts to work on it. This is expressed by

$$W = \frac{\lambda}{\mu(\mu - \lambda)}$$

$$= \frac{2/7}{4/9(4/9 - 2/7)} = 4.1 \text{ days}$$

Since this is probably excessive, quantitative justification for another maintenance crew is now available. The expressions above for L and W are valid only if one crew is available. Other formulas exist for two or more crews. Also, other formulas exist for cases where the input distributions differ from those used in this example. It should be noted that the results for L and W are *average* results over a period of time and seldom, if ever, will apply in individual situations. This is not altogether a disadvantage in that, if justification for another crew is being attempted, the average or long-term utilization is the information being sought.

This discussion was not intended to present an exhaustive coverage of the subject. It was intended to illustrate to the maintenance engineer some of the statistical and operation research methods which are available to him to help in making decisions. These methods, previously confined to the mathematical and statistical, have evolved to the stage where they have become an engineer's tool.

Nondestructive Tests for Inspection[3]

Within the past few years, several nondestructive tests have been developed for the detection of flaws and potential weaknesses in process units. Used judiciously, these tests can save both time and money by helping to prolong plant runs, to reduce downtime, and to increase safety. We will briefly describe the operating principles and possible applications of five of them—those that utilize magnetic particles; dye penetrants; eddy currents; ultrasonics; and radiography.

The first three are mainly for detecting surface flaws and discontinuities in off-stream units. The other two, and possibly the eddy current techniques, have broader applications in that they can also measure wall thicknesses and often detect internal deposits and the effects of corrosion and erosion. Ultrasonics and radiography have the additional advantage of being adaptable for the inspection of on-stream units at temperature up to 1100° F [593° C].

Off-Stream Techniques

Magnetic Particles: Magnetic particle inspection is quick, easy and especially suitable for detecting surface and near-surface flaws in all types of ferromagnetic materials, regardless of size, shape, composition, or state of heat treatment. A magnetizing current is applied to establish a magnetic field in the test object.

A flaw within the field will distort the magnetic flux and create magnetic poles that can be detected by a change in the direction of the needle of a magnetic compass.

Alternatively, magnetic particles can be scattered within the field, where they will be attracted to and held at the magnetic poles, thus indicating the location and size of the defect.

For maximum sensitivity, the surface must be clean, dry, and free of scale. After the test, the surface should be demagnetized to avoid danger of such things as retention of chips or deflection of the arc during subsequent machining or arc welding.

Dye Penetrants: Penetrant inspection will detect flaws that extend to the surface in such varied materials as metals, glass, and ceramics, almost without restriction as to size and shape. The procedure is to clean the surface, apply the penetrant, and allow sufficient time for penetration. Either special dyes or fluorescent liquids are suitable to make the penetrant visible. A developer is needed with the dyes and ultra-violet illumination with the fluorescent liquids.

The penetrant flows into a surface defect and then is drawn from it by the developer. The sensitivity of the developer coating, the type of developer.

Eddy Currents: Eddy currents are useful for detecting and assessing flaws and determining variations in wall thickness of metallic nonmagnetic materials. However, since the inspection technique is complicated, highly trained operators must run the tests and interpret the results.

The test object is placed within a coil that carries an alternating current, thereby inducing so-called eddy currents to flow in closed circulatory paths within the object. These induced currents produce a secondary magnetic field that opposes the primary or exciting field. Thus the currents at any depth within the object are affected by the opposing field of all the currents above them superimposed upon the initial field. The results must be interpreted in the light of all the following variables:

1. The magnitude and frequency of the alternating current.
2. The electrical conductivity, magnetic permeability, and shape of the test object.
3. The relative positions of the coil and the test object.
4. The presence of discontinuities or inhomogeneities in the test object.

Eddy-current inspections are particularly useful for determining the condition of such items as heat exchanger tubes because many can be monitored within a relatively short time without destructive tests.

On-Stream Techniques

Nondestructive tests based on ultrasonics and radiography have been in use for a number of years, but applications have been limited mainly to units that were off-stream or whose temperatures did not exceed 200° F [93° C]. Recently techniques have been developed to permit ultrasonic and radiographic inspections of on-stream units at temperatures up to 1100° F [593° C].

Ultrasonics. The ultrasonic pulse-echo method permits rapid, accurate wall-thickness determinations that are not affected by coke, scale, or liquid in the system. It is also useful for flaw and weld inspection. Only one surface is required, but it must be exposed. Consequently, insulation must be removed wherever measurements are to be made.

The search unit is coupled to the exposed surface by a suitable acoustical couplant, such as motor oil, glycerine, grease, or water so that the acoustical energy from the transducer can be transferred into the test specimen and back again into the transducer. Some locations may have surface contours that prevent proper positioning of the transducer. It is also difficult to measure thicknesses below 0.1 inch with currently available equipment.

The result of the introduction of ultrasound and the reflection of this energy in the material being tested is viewed on a cathode ray tube, which is part of the test instrument. The observed signals are used to determine the thickness of materials and detect and locate flaws.

Special immersion-type search units with water-cooled transducers must be used to inspect high-temperature lines and equipment.

Another type of high-temperature search unit has a contained liquid column between the hot test surface and the transducer. High-temperature silicone grease is used as couplant. It may be used for on-stream measurements up to 750° F [399° C].

Any commercially available ultrasonic pulse-echo flaw detector may be used with both units, and both give the same type of trace.

Radiography. On-stream radiographic inspection requires radioisotopes and film sensitive to gamma radiation. A single exposure can provide a view of large areas of the interior of process equipment. The technique can also be used to determine tower tray position, the amount of fouling deposits in heat exchangers and other equipment, remaining wall thicknesses, the thickness of scale and coke, and the depth of corrosion.

The image of the wall is recorded clearly on the film because the gamma rays must pass through more metal at this section than at any other point. More gamma rays are absorbed by this section, and hence less are available for exposing the film. The image of the wall appears lighter than the remainder of the radiograph. The approximate length of the longest chord that the rays must penetrate is 2½ inches on 6-inch Schedule 40 pipe and 4½ inches on 12-inch Schedule 40 pipe.

Iridium 192 and Cobalt 60 are the radioisotopes most commonly used. Iridium 192 can be used for obtaining thicknesses of pipe of any size up through 6 inches. Cobalt 60 has more latitude and can be used for pipe of larger sizes. However, Iridium 192 is considerably more portable than Cobalt 60 because of the large amount of lead shielding required for most Cobalt 60 storage containers.

Measuring and Analyzing Vibration[4]

The use of vibration measurement and analysis instruments in maintenance applications has grown phenomenally in recent years. The reasons are varied—escalating equipment, energy, and labor costs; greater realization of the cost-effective benefits of preventive and predictive maintenance; and the development of more accurate, compact, lower-cost instruments.

All rotating and reciprocating equipment lends itself to vibration measurement and analysis. Experience has shown that vibration is an important indication of a machine's mechanical condition. During normal operation, properly functioning fans, blowers, motors, pumps, compressors, etc., emit a specific vibration signal, or "signature." If the signature changes, something is wrong.

Excessive vibration can have a destructive effect on piping, tanks, walls, foundations, and other structures near the vibrating equipment. Operating personnel can be adversely influenced too. High noise levels from vibration can exceed OSHA limitations and cause permanent hearing damage. Workers can also experience loss of balance, blurred vision, fatigue, and other discomfort when exposed to excessive vibration.

A Diagnostic Tool

Vibration measurement and analysis provide a quick and relatively inexpensive way to detect and identify minor mechanical problems before they become serious and force costly unscheduled plant shutdowns [table 5-4]. Worn bearings, loose belts, improperly meshed gears, unbalanced shafts, misaligned couplings, etc., are accompanied by specific changes in signatures. Monitoring these changes can permit maintenance to be scheduled well in advance of a major breakdown.

The intervals between regularly scheduled maintenance shutdowns can be optimized if sufficient information about the mechanical condition of a plant's equipment is known. Critical equipment can be monitored continuously by alarm-equipped systems tailored to the machine.

Definitions Vary

Numerous definitions of vibration exist. Most experts agree, however, that vibration can be described as mechanical motion or oscillation about a reference point of equilibrium. The characteristics of vibration can be defined by: (1) amplitude, (2) frequency, and (3) phase.

Amplitude indicates how much vibration is present. Three parameters can be used to measure vibration amplitude (1) displacement, (2) velocity, and (3) acceleration [figure 5-10].

Displacement is an indication of the magnitude of a motion. It shows how much a part is vibrating. Displacement is commonly measured in mils or inches peak to peak (1 mil = 0.001 inch).

Velocity is the rate of change of displacement, and is the first derivative of displacement with respect to time. Velocity is an indication of the destructive energy of a vibrating part. It is usually measured in terms of peak inches per second.

Acceleration, the time rate of change of velocity, is an indication of the

Table 5–4
Likely Causes and General Characteristics of Machine Vibrations

Frequency(cpm or rpm)	*Cause*	*Amplitude*	*Phase*	*Remarks*
Approximately ½ × rotating speed	Oil whip or oil whirl	Often very severe	Erratic	Occurs only on high-speed machines that use pessure-lubricating bearings
1 × rpm	Unbalance	Proportional to unbalance	Single reference mark	Most common cause. If high in vertical direction, check for loose mounting
2 × rpm	Mechanical looseness	Erratic	Two marks	Usually high in vertical direction. Causes: loose mountings, worn bearing housings.
2 × rpm	Misalignment or bent shaft	Should be half that at 1 × rpm. Large in axial direction.	Two marks	Use dial indicator for positive diagnosis.
1, 2, 3, 4 × rpm	Bad drive belts	Erratic	One, two, three or four. Unsteady	Use strobe light to freeze faulty belt or belts.
Synchronous or 2 × synchronous	Electrical	Usually low	Single or double rotating mark	If vibration amplitude drops instantly when power is turned off, cause is electrical.
Many times rpm—usually 8,000 to 25,000 rpm	Bad bearings (antifriction)	Erratic	Erratic-many reference marks	If amplitude exceeds 0.25 mil, suspect faulty bearings.
Rpm × number of blades on fan or pump	Aerodynamic or hydraulic			Uncommon cause.

Source: Ted. F. Meinhold, "Measuring and Analyzing Vibration," *Plant Engineering,* October 4, 1979, p. 83.

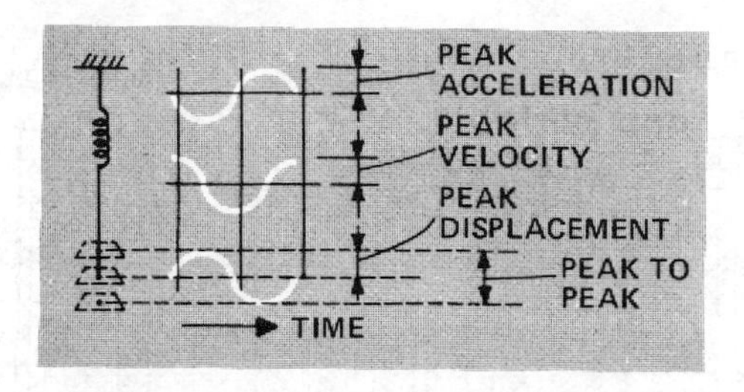

Source: Ted F. Meinhold, "Measuring and Analyzing Vibration," *Plant Engineering,* October 4, 1979, p. 83.

Figure 5–10. Vibration Parameters

forces acting on the part. Acceleration magnitude is proportional to displacement and the square of the frequency of vibration. It is measured in terms of gravity, or G's.

Frequency is the number of vibration cycles in a given period of time. Frequency is usually expressed in cycles per minute (cpm) or cycles per second (Hz).

Phase compares the position of the vibrating part at a given instant relative to a fixed point or another vibrating part. Phase is normally expressed in degrees, where one complete cycle of vibration equals 360 degrees.

Vibration severity is a measure of the destructive energy in a machine. Because the energy of heat and wear is a function of the square of velocity, a measure of vibration velocity is a direct measure of vibration severity. Displacement and acceleration may also be used to measure vibration severity, but when they are used, the frequency of the vibration must be known. Charts showing the relationship between displacement and frequency, and acceleration and frequency [figure 5–11] can be used to approximate vibration severity. Only filtered displacement and acceleration measurements taken at each frequency of interest should be applied to the chart.

Vibration Meters

Peak vibration levels over a wide frequency band are measured primarily with vibration meters. Early vibration meters used spring-loaded probes, oscillating light beams and mirrors, and vibrating reeds. Vibration levels were read directly on a scale or traced by a stylus on pressure-sensitive tape.

In recent years vibration meters have been significantly improved. Most modern meters are electronic; they consist basically of a transducer, inter-

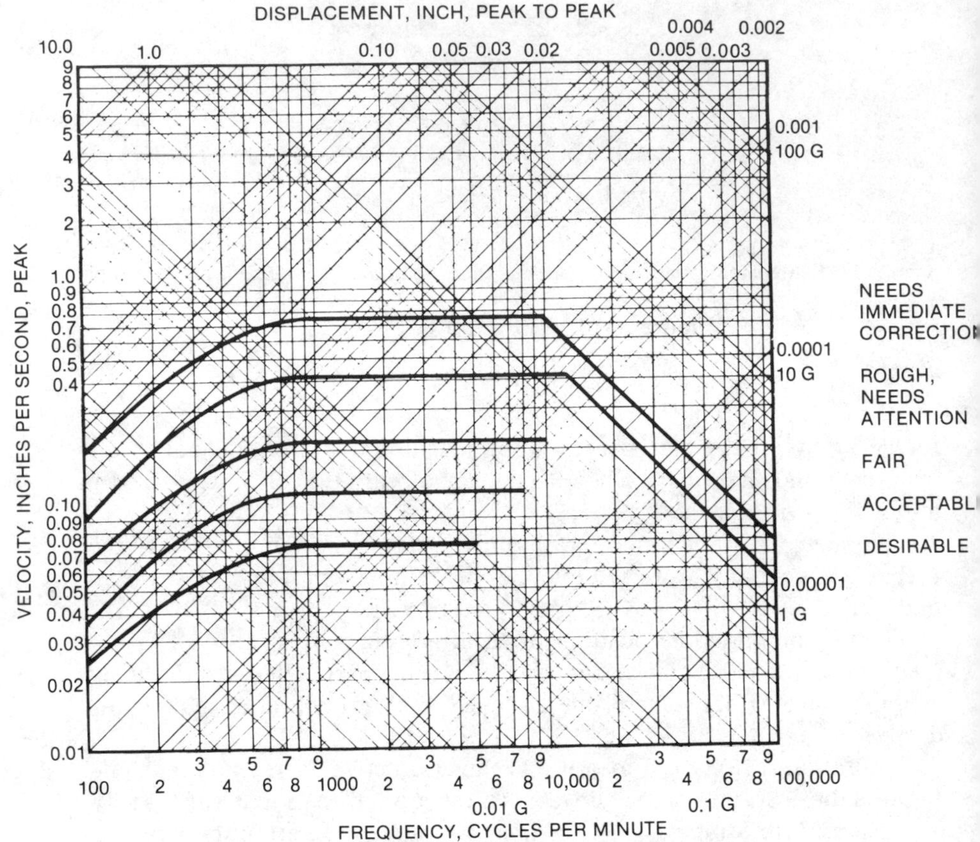

Source: Ted F. Meinhold, "Measuring and Analyzing Vibration," *Plant Engineering*, October 4, 1979, p. 83. Vibration severity levels can be approximated from displacement, velocity, or acceleration measurements by using this nomograph. Only filtered displacement and acceleration measurements should be used. Acceleration values are shown in G's.

Figure 5–11. Vibration Severity Levels

connecting cable, and an a-c voltmeter with suitable rectifiers and damping circuitry. The transducer senses the vibration and converts it to an electrical signal that is measured by the voltmeter. Modern meters are ruggedly constructed, compact, and portable. Power is supplied by rechargeable batteries.

A vibration meter that operates on the laser Doppler principle is also available. Reflected light from the vibrating surface is collected and collimated to a beam by a lens system and mixed with a frequency-shifted reference beam in an optical heterodyne detector. The detector frequency is subsequently downshifted and the resulting frequency shift in combination with signal processing provides an accurate vibration measurement. The meter is designed primarily for difficult applications in which conventional instruments cannot be used.

Transducers

The key component in any vibration measurement system is the vibration-sensing device, or transducer. The three most common types of transducers are: (1) displacement, (2) velocity, and (3) acceleration. No design is best for all applications. In general, displacement transducers are used to detect frequency vibrations below 600 cpm from highspeed, heavy, rotating machinery; velocity transducers are used for 600 to 120,000 cpm vibrations from medium-speed fans and pumps, and accelerometers are used for vibrations above 120,000 cpm from gear boxes and antifriction bearings. However, actual application ranges are much broader [table 5–5].

Displacement transducers are noncontact devices used primarily to measure vibrations in machinery equipped with journal bearings, where the rotor/stator mass ratio is small, or when large vibrations are transmitted to the base from other machinery. Several kinds of displacement transducers are available, but the most commonly used device operates on the eddy current loss principle.

In operation, high-frequency alternating current is passed through a coil embedded in the transducer's tip. The coil generates a magnetic field in front of the tip. When an electrically conductive surface is brought into contact with the field, some of the energy is absorbed. The return voltage decreases as the gap between the tip and the conducting surface is reduced. The meter measures the difference between the output and the return voltages, which is proportional to the gap, or the vibration of the surface. When mounted in pairs 90 degrees apart in a journal bearing, displacement transducers can detect machinery malfunctions with a high degree of accuracy.

Displacement transducers are inexpensive, but an oscillator/demodulator is needed to supply the high-frequency current. The transducers are not affected by lubricating oils and can withstand temperatures to 250 F. Higher temperatures will cause d-c voltage shifts and accuracy losses. Displacement transducers will also sense mechanical and electrical runout (inherent noise), which can cause incorrect readings.

Velocity transducers are the most common of the three major types of transducers and the most useful to maintenance. They are rugged and self-contained and require no amplifiers. In operation, a velocity transducer is somewhat analogous to a reciprocating generator. The transducer's interior consists basically of a fine-wire coil and a permanent magnet.

Depending on the design, either the coil or the magnet is mounted on a spring. When the transducer is attached to a vibrating surface, the housing of the transducer moves with respect to the springmounted element, causing the coil to generate an a-c signal. The output voltage is proportional to the relative velocity between the coil and the magnet. The velocity signal can be integrated to measure displacement. The signal can be transmitted over 1000 ft. through a cable without amplification.

Table 5–5
Transducer Performance and Cost Comparisons

Type	*Frequency Response*	*Temperature Range*	*Advantages*	*Disadvantages*	*Applications*	*Installed Cost*
Displacement	d-c to 3.7 kHz	−60 to 250	Very low frequency; good signals on most journal bearing machines.	Requires mounting to be relatively stationary for accurate measurements. High cost because of probe mounting.	High-speed journal bearings when rotor/stator mass ratio is small	$400
Velocity	10 to 1500 Hz	−60 to 900	Rugged; easy to mount; low impedance; high output; wide temperature range.	Large size; limited frequency range.	Fans; pumps; low- and medium-speed equipment.	$400
Accelerometer	<1 Hz to 100 kHz	−70 to 1400	Easy to mount; wide frequency range; small size; rugged.	Requires charge amplifier; high impedance; low output (especially at low frequencies).	Gear boxes; antifricton bearings; casing measurements; high frequencies.	$500 to $1500
Accelerometer with internal electronic components	<1 Hz to 30 kHz	−60 to 250	Easy to mount; wide frequency range; small size; does not require charge amplifier.	Temperature limitations; some noise generated by integration to velocity and displacement; fragile.	Gear boxes; antifriction bearings; wide range of pumps, fans, and compressors.	$500 to $1500

Source: Ted F. Meinhold, "Measuring and Analyzing Vibration," *Plant Engineering,* October 4, 1979, p. 84. (Based in part on data provided courtesy of Dytronics Co., Inc.)

Velocity transducers are usually fastened to equipment by bolts, cements, or a magnet. These devices resist temperatures to 900 F.

Acceleration transducers are rapidly gaining in popularity because in addition to measuring acceleration, they can measure velocity by integration (electronically) of the signal and can measure displacement by double integration of the signal. These vibration-sensing devices are extremely versatile. However, at low frequencies (less than 10 Hz), acceleration transducers (also called accelerometers) often have difficulty detecting meaningful vibrations because of their low signal-to-noise ratio.

The most common type of accelerometer consists of a mass attached to piezoelectric discs (usually quartz or some proprietary compound). The discs generate an electrical output when they are subjected to vibration. The signal is amplified, converted to a-c voltage, and measured by the meter. Some transducers require a remote amplifier; in others, amplifications is an integral part of the unit.

Accelerometers are small, lightweight, and durable. Normal operating temperature range is −70 to 500 F. Special piezoelectric ceramic discs are available for applications to 1400 F.

Vibration Analyzers

A frequency analysis of a vibration signal is necessary to determine the individual frequency components of the signal. The two major types of vibration analyzers are: (1) swept frequency (tunable filter) and (2) real time. Because identification of specific machinery problems depends on the capabilities of the analyzer, proper selection is important.

Swept-frequency analyzers are usually used for detecting and analyzing signals from machines operating at steady-state conditions with little, if any, variation in speed, load, or power. Most of a plant's frequency-analysis problems can be handled satisfactorily with these instruments.

They have a tunable band-pass filter that separates vibrations on the basis of frequency, so vibration levels can be measured for each frequency. Other similar analyzers use a number of individual fixed-frequency filters that are frequency-scanned sequentially by switching, usually by octave band.

Most of these analyzers operate on rechargeable batteries or a-c power. The instruments are ruggedly constructed and self-contained; they are portable and can be taken to various plant locations. Most can use displacement, velocity, and acceleration transducers. Various readouts, including analog, digital, and direct-tracing charts, are available.

Real-time analyzers (also called spectrum analyzers) are used for detecting and analyzing vibration signals from machines operating at varying conditions. Real-time analyzers are considerably more expensive than swept-frequency instruments.

They detect, measure, and display vibration amplitudes and frequencies instantaneously. The instruments operate on the basis of fast Fourier transforms (or the time compression principle) to increase the data-response speed. Real-time analysis is an effective tool for characterizing a vibration signal, identifying its source, and detecting its change. For example, improper gear meshing, excessive rotational vibration, and bearing noise can be identified by associating mechanical periodicities with specific frequencies. Data readouts are on CRT displays and plotters. Numerous options, including interfacing to recorders, terminals, and computers, are available.

A disadvantage of real-time analyzers is that they require an a-c power source. Long cables are needed if the instruments are to be moved about a plant. The analyzer may present a safety hazard if it is used in an explosive or flammable environment.

One method of escaping the cable problem is to use a battery-powered, instrumentation tape recorder to collect the field data and replay the tape to a real-time analyzer in the laboratory. Replaying the tape at a higher speed reduces analysis time and brings very low frequency signals into the frequency range of less expensive analyzers.

Continuous Monitoring

Numerous continuous-vibration monitoring systems are available. The basic purpose of these systems is to provide constant surveillance of machinery vibration levels at critical points. The system's required degree of sophistication is determined by the importance of the machine or process in a plant. Various alarms are used. Some simply warn the operator of excessive vibration levels; others shut down the equipment.

Computer-based systems account for an increasing percentage of the continuous-monitoring systems in use today. Most of these are used in large power, petrochemical, and chemical plants to monitor compressors, turbines, and generators.

Compact, microprocessor-based continuous-monitoring systems are being installed in some larger process plants. One such system continuously samples data from 48 input channels and prints out information on a 4-inch paper tape. The system performs three basic functions; data logging, trending, and spectrum analysis. In addition, the instrument constantly inspects its own circuitry and responds to any circuit fault signal from its companion vibration monitor.

Deciding what equipment to purchase to launch a vibration-measuring program is not easy. No firm guidelines exist for determining which transducer, meter, or analyzer is best for a particular application. The process of vibration measurement and analysis is an art as well as a science, and on-the-job experience is often the best teacher.

Most vibration experts emphasize the importance of identifying the problem before selecting instrumentation. If the maintenance engineer wants only to get peak readings, then all he needs is a simple vibration meter. However, if he wants to balance a large compressor shaft, he'd better use a vibration analyzer that has a synchronous tracking filter and phase measurement capability.

Future plans for the instruments must also be considered. Analyzers may cost a bit more initially than meters, but they have much greater capabilities. If properly used, analyzers can pay for themselves within a year because of reduced machine downtime and maintenance. Most manufacturers are glad to discuss a potential customer's vibration problem, detail the capabilities of their equipment, and recommend what they believe is needed.

Machinery manufacturers can also recommend what type of transducer to use and where to fasten it to the equipment and can advise users of the mechanical problems they are most likely to experience with the equipment.

Training

Extensive training is not needed to operate a vibration meter. Most plant personnel can be taught to take simple vibration-level measurements in about an hour. Analyzers are more complicated. Skilled technicians are usually required to operate them efficiently. The operator must be familiar with the equipment that is being monitored and understand what is being measured and why. Individuals with engineering training should interpret the collected data.

Some instrument manufacturers offer free training courses to customers. However, these courses are usually geared to the manufacturer's equipment. Courses range in length from 3 to 5 days. Institutions and technical societies also offer occasional courses.

One of the most common mistakes in using vibration instruments is mounting the transducer in the wrong place. Knowing where to place it comes with experience. Always try to get the sensor as close to the source of vibration as possible. The farther away it is, the more complex the transfer function, and the greater the chance for erroneous diagnosis. Velocity transducers should not be used around magnetic fields without proper shielding because they will give false readings. The output signal will act like a vibration signal, but it isn't—it is the effect of the magnetic field. And if day-to-day comparison readings are made, it is important that the transducer be placed in the same position every day. This can be a problem with hand-held instruments.

The surface on which a transducer is mounted should be flat and clean. Often, paint must be scraped off to ensure good contact. If threaded studs are used to attach a transducer, they should not be overly long or overtightened, to prevent damaging the sensor.

Any object (bracket or adapter) used between the transducer and the surface has spring-like qualities and its own natural frequencies. As a result, it will tend to amplify or distort the vibration at frequencies close to its own natural frequency and cause measurement errors.

Hand-held transducers are generally satisfactory for periodic vibration checks. Only enough pressure to keep the sensor from chattering on the surface should be applied. If the sensor is used on curved surfaces, vibration will be measured only in the direction parallel to the sensor's axis. Any unsteadiness of the hand that allows the axis direction to vary may result in unsteady readings.

Maintenance engineers must also remember that vibration is only the indication of a machine's condition. Other indicators must be watched: motor-winding temperature, bearing temperature, inlet and discharge pressures, etc. Vibration can be compared to a human's pulse—the pulse may be normal, but the individual may have a temperature of 105 F and be seriously ill. A pulse measurement alone will not reveal the overall state of his health. The same holds true for vibration—it is only one of several critical parameters and will not reveal the overall condition of a machine.

Outlook

The future appears bright for vibration measurement and analysis instruments. Although prices for meters and swept-frequency analyzers have risen in recent years, the increase has been small compared to the improved quality and capabilities of the instruments. Today's instruments are much more accurate, compact, and reliable than those of the 1960s.

Prices of real-time analyzers have plunged. Instruments that cost $20,000 only 8 or 10 years ago can be purchased for less—and today's lower cost unit has much greater capability and reliability than its predecessor. Battery-operated real-time analyzers are under development and may become standard in the future.

Increased improvements in microprocessors, minicomputers, and computers ensure the market growth for continuous-vibration-monitoring systems. Maintenance engineers are becoming acutely aware of the many benefits that can be reaped by knowing in advance when major maintenance will be required on critical equipment. Computerized systems predict trends and establish the optimum time for a shutdown. Continuous-monitoring systems will continue to be improved. They will be able to do more, and do it more precisely and economically. Mass production of packaged monitoring systems will mean reduced prices. Processing machinery may be delivered to customers with built-in computerized monitoring systems.

Infrared Thermography[5]

Infrared Thermography is a relatively new tool in the inventory of plant-maintenance personnel and energy-conservation coordinators.

Real-time infrared-thermography equipment provides a rapid and accurate inspection of heat loss and heat-radiating areas of on-line process devices that would ordinarily be invisible. In addition to the rapid discovery of these otherwise invisible and unknown anomalies, it is possible to record them by using a simple camera attachment and indicate accurate temperatures and location of these areas.

The equipment operates very much the same as a closed-circuit television system. Instead of receiving incoming light, the infrared camera or scanner receives emitted infrared radiation from the object being observed.

The infrared radiation is directed through special optics, to an infrared detector that has been supercooled to a temperature of approximately −270 °F.

The detector changes voltage outputs as it is impinged with the incoming infrared radiation. These voltages are then amplified and through other electronic circuitry presents a black and white image on a picture tube much the same as a television presentation.

On some equipment the image is framed at a rate of 25 frames per second. Assuming that one point in the thermograph is a known temperature, then the operator can accurately determine the temperatures of other areas in the thermograph. In one frame, or thermograph, it is possible to detect and measure temperatures of as many as 10,000 points the equivalent size of an ordinary thermocouple contact.

The minimum detectable difference of temperatures in the range of 20 °C. to 40 °C. is .2 of a degree Celsius. In the range of 100 °C. to 900 °C. the minimum detectable temperature difference is 2 °C.

When one considers the portability and mobility of the infrared equipment available today, many cost-effective applications in refineries and petrochemical plants come to mind.

The first cost-effective application to industry for this equipment was the inspection of electrical distribution and transmission systems. This inspection evolved some years ago and is still a very important one.

The major advantages of an infrared inspection of any electrical system are:

All common electrical failures announce themselves by beginning to radiate in the infrared spectrum, some times months before they fail due to *overheating*. An infrared inspection of electrical components will detect faulty connections and components rapidly without an interruption of service.

Extreme accuracy and the ability to determine the magnitude of temperature rise permits a high degree of discrimination as to the degree of severity of the problem.

Priorities can be readily assessed for scheduling corrective actions.

Precise pinpointing of problems minimizes the amount of time required for preventive maintenance.

Effort can be concentrated on corrective measures rather than on searching for problems.

By instituting regularly scheduled infrared inspections, unexpected electrical outages can be virtually eliminated.

Many other applications for infrared thermography inspections have surfaced during the past several years in refineries and petrochemical plants as the infrared equipment was improved.

Failure in product flow (erosion, corrosion, thinning, blockages and leakages) often are caused by the product flow itself or external conditions. In both cases, infrared patterns can be recorded, and unlike electrical problems, the data must be recorded by an experienced infrared technologist and carefully evaluated before field repairs are made.

For example, consider a heat exchanger which has both blockages and shell thinning. When operating properly, a continuous flow of hot product enters the main cavity of the exchanger and circulates around cooler product tubes. When inspected with infrared thermography the external shell should show an almost uniform, gradual reduction in temperature from hot-product inlet to cooler outlet sections.

Internal corrosion on baffles or around inlet/outlet sections, and deposits on tube surfaces or internal walls can cause abnormal surface-temperature patterns.

Normal base-temperature profiles should be made of all heat exchangers before the problem develops, to more profitably interpret the infrared thermographs when the problem occurs.

Inspection of refractory and insulation is another cost-effective use for infrared thermography in both petrochemical plants and refineries. These inspections not only determine the location of internal refractory deterioration but also the extent of deterioration by use of computer programs now available.

Vessels with refractory between an internal liner and an external shell can be inspected without removing the outer shell by using external temperature as a guide of internal condition. There infrared data can be used to perform on line repair by pumping additional refractory between the shell and liner. After repair, additional infrared thermographs will indicate if an effective repair has been made.

Infrared thermography, along with other plant data, can be used to inspect for energy losses but also becomes a pictoral and permanent record.

During an inspection of a large plant a number of steam traps were inspected, 3.4 percent were malfunctioning and gave no visible signs that they were defective. A number of steam leaks were found that were not observed previously. Records of all defective traps and their respective locations were recorded and replacement traps installed.

With a conservative estimate of energy loss of $20 per day, per trap, this inspection showed that energy losses totaling $460 per day or $13,800 per month were being consumed by the defective traps and leaks.

Savings from this use of infrared thermography inspection resulted in the reduced use of energy plus the savings in reduced inspection time in location the defective traps and leaks.

With the present availability of hand-held, programmable computers and proper programs to match the applications, infrared and other plant data can be converted to absolute energy losses in thermal units.

These techniques provide on-site energy-loss analysis for both inspection and maintenance personnel and is much more meaningful to management when they know the amount of actual energy dollars lost.

A detailed explanation of all the cost effective uses for infrared inspections in refineries and petrochemical plants would be too lengthy. Following is a list of a few items of process hardware.

Infrared thermography inspections of the following items have proven cost effective for detection of thermal loss patterns:

Cat crackers,
reactors,
flue-gas lines,
slide valves,
c.o. boilers,
furnace walls,
on-line furnace and heater tubes,
heat exchangers,
insulated steam lines,
coke drums,
heat loss in steam boilers,
motor bearings,
pump bearings,

flare lines,
brushed and pigtails on high h.p. motors,
generators,
steam traps,
rotating kilns,
incinerators,
sedimentation in tanks,
liquid level in tanks,
uneven distribution in manifolds,
plugged or restricted flow in product lines,
fin type coolers,
air compressors,
steam turbines,
thermal distribution in cooling towers,
building insulation,
heater and boiler stacks,
cold areas in steam traced product lines,
location of smoldering fires in coke, grain etc., storage tanks,
determine percentage of internal refractory deterioration in vessels and, waste heat boilers.

In the case of a 100,000 b/d, refinery, located in the United States, a competent and experienced infrared inspection team could be expected to complete the following inspection in ten working days.

1. Inspect all insulated steamlines for improper or inadequate insulation and steam leaks.
2. Inspect and record all faulty steam traps.
3. Inspect all relief valves for proper operation, regardless of elevation above grade.
4. Inspect all cat crackers, reactors, and furnaces for refractory deterioration and present an external thermal profile of each vessel.
5. Inspect all heater and boiler stacks for overheating.
6. Inspect furnace tubes for coking in on-line furnaces fired by natural gas.
7. Complete inspection of the electrical distribution and transmission systems giving a final report on the potential trouble spots that would ultimately cause unexpected electrical outage.

After the inspection there would remain time for a general meeting with the concerned supervisory personnel to discuss the reports and other applications that may surface during the initial inspection.

A Survey of Maintenance Diagnostic Instruments[6]

Wider use of maintenance instruments is one of the big remaining untapped areas for cost reduction in the better-run plant maintenance departments. If maintenance management does not emphasize their use, sophisticated PM and troubleshooting testing could pass out of the hands of plant-maintenance departments entirely.

No recognized definition of diagnostic instruments exists. They are portable, temporary-use, analytical instruments, tools, or aids that predict equipment failure, measure the condition of equipment or the quality of its performance, or diagnose the cause of failure or off-normal operation. Note that monitoring, switchboard, process control, and precision laboratory instruments are excluded.

The following instruments are examples of PM aids in use today. The listing is intended to be illustrative, not all inclusive.

A shirt-pocket-size thermistor thermometer about the size of an old-fashioned pocket watch, with batteries and a plug-in probe, reads temperatures within two seconds.

An ultrasonic hardness tester reads the surface hardness in Rockwell C. of rolls, bearing races, shafting, and so on. A lightweight probe is held against the surface; the test takes about three seconds.

An ultrasonic corona detector "hears" corona in voids in cable or splice insulation, in poured potheads, or across insulators before corona can damage insulation.

An analyzer that measures the cathodic potential in millivolts also measures corrosion leakage current between buried tanks or structures and nearby piping or conductors.

A recording volt-ohm milliammeter, with all the flexibility of a standard VOM, can make a graphic record of any scale reading for hours at a time. (Another battery-operated portable VOM has an all-digital lighted readout with decimal point to end reading errors and speed routine circuit checking.)

A laser-beam source and detector readout permits alignment of shafts, fixtures, or structures hundreds of feet apart to a precision of better than 0.001 inch.

A pistol-grip static meter measures from a foot away the electrostatic charge on any surface at which it is aimed.

A portable sonic-resonance tester measures the thickness and soundness of concrete, wool piling, and plywood and even checks the uniformity of fire brick, critical plastics, or metal.

An eddy-current tester with a point probe spots tiny discontinuities on or below metal surfaces without touching the object scanned.

A pencil-probe leak detector with a neon light in the transparent probe flashes whenever the point of the probe gets near a Freon leak.

A five-channel pen recorder will record continuously at pen frequencies well above 60 cycles for over two years without a change in microfilm roll.

A pencil-size tension checker for V-belts.

Small, quick-connect, self-sealing plug-in connectors. Half of the connector screws into any tapped fitting in a hydraulic or pneumatic system while the other half is attached to a pressure gauge. Troubleshooting gauges can then be plugged in much as a voltmeter is in an electric circuit.

A thermopile heat-flow sensor that can be connected to any vacuum tube voltmeter can be calibrated to read heat loss through insulation (in BTU's per square foot per hour) or to check the efficiency of different areas on a heat transfer surface.

A buried cable and cable fault locator allows unskilled crews to trace the course of buried pipe or cable and determine its depth.

A phase-angle meter measures phase-angle difference up to 360° between two about-to-be-connected circuits.

A safety checker for portable electric tools within a few seconds can check for wrong connections, ground leakage under high voltage, adequate grounding of the third wire, and, finally, actual mechanical operation.

There is also a *diagnostic service* that provides an overnight report on the amount of moisture or contaminants in a refrigerant system. The service supplies evacuated cylinders for sampling the gas.

Most Underutilized Instruments

Temperature-sensitive crayons, stickers, or paints that change color permanently, or melt, to indicate when their rated temperature is reached are also useful for ascertaining whether electrical equipment or bearings are overloaded between inspections.

Stethoscopes, both the doctor's kind and the electronic ones, with practice, can be used to hear mechanical trouble in the midst of other noises. Bearing analyzers are actually one form of stethoscope. Pocket-pencil-sized ones come with a pick-up sensitive to phenomena other than noise—to vibration, magnetic fields, and so on.

Smoke bombs, like temperature-sensitive stickers, may not seem to qualify as instruments, but they are highly useful tools in diagnosing air- and heat-flow problems. They show up drafts and dead layers of hot or cold air and permit measurement of actual air velocity in open spaces quickly. The modern ones create a minimum of irritation in occupied areas.

Thermistor thermometers have the temperature-sensing probe at the end of a flexible lead extending from the portable instrument on which the temperature is read directly. Because of the small mass of the sensing tip, temperature reading is so fast that thermistor thermometers make mercury-column and bimetallic dial-type thermometers obsolete for many diagnostic users. Accuracy is good and getting better.

Fiber-optic inspection probes allow examination of the internal mechanism of a closed or inaccessible gear case or other housing. Only an opening large enough to slide in the half-inch (or larger) diameter probe is needed. This is very useful for inspecting gear teeth or finding lost or broken parts.

The stroboscope. The record of diagnostic and trouble-preventing capabilities of this tool keeps growing. Improvements in recent years include light output intense enough to freeze high-speed movement even in today's well-lighted plants; highly accurate speed indication read directly on the dial; photoelectric or proximity pick-off triggering so that no physical contact has to be made with the moving shaft or part; and variable time delay so that nonsymmetry or vibration during any part of the cycle can be studied in detail. This instrument shows up vibration, misalignment, belt slip, hunting, and chatter. If it is used with an optic-fiber probe, one can even see malfunctions in enclosed places. The instrument is especially valuable for maintenance technicians who like to see, rather than mathematically analyze, trouble.

Circuit-breaker testers. It has taken decades of electrical fires, accidents, and near escapes of alert maintenance staffs to the fact that all may not be well with circuit breakers and relays that are hidden away in switchgear, control panels, and substations. Yet the majority of plants still assume that their 600-ampere molded-case breakers will trip with reasonable promptness at 625 amperes. They may not trip at all, in fact. Until solid-state or some inert nonmechanical devices that will unfailingly open power circuits under overload are built, the only safe procedure is routine testing of relays and breakers, both molded-case and power types.

Ultrasonic listening devices. These instruments are known under different names—ultrasonic leak detectors, bearing analyzers, corona detectors, and so on. They "hear" frequencies far above the highest capacity of the human ear (typically, 45,000 cycles) and translate them electronically into ones we can hear. The devices are effective because many trouble noises are far more pronounced in the ultrasonic ranges. Furthermore, much noisy plant equipment does not make any noise in the ultrasonic region. The operator of this instrument, wearing headphones, can walk through a noisy plant and, by scanning with the directional microphone, will hear only trouble-leaking air valves of fittings, unlubricated bearings, corona discharge, and so on.

Torque meters. These bench-mounted instruments indicate the maximum torque output of air, hydraulic, or electric motors. They are particularly useful in checking portable tools. Tool experts say that the chances are better than 50 percent that any given tool in a typical plant is not capable of coming up to full torque within an acceptable period of time. This directly affects production output and product uniformity.

The DC overpotential tester. Cable, insulators, terminations, motors, generators, transformers, and switchgear components rated 2300 volts and above need routine checking at overvoltage to detect signs of impending failure. Within the last ten years, overpotential testing with DC has become the preferred method of PM testing of medium- and high-voltage insulation. The rate at which current leakage rate rises as the insulation is stressed by overvoltage enables one to evaluate insulation condition and even to predict end of life. The method requires properly trained and alert operators. It has not achieved anywhere near the popularity of megohmmeter testing of equipment rated 600 volts and below.

Transient recorders. When plants began experiencing a rash of troubles with early transistors, a new instrument was developed that records the peak transient voltage appearing on the line since the last time the instrument was reset. To the surprise of everyone, including the utilities, voltage spikes of two to five or more times normal were common on many circuits. With that, the transient recorder boom disappeared. These relatively low-priced instruments have a wide range of uses, however, for maintenance departments that have the time and capability to make their own investigations rather than relying on vendors. The devices can be connected to any electrical-quantity-measuring signal. Thus, they can indicate and hold the loudest noise, the peak volume, the greatest volocity, the highest speed, and so on, that occurs overnight or over the next month.

Computing the Number of Craftsmen per Facility[7]

A common problem in any PM program is the optimum manning of a facility. One solution to such a problem lies in the waiting-line approach using the Monte Carlo method to simulate the distributions. Although such procedures readily lend themselves to computers, an illustrative problem is presented which can be solved manually.

We assume a group of 12 similar machines and investigate the number of servicemen required to keep the group operating at minimum cost per unit of output. To help visualize the problem, we prepare a flow chart, as in [figure 5–12].

We need assumptions or estimates of the number of machines, distribution of running times between the end of one service and the beginning of the next on the same machine, cost of downtime, distribution of service times, cost of service channels, availability rule for channels, and priority rule. We start with 12 machines in the upper box [in figure 5–12]. How long does each run before it slides into the lower-left box? How much does it cost per hour for machines to wait for service? How fast (a relative frequency distribution) can a serviceman repair a machine, discharge it to the upper

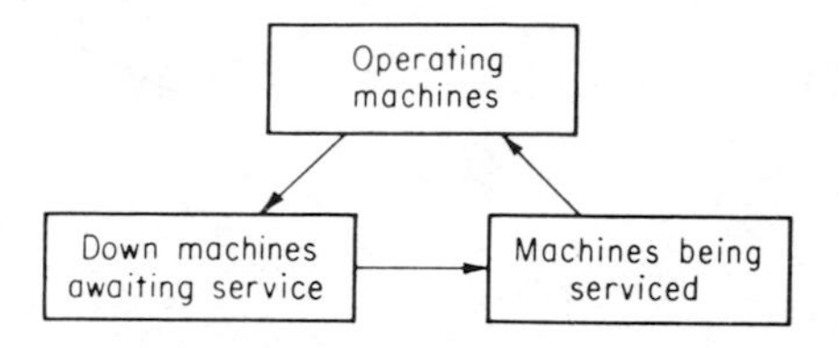

Source: B.E. Goetz, *Quantitative Methods: A Survey and Guide for Managers.* Copyright 1965 by McGraw-Hill, Inc., used with permission of McGraw-Hill Book Company.

Figure 5–12. Flow Chart of Maintenance System

box, and draw another from the queue? How much does it cost to provide a service channel? We shall discuss these questions in the following paragraphs.

Assumptions

We assume 12 similar machines to be serviced by x repair mechanics. Our objective is to determine the optimum number of such servicemen. We investigate the situation and find:

1. The relative frequency of running times from discharge from servicing to reentering the queue awaiting service is as given in the cumulative frequency curve shown in [figure 5–13]. The dotted line shows that 10 percent of the time a machine will run for 24 minutes or less before shutting down for service. The solid line shows that 40 percent of the time a machine will run for 24 minutes or less before shutting down for service. The solid line shows that 40 percent of the time a machine will run for 50 minutes or less before requiring service.

2. The cost of downtime is $5 per hour. Part of our needs for the products of these machines is met by subcontracting at a price for a standard hour's output $5 greater than our out-of-pocket costs to produce a like amount. Every hour of production lost by machine downtime increases the amount subcontracted at a price premium of $5 over our own production cost.

3. These machines are assumed to be subject to six standard types of breakdowns, each of which requires a standard procedure for repair. Standard times have been established for each repair procedure, and little variation occurs above or below the established standards. These six types require the standard times and occur with the relative frequencies given in [table 5–6].

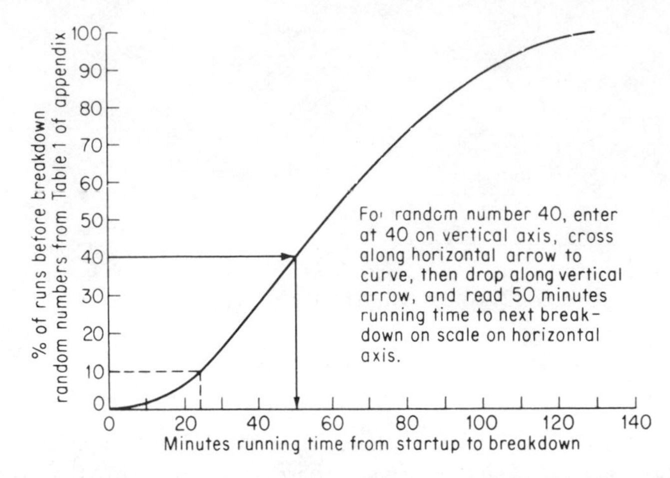

Source: B.E. Goetz, *Quantitative Methods: A Survey and Guide for Managers.* Copyright 1965 by McGraw-Hill, Inc.; used with permission of McGraw-Hill Book Company.

Figure 5-13. Generating Simulated Data as to Running Times

[Figure 5-14] is a graph of the cumulative relative frequencies with which these breakdowns (or needed adjustments) occur. To generate simulated service-time data, we get a two-digit random number from any table of random numbers, enter [figure 5-14] on the vertical axis at the selected figure, cross horizontally to the cumulative frequency line, and then drop vertically to read the simulated data on the scale on the horizontal axis. Thus any random number from 60 to 79, inclusive, indicates a simulated service-time requirement of 80 minutes.

4. Cost per service channel is $16 per serviceman per 8-hour day plus $3 per hour overtime. We assume fractional channels are infeasible (no

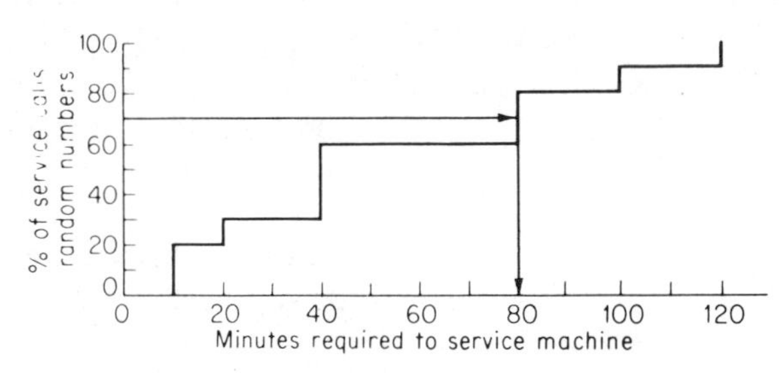

Source: B.E. Goetz, *Quantitative Methods: A Survey and Guide for Managers.* Copyright 1965 by McGraw-Hill, Inc.; used with permission of McGraw-Hill Book Company.

Figure 5-14. Simulated Service Times

Table 5–6
Standard Times for Repair

Type of service required	1	2	3	4	5	6
Standard time to repair	10	20	40	80	100	120
Relative frequency of type	20	10	30	20	10	10

Source: B.E. Goetz, *Quantitative Methods: A Survey and Guide for Managers*. Copyright 1965 by McGraw-Hill, Inc.; used with permission of McGraw-Hill Book Company.

half-time servicemen). Moreover, overtime is added in increments of 1 hour. If a serviceman works an 8½-hour day, he is paid $16 for 8 hours plus $3 for a ninth hour.

5. Servicemen are essentially interchangeable, that is, equally competent on all required types of repair. Any available man can service any of the 12 machines; that is, the service channels are pooled. However, each man works by himself on any given job. A machine cannot be put into operation sooner by two or three men working on it. All servicemen come to work an hour after the machines begin operating in the morning. All work a minimum of 8 hours per day. (Some hours may be idle, but the man is paid his usual rate for standby time.) All machines are put in running order before servicemen leave at night, although each may leave when he finishes the job on which he is working unless there are machines waiting for service. All machines must be ready for operation at the beginning of the following day.

6. Since all 12 machines are alike, as servicemen become available they draw from the queue the machine that needs least work to return it to production. However, once started on a job, the serviceman finishes it. He does not interrupt it to fix another machine, no matter how little attention it needs.

Monte Carlo Procedure

Since the machines average about 60 minutes of running time before breakdown and since service time averages around 60 minutes per breakdown, we would expect about half the machines to be down even if there were 12 servicemen. With fewer than six servicemen, we would expect all to be busy most of the time, with a queue of machines waiting for attention most of the time. With downtime at $5 per hour and servicemen at $16 per 8-hour day, we would surely need six or more but not more than 12. Consequently, we should try alternative plans for supplying 6, 7, 8, . . . channels,

Table 5–7
Simulated Data

M	*RN*	*Ser*	*RN*	*Run*	*M*	*RN*	*Ser*	*RN*	*Run*	*M*	*RN*	*Ser*	*RN*	*Run*	*M*	*RN*	*Ser*	*RN*	*Run*	*M*	*RN*	*Ser*	*RN*	*Run*
1			26	38	7	88	100^{D}	37	48	1	98	120^{F}	20	33	7	19	10^{F}	35	46		94	120	66	73
2			09	23	8	09	10^{E}	96	113	2	65	80^{E}	64	71	8	27	20^{F}	14	28		25	20	66	73
3			28	40	9	65	80^{B}	81	90	3	53	40^{A}	69	76	9	25	20^{F}	70	77		55	40	40	50
4			48	57	10	10	10^{C}	63	70	4	14	10^{C}	53	61	10	43	40^{F}	97	115		63	80	73	80
5			09	23	11	19	10^{B}	75	82	5	52	40^{B}	27	39	11	52	40	76	84		16	10	57	64
6			86	96	12	63	80^{C}	59	66	6	79	80^{D}	91	104	12	21	20	77	85		56	40	35	46
7			77	85	1	52	40^{A}	33	44	7	31	40^{D}	31	43	1	99	120	58	65		73	80	65	72
8			86	96	2	60	80^{B}	88	99	8	73	80^{C}	19	33	2	12	10	48	57		46	40	58	65
9			42	52	3	63	80^{A}	53	61	9	08	10^{E}	73	80	3	66	80	45	54		26	20	44	53
10			11	25	4	31	40^{F}	68	75	10	06	10^{D}	47	56	4	56	40^{B}	78	86		57	40	28	40
11			27	39	5	35	40^{C}	95	109	11	93	120^{C}	26	38	5	68	80^{E}	07	20		84	100	59	66
12			50	58	6	31	40^{C}	66	73	12	25	20^{B}	93	107	6	13	10	60	67		41	40	84	94
1	47	40^{F}	38	48	7	30	40^{C}	45	54	1	40	40^{A}	82	91	7	33	40^{B}	14	28		64	80	34	45
2	75	80^{A}	04	15	8	57	40^{D}	49	57	2	53	40	88	99	8	71	80^{F}	51	59		22	20	01	7
3	21	20^{E}	98	118	9	21	20^{E}	72	79	3	58	40^{A}	46	55	9	62	80	71	78		22	20	62	69
4	02	10^{A}	14	28	10	41	40^{F}	92	115	4	51	40^{A}	81	90	10	28	20	18	32		21	20	05	17
5	17	10^{B}	85	95	11	50	40^{B}	95	109	5	20	20^{B}	28	40	11	10	10	03	13		86	100	22	35
6	69	80^{E}	80	88	12	50	40^{E}	86	96	6	82	100	34	44	12	26	20	81	90		98	120	49	58

Source: B.E. Goetz, *Quantitative Methods: A Survey and Guide for Managers*. Copyright 1965 by McGraw-Hill, Inc.; used with permission of McGraw-Hill Book Company.

M = machine number

RN = random number from random numbers table

Ser = simulated service time

Run = simulated running time

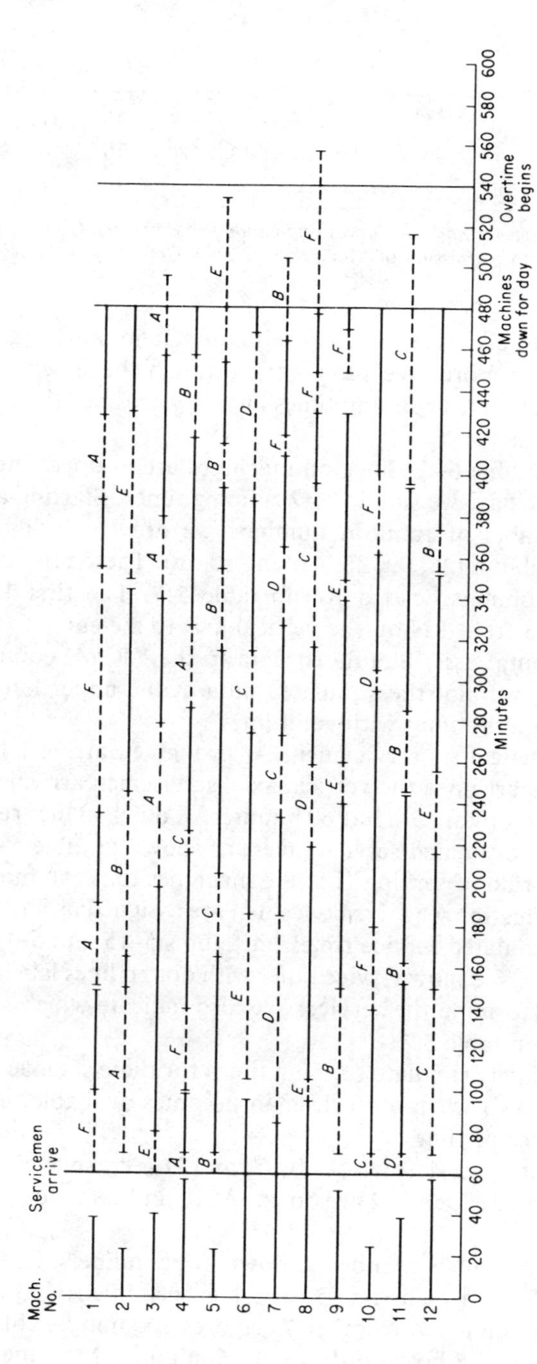

Source: B.E. Goetz, *Quantitative Methods: A Survey and Guide for Managers.* Copyright 1965 by McGraw-Hill, Inc.; used with permission of McGraw-Hill Book Company.

Figure 5–15. Gantt Chart on Twelve Machines

Table 5-8
Estimated Service Time

Machine number	1	2	3	4	5	9	10	11	12
Service time required	40	80	20	10	10	80	10	10	80
Serviceman assigned	F		E	A	B		C	D	

Source: B.E. Goetz, *Quantitative Methods: A Survey and Guide for Managers*. Copyright 1965 by McGraw-Hill, Inc.; used with permission of McGraw-Hill Book Company.

watching total costs decrease, reach a minimum, and start up, at which point we need try no more. We shall work through the six-channel alternative. The same simulated running times and service times should be used for the other alternatives.

1. We prepare table 5-7. Since all machines are in operating order at the beginning of the day, we simulate 12 running times. Starting arbitrarily anywhere in our table of random numbers we draw two-digit random numbers down a column (26, 09, 28, 48, and so on). These provide random numbers for the columns headed RN in table 5-7. The first 12 are run through figure 5-13 from RN on the vertical axis to the curve to simulated data on the horizontal axis. Simulated data in the "Run" column are so generated. Finally, we plot the simulated data at the upper left of figure 5-15. We show running times with solid lines.

2. Our hypothetical six servicemen, A through F, arrive 1 hour after operations begin. Start-up is the vertical axis; servicemen arrival is marked by the vertical line on figure 5-15 at 60 minutes. A quick glance reveals nine machines down for estimated service times, as shown in table 5-8.

Following the rule of serving first the machines that can most quickly be returned to production, the six servicemen are assigned as indicated; and we enter the six simulated service times on figures 5-15 and 5-16, starting all six at 60 minutes. We show service times with dotted lines labeled according to the man performing the service. We also indicate which serviceman performed service in table 5-7.

3. We immediately simulate running times for these six machines. One or more may be down when a serviceman becomes available, and it may have the shortest repair time.

4. At 70 minutes, serviceman A, B, C, and D become available. Since only three machines, 2, 9, and 12 are down, A, B, and C are assigned and D is idle.

5. The next machine is number 7, down at 85 minutes. Serviceman D and E are available, and we assign D, who has been idle for 15 minutes (E has been idle for 5 minutes). Number 7 requires 100 minutes of service.

6. Machines 6 and 8 both go down at 96 minutes. Machine 6 requires

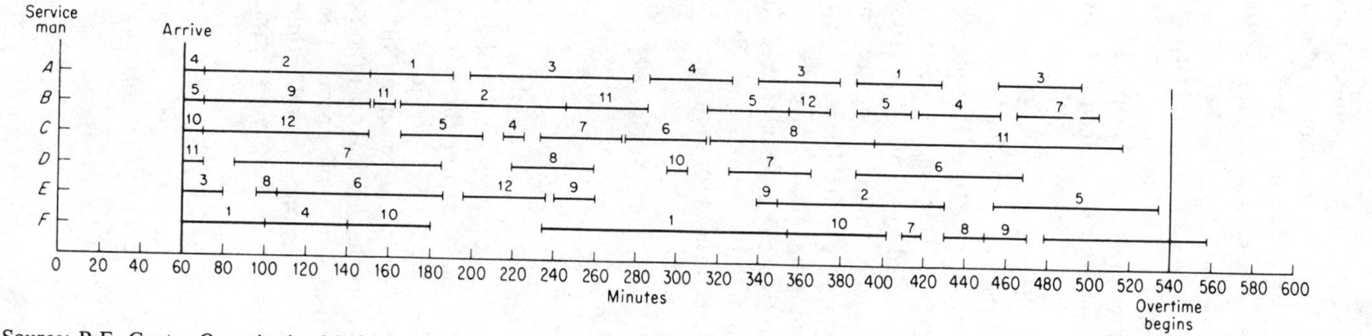

Source: B.E. Goetz, *Quantitative Methods: A Survey and Guide for Managers*. Copyright 1965 by McGraw-Hill, Inc.; used with permission of McGraw-Hill Book Company.

Figure 5–16. Gantt Chart on Six Servicemen

80 minutes of service time; machine 8, only 10 minutes. Consequently, we assign the only available serviceman E to machine 8. Machine 8 now runs for 113 minutes before requiring service again.

7. When the next serviceman F becomes available at 100 minutes, two machines are down: 4 and 6. Machine 4 requires 40 minutes service, and 6 requires 80 minutes; therefore F is assigned to machine 4. Notc that machine 4 was waiting for service for 2 minutes.

8. The above processes continue until, at 480 minutes, the machines are shut down for the night. The load on the servicemen is given in detail in figure 5–16.

9. Unexpired running times should be carried forward to the next day since the historic frequency distribution should, and presumably does, report how many minutes machines run from the end of one service to entering a queue for the next service. This implies a bias in starting up the Monte Carlo, which can be eliminated by ignoring the first 2 or 3 hours or by dilution via a sufficiently long run. In any event, the simulation should be run long enough to exhibit stability in its answer.

10. The same simulated running and service times should be run through Gantt charts, similar to figure 5–15, constructed for 7, 8, . . . servicemen, enough to find the minimum-cost solution. Obviously, the same numbers of days' operation should be used for each such simulation to yield comparable results.

Comparing Alternative Plans

Each Gantt chart should be costed out and the totals compared. While the single-day sample in figure 5–15 is not long enough and is cursed by initial bias, it can serve as a demonstration of costing the plan. The waiting time for the 12 machines is shown in table 5–9.

All alternative plans will be identical as to service times, since the differences between plans do not affect the physical operation of the machines, which determines both running times and service times. The plans will differ in only three rows of the general master budget model: time waiting for service, number, and cost of service channels, and amount of overtime premium paid. For six operators, these data are

Cost of waiting time (260 minutes)($5/60)	=	$21.67
Cost of providing 6 service channels 6($16)	=	96.00
Cost of overtime, 1 hour at $3	=	3.00
Total cost of plan		$120.67

ble 5-9
aiting Time for Twelve Machines

chine Number	*1*	*2*	*3*	*4*	*5*	*6*	*7*	*8*	*9*	*10*	*11*	*12*	*Total*
st hour	22	37	20	3	37	0	0	0	8	35	21	2	185
bsequent	2	15	0	2	0	10	0	0	30	0	4	12	75
tal waiting time	24	52	20	5	37	10	0	0	38	35	25	14	260

urce: B.E. Goetz, *Quantitative Methods: A Survey and Guide for Manager.* Copyright 1965 by Graw-Hill Inc.; used with permission of McGraw-Hill Book Company.

If unexpired simulated running times are carried forward, presumably there will be even more waiting time during the first hour, and so this $120.67 is probably understated. Since nearly two-thirds of all waiting time is in the first hour, in spite of the favorable bias, probably the hour lag in the working day for servicemen is not a good idea. We could experiment, by simulation, with one-half-hour lag and/or with no lag. Experimentation by simulation is much quicker and far cheaper than real experiments in the plant. And we should have correct, tested answers *before* making changes. The $21.67, whether or not the $3 overtime pay is included, is enough larger than the $16 cost of another service channel and, unless the suggested change in hours worked reduces drastically the 185 minutes lost in queue during the first hour, we should surely try a simulation run on seven channels.

The six servicemen have been idle for intervals given in the following table. (A really serious simulation presumably would introduce allowances for personal time and fatigue and would balance loads more evenly among the men so that most of their rest time would coincide with the time they would be idle for lack of work.) This is more than 10 hours of idle time; thus, if we could schedule service time, by preventive maintenance, five servicemen might be just sufficient. [See table 5-10.]

All such questions can be answered by formulating each alternative with sufficient precision to enable simulation and with sufficient attention to conditions in the real world to achieve accurate representation and by running the simulation sufficiently long to achieve stable results, i.e., long enough for some one plan to emerge as best and to persist in the status. But the longer a simulation is run, the more it costs; therefore it should be run just long enough to be reasonably sure of the answer.

If you enter the unused data in [table 5-9] and in [figures 5-15 and 5-16], it will soon be obvious that this is properly a job for a computer, especially since they represent so short a period and an adequate run should

Table 5–10
Table of Idle Time

	Servicemen						
	A	*B*	*C*	*D*	*E*	*F*	*Total*
Idle time in last hour	45	36	24	73	7	0	185
Idle time in earlier hours	51	37	36	85	133	80	422
Total idle time	96	73	60	158	140	80	607

Source: B.E. Goetz, *Quantitative Methods: A Survey and Guide for Managers.* Copyright 1965 by McGraw-Hill, Inc.; used with permission of McGraw-Hill Book Company.

cover several days. And, if you work up a second page similar to [table 5–9] in order to extend the simulation run to three days, you will conclude that that too is work for a computer. Finally, identifying, summarizing, and costing waiting time is a meticulous, tedious job ideally suited to a digital computer.

Predict Failures and PM Frequency[8]

Maintenance management can increase equipment availability by mathematically predicting failures. Based on actual shutdown data, a mathematical model can be constructed which will tell you when to repair or replace equipment to minimize downtime. The model can be reduced to a simple line on a nomograph.

Two steps are required to determine optimum maintenance for any specific equipment item.

Step 1. From equipment failures and shutdown records, draw a Reliability Curve and obtain the mean downtime for failures and the mean downtime for scheduled repairs or replacement.

Step 2. From Step 1, determine whether there is an optimum time interval between equipment overhauls.

Reliability Curve

Use the following procedure to draw the Reliability Curve.

1. From equipment history records, tabulate the Time Between Failures (TBF). Be careful not to include other than failure data; including scheduled shutdowns will distort the sample.

2. List the Time Between Failures (TBF) from the shortest to the long-

est and at the same time number each failure interval starting with 1 as the shortest time, i.e., $n = 1, 2, 3, \ldots, N$ where N is the total number of failure intervals tabulated. [See table 5–11, left two columns.]

3. Calculate the probability of obtaining a time between failure greater than each of the failure intervals tabulated using the following equation:

$$R(t) = [(N - n) + 1]/(N + 1)$$

4. The probability of failure at t hours or less is $F(t)$ and is calculated using the equation:

$$F(t) = 1 - R(t)$$

5. On [figure 5–17] plot the values of $R(t)$ versus time between failures (TBF). Draw a straight line that visually fits the plotted points. It is the Reliability Curve.

6. Where the Reliability Curve crosses Line B on [figure 5–17] draw a

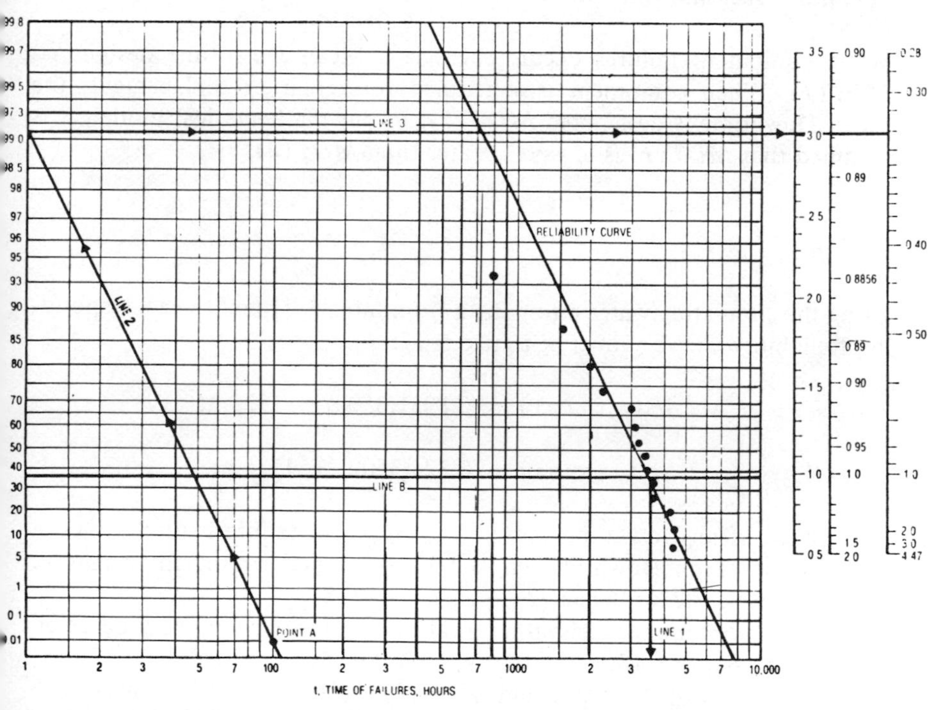

Source: Luis E. Niño, "Predict Failures and Overhaul Interval to Minimize Downtime," *Hydrocarbon Processing,* January 1974, p. 108.

Figure 5–17. Nomograph to Plot Reliability Curve and to Find Values for Overhaul Interval

vertical line down to the Time Between Failures scale (see Line 1) to obtain the value of η, the scale parameter used in later calculations.

7. Starting at Point A, draw Line 2 parallel to the Reliability Curve. At the intersection of Line 2 and the ordinate or $R(t)$ scale, draw horizontal Line 3 to intersect the auxiliary scales β, μ/η and σ/η. The shape parameter, β, is read directly from the first scale and the second scale value is used to calculate the Mean Time Between Failures ($MTBF$).

$$MTBF = (\mu/\eta)\eta$$

8. Compare the calculated $MTBF$ with the arithmetic mean of the Time Between Failures, $\bar{X}$. If the values differ significantly from each other, adjust the Reliability Curve until the difference is reasonably small (less than 20 percent).

9. The equation for the adjusted Reliability Curve is the reliability function of the equipment and is shown in table 5–11.

Optimum Overhaul Interval

When equipment failures occur, there is a Mean Downtime designated $MDT(1)$. When equipment is scheduled for repair or replacement, the Mean Downtime is designated $MDT(2)$. In our mathematical model, it is assumed that $MDT(1)$ is always greater than $MDT(2)$.

Example

Using the circulating water pump data from table 5–11 and the corresponding Reliability Curve shown in figure 5–17:

$$MTBF = (\mu/\eta)\eta = 0.8931(3500) = 3125 \text{ hr}$$

$$\sigma = (\sigma/\eta)\eta = 0.324(3500) = 1134 \text{ hr}$$

Given: Mean Downtime caused by equipment failure is three times as long as the Mean Downtime caused by scheduled equipment repair or replacement or $MDT(2)/(MDT(1) = 1/3 = 0.33$.

Does an optimum overhaul interval exist? This question may be determined using figure 5–18. From figure 5–17, it was determined that $\beta = 3.0$. The intersection of this value and $MDT(2)/MDT(1) = 0.33$ is at a point above the curve on figure 5–18, which confirms that an optimum overhaul exists. If the intersection occurred below the figure 5–18 curve, the policy should be to repair or replace the equipment at failure.

Table 5–11
Analysis of Failures

n	TBF	$F(t)$	$R(t)$	*Observations*
1	810	0.067	0.933	Sample statistics:
2	1,530	0.133	0.867	$\bar{X}$ = 3,127 hrs.
3	2,000	0.200	0.800	s = 1,102 hrs.
4	2,470	0.267	0.733	
5	3,000	0.333	0.667	Theoretical distribution
6	3,050	0.400	0.600	parameters:
7	3,230	0.467	0.533	η = 3,500 hrs
8	3,380	0.533	0.474	β = 3.00
9	3,516	0.600	0.400	μ/η = 0.893
10	3,708	0.667	0.333	α/η = 0.324
11	3,786	0.733	0.267	μ = $MTBF$
12	4,368	0.800	0.200	= 0.893 (3,500)
13	4,400	0.867	0.133	= 3,125 ~ 3,127 hr.
14	4,532	0.933	0.067	α = 0.324 (3,500)
				= 1,134 ~ 1,102 hr.
				Reliability function
				$R(t) = \exp(-t/3{,}500)^{3.0}$

Source: Luis E. Niño, "Predict Failures and Overhaul Interval to Minimize Downtime," *Hydrocarbon Processing,* January 1974, p. 108.

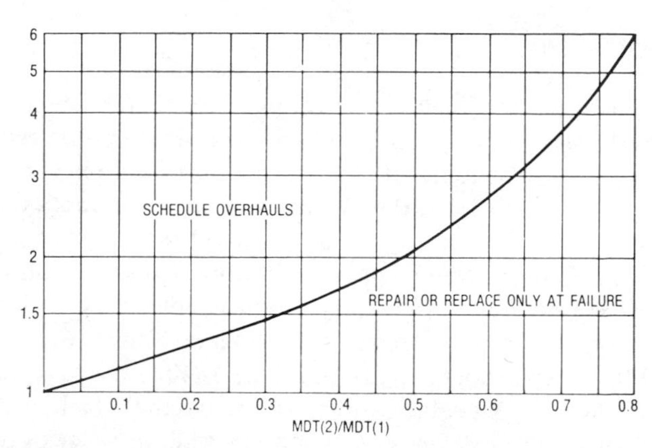

Source: Luis E. Niño, "Predict Failures and Overhaul Interval to Minimize Downtime," *Hydrocarbon Processing,* January 1974, p. 108.

Figure 5–18. Curve to Determine if an Optimal Overhaul Interval Exists

Optimum overhaul interval is designated *To*. Enter [figure 5–19] at $MDT(2)/MDT(1) = 0.33$ and read from the curve,

$$Z = (To - MTBF)/\sigma = -1.0$$

$$To = 3125 + (-1.0)1134 = 1991 \text{ hr}$$

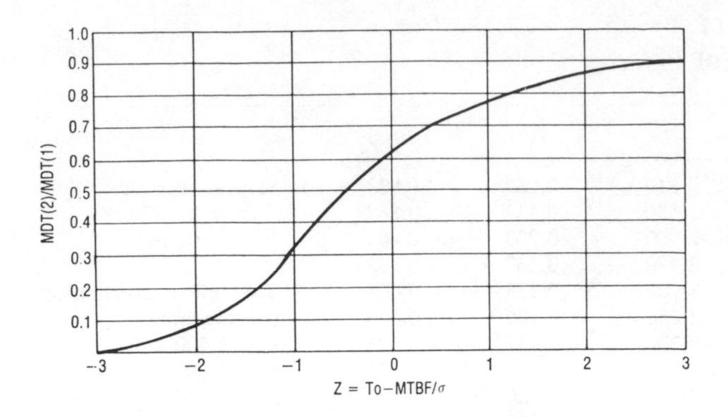

Source: Luis E. Niño, "Predict Failures and Overhaul Interval to Minimize Downtime," *Hydrocarbon Processing,* January 1974, p. 108.

Figure 5–19. Curve to Find Optimal Overhaul Interval

Assuming that the pumps are continuously operating, they should be overhauled every 83 days or, say, every 3 months.

Practical Limitations

The overhaul interval calculated through these replacement models, however, need not always be consistent with the other system requirements. In our example, the probability that a pump does not fail before 1991 hours is 0.83. This is obtained from the Weibull Reliability Function, $R(t)$, column in [table 5–11]. If operating conditions required a minimum reliability of 0.90 for these pumps, the *To* previously obtained would optimize availability but would not fulfill reliability requirements. In such instances, an overhaul interval must be selected using the reliability model such that $R(To) \geq 0.90$. A convenient figure might be 1440 hours (2 months) which gives the pumps a 0.93 probability of surviving the overhaul interval. It must be kept in mind, however, that any decision to increase or decrease the optimum overhaul intervals is made at the cost of an increment in the equipment downtime rate and consequently a reduction in the system's effective capacity.

The concepts presented, as well as the mathematical models and the procedures to derive optimized solutions, are rather straightforward and simple in their application. The practical obstacles usually encountered are those associated with obtaining the shutdown data. Equipment history records, well maintained through an adequate shutdown report system, can easily and rapidly pay for themselves with the benefits of the additional capacity obtained through the application of optimized maintenance policies. The theory underlying these procedures can be found in most reliability engineering books and handbooks.

Design for Maintenance

As mentined earlier in this chapter, it is essential that there be a maintenance and new-construction coordinator to work with plant forces or with contractors. This coordinator will attempt to oversee the design of the new construction so that the new design facilitates maintenance activities.

The following are typical of the items with which this coordinator is concerned:

1. assurance that the purchasing agent of the new construction concentrates his request for bids with manufacturers that are represented by equipment already in the plant, thus ensuring that the maintenance craftsman is trained in maintaining this type of equipment and minimizes the spares that must be maintained in the storehouse;
2. assurance of access to equipment such as truck lanes, pumps concentrated at battery-limits pump pit, and trolley beams to assist in putting large items of equipment onto trucks;
3. sufficient flanges to assist in cleanout and inspection of pipelines;
4. valves on both sides of equipment so that equipment may be removed without danger of product leakage and blocks and bypass valves for control valves so that they may be removed with the unit continuing on manual control;
5. assurance that there is sufficient unit lighting to assist in maintenance activities after dark;
6. assurance that there are sufficient water and air-hose connections so that large numbers of hoses are not necessary;
7. location of heat exchangers where tube bundles can be easily removed; and
8. assurance that overhead-pipe-rack clearance is sufficient to allow mechanical equipment to enter the unit.

Notes

1. This section is reprinted from J.C. Andrica, "Establishing a PM Inspection Program," *Plant Engineering,* August 9, 1979, pp. 109–112.
2. This section is reprinted from Lawrence Mann, Jr., "Statistical

Methods Facilitate Maintenance Planning and Scheduling,'' American Society of Mechanical Engineers Report 66-PET-12, 1966.

3. This section is reprinted from Bernard Ostrofsky, ''Nondestructive Tests for Inspection of Refinery Units,'' *Techniques of Plant Engineering and Maintenance* 19:35-139. Copyright 1968, Clapp and Poliak, Inc.

4. This section is reprinted from Ted F. Meinhold, ''Measuring and Analyzing Vibration,'' *Plant Engineering,* October 4, 1979m pp. 82-91.

5. This section is reprinted from Ken Faulkner, ''Infrared Thermography'' *Proceedings of the 1978 National Petroleumn Refiners Association Refinery and Petrochemical Plant Maintenance Conference,* pp. 125-127.

6. This section is based on G.C. Quinn, ''Survey of Maintenance Diagnostic Instruments,'' *Techniques of Plant Engineering and Maintenance* 19:217-221. Copyright 1968, Clapp and Poliak, Inc.

7. This section is reprinted from B.E. Goetz, *Quantitative Methods: A Survey and Guide for Managers,* pp. 432-439. Copyright 1965 by McGraw-Hill, Inc.; used with permission of McGraw-Hill Book Company.

8. This section is reprinted from Luis E. Niño, ''Predict Failures and Overhaul Interval to Minimize Downtime,'' *Hydrocarbon Processing,* January 1974, pp. 108-110.

6 Measuring and Appraising Maintenance Performance

Many indexes and some techniques, such as work sampling, will measure the performance of the maintenance work force. Also necessary for any complete maintenance-management system are indicators that show how well each level of maintenance management is performing. This chapter will consider both measurement and appraisal of total maintenance performance at three levels: top management, maintenance management, and the individual maintenance craftsmen.

Maintenance management competes with production for use of the firm's resources. The budget presented by production is based on percentage return on investment. The budget normally submitted by maintenance management is based on demand for its services. To compete successfully with production for funds, maintenance management must therefore reform its budget so that it also reflects percentage return on investment. To some extent, this forces maintenance management into program budgeting. A program budget requires that the cost of maintenance be expressed in terms of the end use of the dollars given to it rather than in the traditional terms of salaries, materials, and equipment. It is easy for production to show the profit or loss for each of its production units. Because of the dynamics of plant maintenance, however, it is seldom possible to isolate the effects of any one maintenance program, whether it is designed to reduce maintenance costs or increase maintenance productivity. Economics is the primary consideration, however, and maintenance must attempt to isolate the effects of specific programs in order to justify capital investment in them. There is no reason to believe that such estimates would be less reliable than the production estimates offered by operations.

Although there are many indicators on all three of the levels mentioned, there are some that appear to be used by a number of organizations.

Some indicators used by *top management* to judge the performance of the maintenance work force are as follows:

$$\frac{\text{Maintenance material dollars}}{\text{Maintenance labor dollars}} \times 100 = \text{Percentage}$$

$$\frac{\text{Total maintenance dollars}}{\text{Company employees}} = \text{Dollars per employee}$$

$$\frac{\text{Spare parts} - \text{Dollar inventory}}{\text{Plant-replacement investment}} \times 100 = \text{Percentage}$$

$$\frac{\text{Maintenance cost in dollars}}{\text{Units of Production}} = \text{Dollars per unit}$$

$$\frac{\text{Maintenance people}}{\text{Plant-direct people}} \times 100 = \text{Percentage}$$

$$\frac{\text{Maintenance payroll}}{\text{Plant payroll}} \times 100 = \text{Percentage}$$

$$\frac{\text{Outside contract labor}}{\text{Internal maintenance labor}} \times 100 = \text{Percentage}$$

Number of maintenance employees =

Cost of maintenance per time period =

Some indicators used by *maintenance management* to judge its performance are as follows:

$$\frac{\text{Standard labor hours}}{\text{Actual labor hours}} \times 100 = \text{Percentage}$$

$$\frac{\text{Preventive-maintenance hours}}{\text{Total maintenance hours}} \times 100 = \text{Percentage}$$

$$\frac{\text{Overtime hours}}{\text{Straight-time hours}} \times 100 = \text{Percentage}$$

$$\frac{\text{Backlog of labor hours}}{\text{Total hours available per period}} = \text{Number of periods of backlog}$$

$$\frac{\text{Craft hours planned by planner}}{\text{Hours worked on plan}} = \text{Ratio of planner effectiveness}$$

$$\frac{\text{Man-hours required for emergency}}{\text{Total man-hours required}} \times 100 = \text{Percentage}$$

$$\frac{\text{Man-hours worked on schedule}}{\text{Total man-hours worked}} \times 100 = \text{Percentage}$$

$$\frac{\text{Man-hours on late starts and early quits}}{\text{Total man-hours paid}} \times 100 = \text{Percentage}$$

$$\frac{\text{Work orders executed as scheduled}}{\text{Total work orders scheduled}} \times 100 = \text{Percentage}$$

$$\frac{\text{Man-hours spent on breakdown repairs}}{\text{Total man-hours}} \times 100 = \text{Percentage}$$

$$\frac{\text{Jobs executed at or within } \pm 15\% \text{ of estimated cost}}{\text{Total estimated jobs executed}} \times 100 = \text{Percentage}$$

$$\frac{\text{Man-hours of work entered in log books}}{\text{Total direct-maintenance man-hours}} \times 100 = \text{Percentage}$$

$$\frac{\text{Equipment running time}}{\text{Equipment running time + downtime}} \times 100 = \text{Percentage}$$

$$\frac{\text{Number of breakdowns caused by low-quality maintenance}}{\text{Total number of breakdowns}} \times 100 = \text{Percentage}$$

$$\frac{\text{Inspections incomplete}}{\text{Inspections scheduled}} \times 100 = \text{Percentage}$$

$$\frac{\text{Jobs resulting from inspections}}{\text{Inspections completed}} \times 100 = \text{Percentage}$$

Indicators used to determine *crafts productivity* are as follows:

$$\frac{\text{Actual man-hours worked per unit of time}}{\text{Estimated man-hours worked per unit of time}} \times 100 = \text{Percentage}$$

Work-sampling studies:

Productive-time craftsmen = Percentage

Travel = Percentage

Avoidable delay = Percentage

Unavoidable delay = Percentage

Other benchmarks used to compare maintenance performance are as follows:

Typical ratios of maintenance cost to total production costs:

Pharmaceutical = 2%

Metal working = 1–3%

Textiles = 3–5%

Paper = 8–10%

Chemical = 10–15%

Mining = 15–25%

For the storehouse:

$$\frac{\text{Number of requisitions for which items not in stock}}{\text{Total requisitions received}} \times 100 = \text{Percentage}$$

Number of items issued per time period =

Turnover of complete inventory =

Dollar value of items in storehouse =

$$\frac{\text{Storehouse-indirect costs}}{\text{Number of items issued}} = \text{Dollars per item}$$

$$\frac{\text{Storehouse-indirect costs}}{\text{Cost of issued items}} \times 100 = \text{Percentage}$$

Miscellaneous:

$12/h.p./yr. for reciprocating equipment (1977 dollars)

$4/h.p./yr. for centrifugal equipment (1977 dollars)

Maintenance cost as a percentage of replacement cost:

Refinery = 2%

Chemical plant = 6%

Maintenance staff as a percentage of total employees

Paper and pulp = 10.5%

Chemical = 12.5%

Petroleum = 17.8%

Primary metals = 10.4%

Maintenance hourly employees per maintenance supervisor

Paper and pulp = 8

Chemical = 8

Primary metals = 10

No matter which of these indexes are used, it is important to remember that the index must be proved by applying, for example, the last three years' data to it to determine if it truly tracks what you believe you are measuring. Furthermore, it is important to remember that the trend of the index is as important as or more important than its value or reading at any one time. Trends are much more creditable than instantaneous readings.

Analysis of Labor Productivity

Of the three elements of maintenance resources—manpower, material, and equipment—by far the most variable and most difficult to control is manpower. Maintenance managers have little control over wage rates, which are most often established by negotiation and labor contracts. Maintenance management *can* control the how, when, and where of the work performed, however, that is, the productivity of its work force. Continuing attention to labor productivity is a necessary element of any comprehensive maintenance-management information system. Control reports should be considered on a daily, weekly, monthly, and yearly basis.

Utilization of maintenance manpower is monitored and reported in a number of ways—most commonly as the ratio of man-hours actually worked to man-hours estimated in accordance with standards. Deviations from the standards estimates should be investigated. They might indicate a real change in productivity, a need for a training program, a need for increased attention to industrial relations, labor unrest, outdated standards, or failure of field supervisors to make use of the standards.

The format of the reporting mechanisms, the degree of detail, and the frequency of issuing reports all vary widely among process plants—and also over time, since fewer and less frequent reports are necessary as the program becomes a developed operation. Decisions about what reports should be used and what is to be included should be worked out with maintenance management and the field forces. The report formats should be designed with the entire system in mind. If these reports do not result in corrective action, they add nothing to the system; indeed, they detract from the system in that they collect and circulate data for which there is no use.

The following four reports are the most typical:

1. weekly labor-analysis report,
2. maintenance weekly performance report,
3. jobs-completed summary, and
4. maintenance-performance-trend report

Sources of Reported Data

All information having to do with craft assignments, number of people assigned to the crew, and location in the plant is reported by maintenance planning and scheduling.

The area supervisor or maintenance foreman reports the time distribution for all individuals assigned to him during any particular shift. This report can be a separate time card or a portion of the maintenance work order.

The total hours actually worked, the number of planned man-hours, and the percentage of man-hours on standards can be obtained from the maintenance-planning office after the work order has been closed out.

Responsibility for the continued and up-to-date reporting of data rests with the maintenance-planning function.

The Four Reports

The weekly labor-analysis report, shown in figure 6–1, is used to measure productivity and to calculate the performance index. It also permits comparison of performance with standards.

The productivity index is calculated by dividing the total number of man-hours estimated for all of the work orders by the number of man-hours

WEEKLY LABOR ANALYSIS

______ Half of 19 __ Shift

Area No. ____________

Week Of	No. Craftsmen Assigned	No. M-H Planned	No. M-H Not Planned	% Planned M-H Actually Worked	% Work On Standards	Productivity Index	Performance Index

Figure 6–1. Weekly Labor-Analysis Form

actually required to perform those work orders. The performance index is calculated by dividing the total man-hours estimated for work orders by the total maintenance man-hours accumulated from the time cards. This report is used by maintenance management to monitor the performance of maintenance personnel; a copy goes to plant management each week.

The maintenance weekly performance report, shown in figure 6–2, analyzes the weekly performance of each group of maintenance craftsmen. The input is the same as for the labor-analysis report, and so are the calculations for productivity and the percentage of work on standards. The crafts cate-

MAINTENANCE WEEKLY PERFORMANCE REPORT							
						Week ending	
Crafts	No. Men Assigned	Productivity, %			% Measured Or On Standards		
		Average Last Year	Last Week	This Week	Average Last Year	Last Week	This Week

Figure 6–2. Maintenance Weekly-Performance Report

gory could be either the traditional AFL-type craft breakdown or it could be geographical areas of the plant or groups of craftsmen. This report compares this week's maintenance performance with that of last week and the same week last year.

The jobs-completed summary is a weekly report showing that work orders have been completed, together with the performance on each. This report assists maintenance planning, as well as maintenance management, by indicating the degree to which maintenance has the ability to foresee what work will be required. The existence of a large number of unplanned work orders is reflected by a large number of jobs that stretch out and do not get completed.

The maintenance-performance-trend report is divided into crafts or groups. This report is most often a graph, with percentage performance as the ordinate and the months of the year as the abscissa (see figure 6–3).

Figure 6–4 illustrates an optional comprehensive monthly planning and scheduling report. This document is useful when the maintenance management information system is perfected. It is self-explanatory.

Performance-Improvement Measures

It is not sufficient merely to establish performance and productivity indexes. It is necessary to be able to determine whether the program is

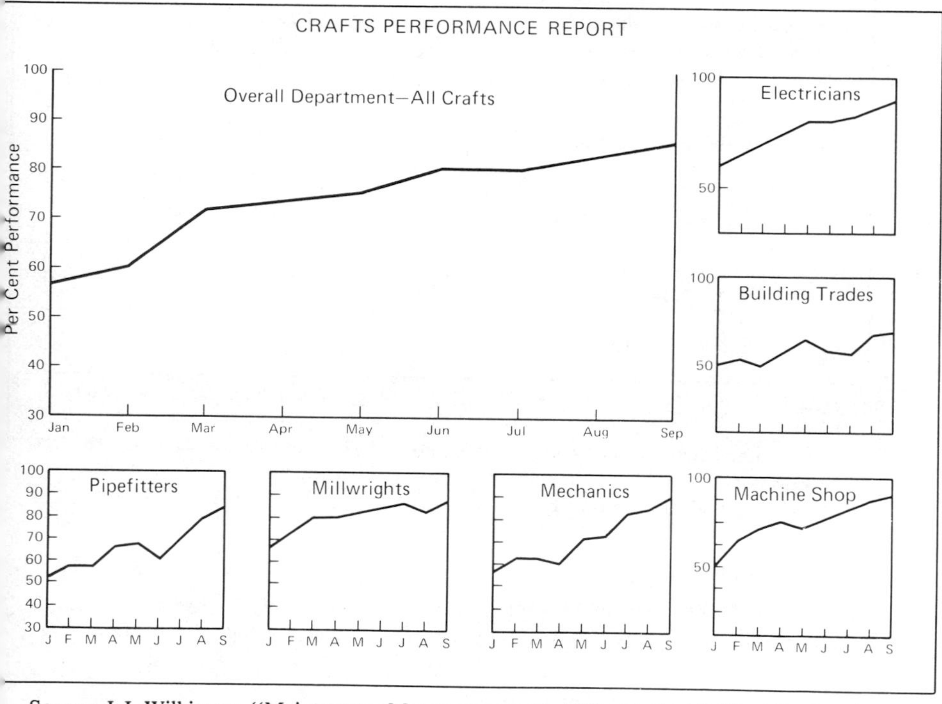

Source: J.J. Wilkinson, "Maintenance Management—Part III—Control," *Plant Engineering,* June 12, 1969, p. 61.

Figure 6–3. Crafts-Performance Report

MAINTENANCE PLANNING–SCHEDULING MONTHLY REPORT

Report Period – May 3 – May 30

	BOILERMAKER		
Planning–Scheduling Savings	This Month	Last Month	this Year
a. Total no. of work orders issued (includes all immediate action, today and cross orders issued).	2324	2255	12,235
b. Total craft work orders issued	354	316	1,856
1. Regular scheduled jobs	231	219	1,240
2. Today	9	12	283
3. Immediate action	114	85	333
c. Percent jobs I.A.–today (b2 + b3) + (b)	35%	31%	33%
d. Number jobs completed	368	400	1,911
e. Average length of completed job (actual hrs.)	11.6	8.8	9.4
Planning–Scheduling Effectiveness			
f. Total manhours on *all* craft work	4061	4128	20,526
g. Total manhours on planned work	3858	3856	18,952
h. Coverage g/f × 100	95%	93%	92%
i. Percent Ttl. Hrs. I.A.–today work	18%	15%	14%
j. Total Std. Hrs. on completed planned work	3824	2802	15,661
k. Total actual hrs. on completed planned work	4309	3181	17,876
l. Productivity of completed planned work (j/k × 100)	89%	88%	88%
m. Base period productivity index	43%	43%	43%

Planning–Scheduling Savings	Last Year	This Year
n. Net savings in manhours (planned crafts) $\frac{k \times l}{m}$ – (k + 1/9 planning–scheduling manhours)	37,153	17,653
p. Net savings in manhours (Total this year) (Total Hrs. saved–Total planning Hrs. expended)	90,289	
q. Increased craft output (Total maintenance) p/f × 100 f = Total of craft cum.	59%	

TOTAL PLANNING HOURS TO DATE: 10,080

Source: G.W. Smith, "Maintenance Planning/Scheduling," *Plant Engineering,* February 6, 1969, p. 58.

Notes: I.A. = immediate action. Report is a measure of the effectiveness of planning, scheduling, and work measurement. It is a reminder to management that planning pays its way.

Figure 6–4. Maintenance Planning and Scheduling Monthly Report

improving and the rate of that improvement (or deterioration). This requires that some base or reference level first be established.

The most frequently used measures of maintenance performance are

1. number of individuals in maintenance work force,
2. size of maintenance-labor payroll (dollars),

3. labor-productivity index, and
4. result of work-sampling studies

The number of individuals in the maintenance work force is a popular index since it is so easy to measure. (One study reports that, in the chemical and allied products industry, approximately 12 percent of the employees are in maintenance; in the primary metals industry, this figure increases to 20 percent; and in the paper and allied products industry, the figure is 11.1 percent.) This index may be advantageous in industries where technology is stable and production does not vary from one time period to another; where technology is moving rapidly and production is variable, however, the number of individuals in maintenance would fluctuate for these reasons alone.

Another disadvantage to using the size of the work force as a measure is that it assumes that the level of maintenance would remain unchanged from one reporting period to another.

The same objections apply to using the dollar size of the maintenance payroll as an indicator. It must be adjusted constantly for changes in wage rates, the mix of experienced workers and apprentices, and premiums paid for turnaround and overtime work.

Perhaps the most reliable indicator is one of the labor-productivity indexes, which measure the effort of the work force against standards or some predetermined target. The most popular of these indexes compares the actual hours expended on the work order to the standard hours that should be required to complete the work. Current production is then expressed as a percentage of a base productivity level determined by using standards developed in the plant or obtained from other industries in the same type of production or from consultants. This can only be done when the jobs are planned in sufficient detail to make them comparable.

Before installing any labor-productivity-improvement program, most process plants report that their productivity is between 35 and 50 percent. This figure can be expected to increase approximately 25 percent with the successful implementation of a productivity-improvement program. It should be emphasized that most of the inefficiencies that exist in maintenance programs are the result of some management-controlled variable rather than an employee-controlled variable. It appears that approximately 80 percent of the improvement gained can be attributable to improved procedures, improved organization, and other management-controlled improvements and approximately 20 percent to the work force becoming aware that management is interested in increasing productivity.

There are a number of ways in which productivity can be increased. Ways of increasing productivity can be discussed in meetings of foremen and craft groups. The mere indication that the productivity of some crafts is below that of others is a powerful incentive for increased productivity. The posting of productivity levels on an appropriate bulletin board will indicate to the craftsmen where they stand relative to the remainder of the work force. The productivity-increase program should include encouragement of the individual craftsmen, and the foremen should be subjected to discussions indicating the means by which this would be accomplished. Awards of certificates or merchandise could increase interest in the program. Attention to training programs will help identify craftsmen whose training is inadequate and will give management an opportunity to respond to their needs. Method-studies programs would indicate more productive ways of performing maintenance jobs. Work-sampling programs will also indicate ways of increasing productivity.

Maintenance-Work Sampling and Analysis[1]

The immediate purpose here is to explain the practical application of work sampling as a tool to improve the efficiency of maintenance operation. Work sampling is analogous to taking a series of instantaneous snapshots of maintenance activities, so that, when the frames are arranged in the proper sequence, they present a comprehensive and reliable movie of each activity—which can then be studied in detail. We will discuss here the mechanics of collecting the survey data and the methods of analyzing, interpreting, and applying those data.

Although we can safely take it for granted that the principle of sample taking is mathematically valid, the value of its practical application to maintenance must be determined before time, money, and manpower are invested in the technique.

How Sampling Pays Off

Work sampling permits us to attach strict numerical values to maintenance performance in such a way that progress or its absence can be traced and measured from one time to another. Thus, the dollars it might take to improve maintenance performance—through the introduction of new tooling, the addition of new facilities, the revision of existing facilities, or the addition of more supervision—can be justified in concrete terms.

In addition, the alert supervision of the work-sampling process provides eye-opening clues to specific weaknesses in departmental methods and procedures; to deficiencies in supervisory coverage; and to any inadequacies in departmental skills, supplies, tooling, equipment, or facilities.

Thus, if it is desirable to trace maintenance performance on a definite scale of values, to find and exploit the most fertile areas of improvement on a priority basis, or to establish sound economic justification for possible expenditures for maintenance supplies, tooling, equipment, facilities, or personnel, then there is reason to make an investment in a work-sampling survey.

Education in the Technique

Unless we have the benefit of previous knowledge and experience, the first necessary step in conducting a survey is education. This calls for the study of a good book and available reprints on the subject. The next step is to pass this education along to the others who will be concerned with and involved in the survey. This can probably be best accomplished through a series of informative small-group meetings at which the general idea of a work-sampling survey can be explained. (Training films on the subject—which can be rented or purchased—will be a big help here.) At such meetings, the reasons for conducting the survey should also be outlined—and they may well be summarized as the need to measure management and supervisor effectiveness. Whenever possible, the craft group that is to be covered by the survey should be represented at all the presurvey orientation meetings. The more the hourly men understand the subject, the more easily they will accept the survey.

Training of the survey observers is of basic importance. They must understand the definitions of the activity categories, the sample-taking techniques, and the meaning of statistical bias.

If several observers are to be used, it is very important that they all see alike and talk the same language. This is made possible by careful, repeated review of the definitions of the categories and by presurvey practice field runs.

At this point let us stop to consider what the survey will consist of from the sample takers' viewpoint. First, there must be random selection of the time of each sample and of the employee to be observed (and, if necessary, interviewed to determine his precise activity at the instant of observation). This is, in effect, the "snapshot" that has to be labeled, and the label is the activity description. The data collected in each sample should include (1) the date and time of the observation, (2) the observer's identification, (3) the craft or labor grouping of the observee, (4) the name of the observee's supervisor, and (5) the activity of the observee at the instant of observation. Other items, such as the location of the observation, may also be included. The data may be recorded in any one of a number of ways—keysort cards, for example, which can be punched to show the desired information.

The Categories

Generally, the activity categories used are (1) work, (2) preparation, (3) travel, (4) delay, (5) cleanup, and (6) idleness.

Craftwork is defined as doing the job itself. *Attention work* is something that can be very tricky to identify. At times it may be hard to distinguish from idleness without a close look, a full understanding of the situation, or an interview with the observee; for example, a standby employee may actually be idle and merely watching the job at the time he is observed, but if his assignment is to be there in a standby capacity, he is to be classified as being on attention work. In other cases, an attention worker may be a ladder holder, he may be observing the results of work he has done to see if it has produced the desired results, or he may be observing conditions before beginning a job to determine the exact situation and its requirements.

Job preparation can be preparation of the site, of the people concerned, or of the plans. This activity may occur either at the job site or in another location. It can consist of moving tools and equipment into the job area, hanging or placing a scaffold, clearing the work area, erecting chain falls or barricades, or anything else that it may be necessary to do before starting the job but that is not really part of the job. Preparing people includes any contact with engineers, supervisors, or others who may provide advice or instructions. Preparing plans is any study of related written instructions, drawings, or sketches.

Travel can be either to the job or from the job, and it may be either empty or loaded. Carrying a normal kit of tools is not considered loaded *travel,* whereas conveying tools or supplies needed to perform a particular job is.

Delay is a category that can have many subdivisions. Typical delays are waiting for help on the job, waiting for instructions, waiting for tools or equipment, waiting for the equipment to be available, waiting for job permits, waiting for materials (including waiting at the stores window), waiting for transportation, and any other waiting, however necessary, that delays the progress of the job.

Cleanup consists of pickup and cleanup on the job, replacement of tools and equipment, job-site restoration, and so on.

Idleness can be broken down into idleness on the job and idleness off the job. It is a technicality, perhaps, but if an observee is found traveling to or from a point of idleness—such as a break area, a restroom, or a first-aid station—he should be scored as idle off the job rather than as traveling.

A short interview with the observee is often essential to accurate classification. One problem is how to categorize the scheduled craftsmen that are not seen when the observer visits the site. The two approaches commonly used in this instance are (1) to assume that the missing craftsmen are representative of those working and to ignore them or (2) to question the crew foreman and determine where these craftsmen are. The latter course of action will result in more observations.

Once the survey observers have a common understanding of the activity classifications, the major part of their preparation will have been accomplished.

Bias

It is very important that all survey participants understand the term *bias*. Any practice that tends to influence the samples from the normal introduces bias, and any departure from randomness or any survey tactic that tends to be fixed, patterned, or otherwise predictable will do so. Thus, if an observer follows an unchanging route, his sampling will be biased. Taking too many samples on any one day, on any one day of the week, or in any one time period in one day will also introduce some form of bias. Whenever bias is found, the samples taken during the period in which it existed should be discounted.

Selection of Observers

Selection of the observers is an important consideration. They may come from any plant-employee group that can spare the time, but it is preferable to use people who know the plant and the men well, if they can be found. Although using the maintenance foremen as observers may cause problems, it is highly recommended. The experience gives the foremen an entirely different viewpoint on what is going on in the department, enhances his appreciation of the survey, and ensures that he will accept the findings. This is of paramount importance, for most of the survey's success will depend on postsurvey actions that will require the foremen's understanding and cooperation.

The Pace of the Survey

The next step is to set up the survey mechanisms, which will require a decision on the pace of the survey. Normally, a survey should include at least 2,400 observations to ensure that the results are reasonably reliable. Thus, the duration of the survey will be determined by the number of surveyor man-hours that can be allotted to the sampling daily.

Since maintenance tends to be cyclic in nature, and the most common cycle period is probably one month, one month is reasonable as the minimum span. This calls for about 100 to 150 samples per day, which can be allotted on the basis of the time the observers have available. It should be remembered that an observer looking at a crew of eight craftsmen makes eight observations. In a process plant, the average observer working on an intermittent basis can take about eight samples in a half-hour. Saturation sample taking by a single full-time observer can result in many more, primarily because the saturation sample taker becomes very familiar with the jobs going on at a particular time and has a good idea of where his observee targets are to be found.

Ensuring Randomness

Since the samples must be taken at random times on randomly selected people, another presurvey step is to assign random numbers to the observee group and to the times of the work day, which is divided into ten-minute segments. Ten-minute segments provide 48 observation periods in the eight-hour work day and are small enough to permit representative observation of activities that tend to occur at a particular time of day. The completed survey should have a nearly equal distribution of samples in each of the 48 time periods; for example, a fifteen-minute segment at the start of the day, before and after a break period, or before the end of the day can yield very misleading results, depending on whether the samples are taken in the first, middle, or last five minutes of the fifteen-minute period. This effect tends to be greatly reduced when ten-minute time intervals are used.

It is advisable to have the observers contact the maintenance clerk, and the maintenance foreman if possible, before starting a sample-taking trip in order to get some idea of where to find the observees. It is also advisable to have observers follow varying sample-taking routes in order to avoid telegraphing their plans or otherwise introducing bias. We also found it best not to take more than one sample at any one time when a group of two or more designated observees was working in a localized area.

Many process plants are geographically spread out, and the possibility of making many observations in a short period of time does not exist. One approach to this problem is to randomize the sequence in which areas are sampled. Obviously, there will be much backtracking in this system, but it will go a long way toward ensuring randomness in the observations. The random-time approach is not practical in these cases because of the difficulty in judging the exact arrival time at a particular location.

Until a new observer has taken about 100 samples, it is desirable to monitor his scoring for any significant deviations from the general average that might indicate errors of activity definition or introduction of some form of bias.

Presenting the Data

Graphical presentation of the data may make analysis easier. Such general items as 45 percent direct work, 5 percent attention work, 5 percent preparation, 10 percent delay, 25 percent travel, and 10 percent idle are not sufficient for analysis. We therefore graphed the day into ten-minute intervals and drew the various activity percentages from ten-minute period to ten-minute period across the graph. This gave us an activity profile of the day.

The graph showed how quickly or slowly work started in the morning. It showed the times when average and maximum work outputs were attained. It showed the time when travel was highest, the pockets of idleness, and the effect of the coffee break. It showed work slackage before lunchtime and before the end of the day, and a lack of cleanup time. It also showed the need for more preinstruction time for the next day's work.

Taking Action

Our graphs made it possible to pinpoint the times of the day when activity distribution was not what we felt it ought to be, and we were able to recommend steps that would improve efficiency.

We reasoned, for example, that there was excessive travel early in the day because the mechanics were surveying the jobs before obtaining the needed tools and supplies—something that could be avoided in many cases by preinstruction from the foremen the day before. We also obtained a job-supply cart to be positioned near concentrated work centers to reduce the need for frequent trips to stores.

Although we may have suspected that work tended to fall off too early before lunch, to pick up too slowly after lunch, and to fall off again too early at the end of the day, it was not a hard fact that stimulated action until the foremen saw it on the graph.

Work sampling can be applied as successfully to the maintenance foremen as to the hourly workers. It will reveal that many nonsupervisory activities that often take up the greater part of a maintenance foreman's day and will emphasize the fact that an increase in the proportion of time he spends on supervision will improve the efficiency of his entire crew.

In conclusion, it must be emphasized that work sampling is useless unless the results are intelligently analyzed and the findings are used to set in motion the improvements suggested by the survey.

Maintenance-Control Indicators

For evaluating the performance of maintenance management, each plant must seek out its own evaluators that are designed to measure what management wants to have measured. Far too many plants make decisions based on indexes that, in fact, have little relationship to the performance of their maintenance systems.

The trend of the performance indexes is more important than their value at any one time, and most plants in the process industry find that no one indicator can suffice for decision-making purposes. Performance indexes must be reviewed constantly to ascertain whether they are still valid. In plants with maintenance-management information systems, the data are grouped in such a manner that each decision maker throughout the management process receives the information he needs for making his decisions and receives no more information than he needs.

The indexes reviewed here are in addition to those listed earlier and are representative of those in common usage today. They will be divided into groups, but occasionally an index in one group may serve the purposes of another group.

Status-of-Load Indicators

These indicators show the status of maintenance work that has been planned, scheduled, or dispatched and for which all resources, including materials, are available.

Backlog. The number of man-hours of backlog indicates, both for total crafts and within each craft, the number of planned and engineered man-hours of work for which there had been no craftsmen to perform that work (see figure 6–5). The trend in the backlog value indicates whether or not the plant has adequate resources with which to perform necessary work. An ever-extending backlog might indicate an increasing level of maintenance, or it might indicate that the work is increasing faster than resources are being allocated to the maintenance department. One way of dealing with such a situation, particularly if it is a short-term one, is to package the work so that portions of it can be subcontracted. This can only be done when the collective-bargaining agreement allows for selective subcontracting.

Some level of backlog is a healthy situation, of course, in that maintenance management has an alternative of jobs. A popular argument centers on what constitutes a satisfactory backlog. Some plants operate with a two-week backlog of work and other plants are satisfied to operate with a two-month backlog. Most plants consider a work order in the backlog only

MAINTENANCE BACKLOG REPORT

Craft	Avg. Weekly Available Manhours	Standard Hours Backlog	Actual Hr. Backlog at 80%	No. of Days Backlog	Backlog Days Less 15% I.A.	Act. Hrs. Backlog for Prev. Months Based on 80% Productivity					
						June	May	Apr.	Feb.	Jan.	Dec.
B	885	1048	1310	.4	8.7	1338	2193	1945	2046	2119	1610
C	570	845	1056	9.3	10.9	1539	1694	999	1042	1365	1196
E	440	1121	1401	15.9	18.7	1923	2289	2670	2021	1910	2061
M	910	806	1008	5.5	6.5	664	2669	2315	2668	2465	2323
P	355	1803	2254	31.7	37.3	2445	1985	1930	1738	2010	1965
PC	275	516	645	11.7	13.8	493	1266	549	666	603	623
PF	865	1506	1883	10.9	12.8	2390	3300	2846	3416	2185	950
R	475	684	855	9.0	10.6	841	1658	1236	1174	1024	1528

Source: *Plant Engineering,* February 6, 1969, p. 58.

Note: Backlog report, broken down by craft, indicates manpower loading, permits management to make employment decisions. By listing the information for the past six months, important trends can be identified.

Figure 6–5. Maintenance Backlog Report

when all materials are available for that job. Some plants consider a work order in the backlog when the material is scheduled to be in the plant.

Open Work Orders. The number of open work orders each day indicates whether the maintenance work load is increasing or decreasing. If the plant writes a large number of work orders and they are all for approximately the same number of hours, then the number of work orders open each day (or, conversely, the number closed out each day) is an indication of the maintenance work load and an indication of the progress in processing work orders.

Planning Indicators

Planning indexes are used for monitoring the efficiency of the maintenance-planning function. The ratio of man-hours spent on planned work to total maintenance man-hours indicates the amount of planned work relative to total work (see figure 6–6). Most process industries appear to be satisfied when approximately 80 percent of their maintenance man-hours are spent on planned work and the remaining 20 percent on emergency work. Normally, it is uneconomical to try to plan all maintenance work. Another way of reporting the same index is as the ratio of man-hours of immediate-action work to total manhours. Immediate-action work is defined as work that must be attended to on the same shift on which it is reported.

The ability to forecast is an indicator of good planning, and the jobs-completed-on-schedule statistic indicates how well maintenance management arranges and controls the work orders. One method of deriving this index is shown in figure 6–6.

Most maintenance management is very conscientious about overtime. An abnormal or unusual amount of overtime often indicates a lack of planning or a deficiency in the ability to forecast. Most plants restrict overtime for the maintenance work force to some percentage—perhaps 5 percent—above which permission must be obtained from higher management.

In plants with planned maintenance systems, the percentage of man-hours worked on work orders planned from standards is a good indicator of maintenance's ability to plan.

Downtime has been mentioned previously as an indicator. If the amount of downtime is at a minimum or is satisfactory to plant management, it can be assumed that maintenance is planning satisfactorily or at least to the satisfaction of the plant management. Overmaintenance must be guarded against if this index is used.

SCHEDULE EVALUATION

	CRAFT	BOIL.	CA	PERS.	ALL CRAFTS
(A)	ACTUAL HOURS WORKED From Time Cards	158		12%	933
(B)	HOURS WORKED ON SCHEDULE (From Daily Job Cards)	126		08%	753
(C)	% HOURS ON SCHEDULE (B ÷ A) × 100	80%		%	80%
(D)	HOURS WORKED NOT ON SCHEDULE (From Daily Job Cards)	13		3%	68
(E)	% NOT ON SCHEDULE (D ÷ A) × 100	8%		3%	7%
(F)	RUSH HOURS (From Daily Job Cards	19		0%	121
(G)	% RUSH HOURS (F ÷ A) × 100	12%		%	13%
(H)	JOBS SCHEDULED (From Daily Schedule)	13		0%	90
(I)	JOBS WORKED ON TODAY . . . (From Daily Job Cards)	9		%	72
(J)	% SCHEDULE WORKED ON (I ÷ H) × 100	69%		9%	80%
(K)	SCHEDULE RATING, INDEX . . . AVERAGE OF WORK ON SCHEDULE AND JOBS WORKED (C + J) ÷ 2	74%		8%	80%

Source: *Plant Engineering,* February 6, 1969, p. 59.

Figure 6–6. Schedule Evaluation

Other Productivity Indicators

Perhaps the productivity indicator receiving the widest attention today is work sampling. Although it is used for a number of purposes, work sampling is discussed here merely as a means of productivity assessment. Therefore, we will concentrate on the percentage-productive category to the exclusion of the other categories. What is considered productive must be defined in the write-up of the work-sampling program. What one company considers as productive work another might consider in some other category. In the chemical and process industries, 40 percent productive appears to be on the low side, and 60 to 65 percent productive appears to be satisfactory.

The previously mentioned ratio of maintenance-labor cost to maintenance-material cost, to a certain extent, indicates the productivity of the work force. As emphasized before, this index is most relevant when the cost of material is relatively constant.

If it is assumed that the technology of the plant remains relatively stable, productive units are neither added to nor subtracted from the plant, and the level of maintenance remains the same, then the maintenance cost per productive unit would also be an indicator of productivity. The base periods used would be periods of relative stability in product demand, inflation, and so on.

Another productivity index can be obtained by comparing actual maintenance cost to budgeted maintenance cost. This index assumes that maintenance management has the ability to forecast and project what maintenance needs will be and to apply a dollar figure to these needs before the beginning of the financial period. When the actual maintenance costs vary from the forecasted costs, it may be assumed that productivity has either increased or decreased, according to the direction of the difference. Many aspects of the modern plant other than maintenance could cause this index to change, however, and management must investigate the problem to find the source of the change.

Several rule-of-thumb indexes are used in the process industries. One such index stipulates that two maintenance people are necessary for every million dollars of investment. Another is based on the premise that total maintenance cost should be 7 percent of replacement investment, and still another is based on the tenet that total maintenance cost should be 6 percent of sales.

Periodically, Albert Ramond and Associates publish a profile of maintenance management. The most recent profile includes 61 companies classified as chemical and allied products. This type of industry is considered fairly representative of the process industries in general. The indicators found for these 61 process industries are shown in table 6–1.

Productivity versus Manpower Requirements[2]

A 20 percent improvement in the productivity of your maintenance force is quite possible, and this can mean the accomplishment of 100 percent more work. You can actually double the output of your maintenance work force or get the same amount of work done with half as much manpower, unless you are already high on the productivity ladder or if you are taking the easy way out by not considering all aspects of productivity.

Table 6–1
Profile of 61 Process Industries

Number of total plant employees per maintenance employee	10
Number of maintenance hourly employees per maintenance supervisor	10
Number of maintenance hourly employees per maintenance staff planner	38
Percentage of maintenance hours planned	73
Number of maintenance hourly employees per industrial engineer	94
Number of maintenance employees per maintenance stores clerk	21
Number of maintenance hourly employees per maintenance staff employee	14

Figure 6–7 shows the volume of work accomplished at different levels of productivity with the same amount of maintenance manpower. Note that an increase in productivity from 20 to 40 percent results in the accomplishment of four volumes of work instead of two, and that an increase from 30 to 60 percent in productivity produces a similar 100 percent improvement. An increase from 40 to 50 percent produces 25 percent more work, whereas an increase from 80 to 90 percent produces 12.5 percent additional work output. This points up the need to know where you stand in order to analyze past accomplishments or to predict how contemplated charges will affect work output.

Productivity

As used in this section, productivity is a measure of essential direct-work activity modified by the effectiveness with which that activity is being performed.

Effectiveness

Effectiveness is a measurement of the overall competence with which a job is being performed. Is the job being accomplished by thoroughly trained workmen in the best manner with proper tools and equipment? Can the repair procedure be upgraded by using different materials or components that will increase the rapidity of the repair or possibly eliminate the requirements for periodic repairs completely? Since work sampling measures only

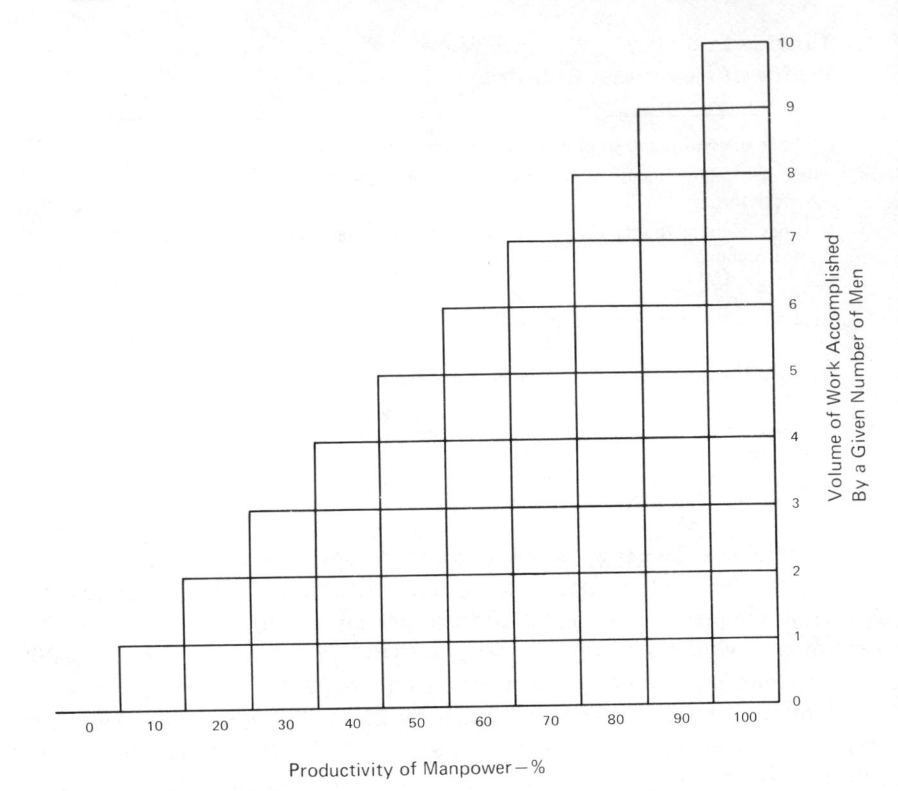

Source: Russell W. Lortz, "Productivity vs. Manpower Requirements," *Techniques of Plant Engineering and Maintenance* 20:177–181. Copyright 1969, Clapp and Poliak, Inc. Variation in volume of work is accomplished by change in the percentage-productivity basis—constant level of manpower.

Figure 6–7. The Volume-of-Work Chart

the activity level and gives no indication of whether the work is essential or is being done in the best way, effectiveness information must be developed by maintenance analysis—a complete study of each of the larger and repetitive jobs. The following are some of the questions that must be asked and answered if a gain in effectiveness is to be made:

1. Is this repeat job a result of (a) poor quality or sloppy work, (b) unsatisfactory material, (c) improper design of a component or the entire unit, or (d) inadequate training of the workmen?
2. Could the time required for repairs have been decreased by (a) methods improvement, (b) improved tools or more tools, (c) additional training of workmen?

3. Could an adequate PM program eliminate the cause of the breakdown?
4. If this is a PM job, is the frequency too high?
5. Are the men motivated to try to do the best possible job?

Maintenance analysis is a time-consuming procedure, but it is essential for approaching the optimum in maintenance effectiveness. Nevertheless, much can be done to improve effectiveness without detailed studies. Supervisors are in a position to analyze and eliminate the causes of many repetitive jobs. Electrical overloading and corrosion, for example, are easily identified and corrected. In the process industries, corrosion is a big contributor to maintenance costs. This can frequently be eliminated by a change in metallurgy or possibly by a change in processing or inhibitor injection. We must remember that repair jobs can be eliminated by eliminating the cause; thereby, overall effectiveness has been increased and there is no longer need to be concerned about productivity on the jobs in question.

Direct Time and Effectiveness

It is now possible to define productivity in terms of direct-maintenance time and effectiveness. Productivity equals direct maintenance time multiplied by effectiveness; for example, if direct-maintenance time is 30 percent and effectiveness is 66.6 percent, then productivity equals 30 times 0.666, or 20 percent. The percentage of direct-maintenance time can best be determined by work sampling. Effectiveness can best be determined by detailed maintenance-analysis studies, although it also can be estimated from a supervisor's observations and knowledge of major repetitive jobs.

Manpower

Figure 6–8 shows a way to determine the effect of changes in effectiveness and direct-maintenance time on the relative size of a work force. As can be seen in the figure, a work force that is spending 40 percent of its time on direct maintenance at 75 percent effectiveness must be 3.34 times as large as the theoretical work force on 100 percent direct maintenance at 100 percent effectiveness.

What will the effect on the size of the work force be if the effectiveness remains at 75 percent and the direct maintenance time reaches 50 percent? The lines at the top of the figure show that this increase in direct-maintenance time at the same effectiveness will mean that only 80 percent of the previous work force, or 2.68 times the theoretical, can do the same amount

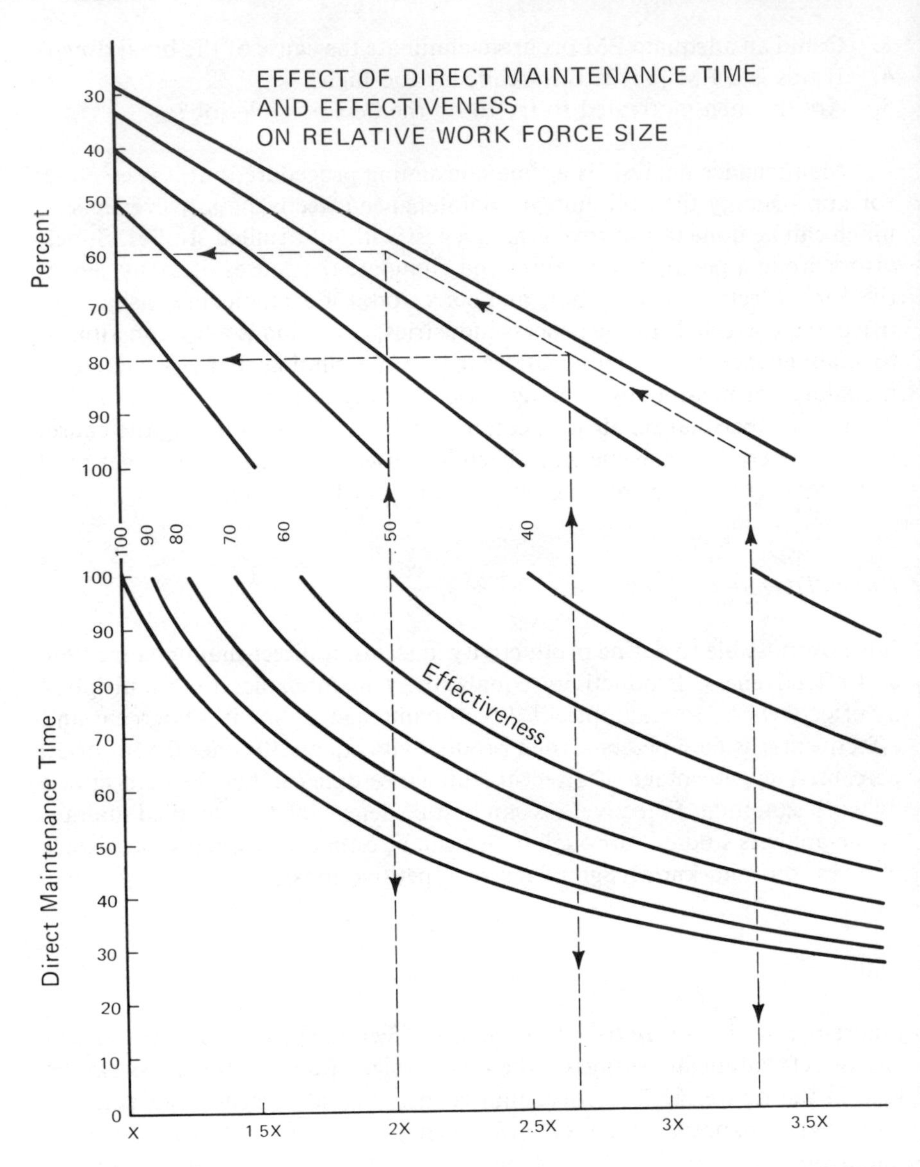

Source: Russell W. Lortz, "Productivity vs. Manpower Requirements," *Techniques of Plant Engineering and Maintenance* 20:177–181. Copyright 1969, Clapp and Poliak, Inc.

Figure 6–8. Effect of Direct Maintenance Time and Effectiveness on Relative Work-Force Size

of work. If effectiveness could be increased to 100 percent at 50 percent direct-maintenance time, the new work-force requirement would be only 60 percent of the original number, or twice the theoretical. This means that, if we started with a work force of 334 men and the work load remained constant, we could reduce the work force to 268 men by increasing direct-maintenance time from 40 to 50 percent at the same effectiveness. By increasing effectiveness from 75 to 100 percent with a direct-maintenance time of 50 percent, we could get the same amount of work done with a work force of 200 men instead of the original 334. This figure highlights the fact that two types of improvements are available to increase productivity. Maintenance management should always strive not only to eliminate all the things that detract from direct-maintenance time but also to eliminate unnecessary work and to ensure use of the best procedures, equipment, and tools for the work that has to be done.

A practical example is supplied by a manufacturing plant in which the instrument-electrical force of nineteen hourly workers had a direct-maintenance time, as measured by work sampling, of 43 percent. As a result of a management decision, in the light of circumstances that had nothing to do with their work load, the number of hourly workers was reduced to twelve. Although the reduced force had the same amount of equipment to maintain, their direct-maintenance time rose to 50 percent. Since there was a force reduction of 36.8 percent, while the direct-maintenance time increased by only 16.3 percent, it is obvious that the effectiveness with which the work was being done must have increased. How much it increased can be determined by the following simple equation. Assuming 60 percent effectiveness,

$$19 \text{ men} \times 43\% \text{ DM} \times 60\% \text{ eff.} = 12 \text{ men} \times 50\% \text{ DM} \times \text{eff.}$$

$$\text{Eff.} = \frac{19 \times 43 \times 60}{12 \times 50} = 82\%$$

Thus, the effectiveness had increased by 22 percentage points. If we now divide the increase in percentage points of effectiveness, 22, by the original effectiveness of 60, we will find that we have increased the effectiveness of the essential work being performed by 36.7 percent. This illustration shows that effectiveness is very real. Although it is difficult to measure, its presence should never be ignored.

Suggested Composite Index[3]

The first step in upgrading a maintenance operation should be a meeting to present the concepts to be used, to substantiate their validity, and to review the expected benefits. This orientation should be attended by all maintenance supervisors and by representatives of production supervision, plant engineering, and top management. Involvement of top management is especially important to convince maintenance personnel of management's interest in, and commitment to improving the maintenance operation.

Companies use different methods to evaluate their maintenance operations. As has been seen, some use the ratio of maintenance cost to total operating cost, and some compare maintenance cost with the dollar value of product shipped. Other companies measure maintenance performance as current maintenance cost compared to previous year's cost.

A better assessment of maintenance operation is the measure of its composite service to production. This evaluation can be used to make comparisons with other maintenance operations in similar or different industries. The composite-service method of evaluation comprises 16 basic measures relating to maintenance effectiveness. Values are easily obtained without special technical training or extensive investigation.

Service Capability Evaluation

The four values required to evaluate service capability [figure 6–9] are compiled as follows:

1. The percentage of maintenance absenteeism is obtained from personnel records. For illustrative purposes, the records show absenteeism to be 4 percent.
2. Crew-activity level can be measured by conducting a simple, 10 day sampling study. The number of maintenance personnel observed and the number actively engaged in actual maintenance work should be recorded. Dividing the number of people actively engaged in maintenance work by the total number observed will give a crew-activity percentage for use on the chart. In this case, it is 60 percent.
3. The average work-order-completion time (in days) can be computed after a review of a number of completed work orders. The completion dates are compared with the issue dates, and the average of these times is plotted on the chart. It took an average of nine days to complete work orders.
4. Forecasting effectiveness can be determined by reviewing maintenance job schedules and noting those jobs completed on the day scheduled. Dividing the number of jobs completed on schedule by the total number of jobs studied will yield the percentage forecasting accuracy for use on the chart. In this case, it is 43 percent.

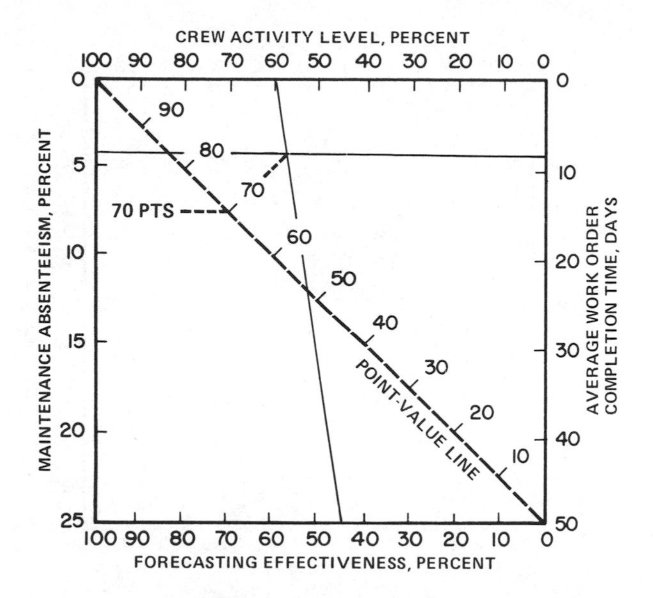

Source: Frank V. Claire, "Evaluating Maintenance Operations," *Plant Engineering,* June 23, 1977, p. 125.

Figure 6–9. Service-Capability Evaluation

After these four values have been determined and located on the chart axis, the values on opposite sides of the chart are connected by straight lines. When both lines have been drawn, a dotted line is drawn from the intersection point perpendicular to the diagonal point-value line. The point value for service capability is read where the dotted line crosses the point-value line. In the example, the point value is 70.

Expenditures Evaluation

The four values required for this chart [figure 6–10] can be obtained as follows:

1. Percentage of maintenance wages for direct maintenance labor is based on accounting records of hourly wages paid for direct maintenance labor for a given period. Wages paid for maintenance clerical help, tool crib attendants, and maintenance supervision are excluded. Direct maintenance labor wages divided by the total maintenance labor cost

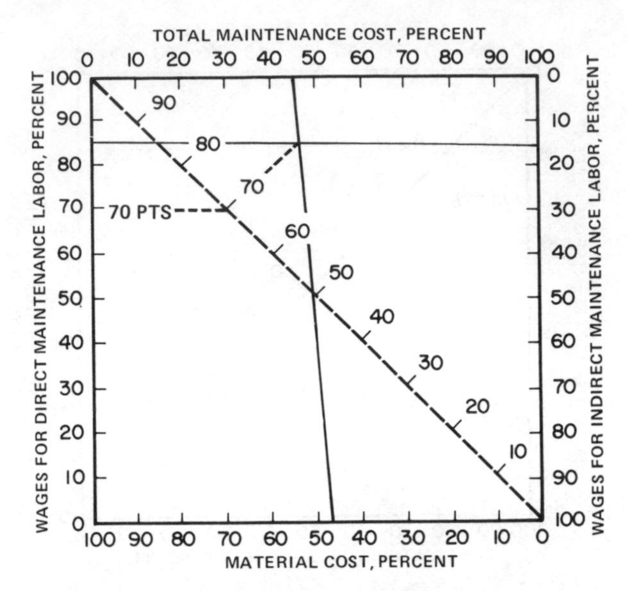

Source: Frank V. Claire, "Evaluating Maintenance Operations," *Plant Engineering,* June 23, 1977, p. 125.

Figure 6–10. Expenditures Evaluation

for the same period will provide the required percentage figure. This value is 85 percent in the example.

2. Total maintenance labor cost as a percentage of direct labor cost can also be obtained from accounting records. Dividing total maintenance labor cost for a given period by total direct labor cost for the same period will yield this percent. In this example, it is 45 percent.
3. The percentage of maintenance wages for indirect maintenance labor includes the wages paid for maintenance clerical help, tool crib attendants, and maintenance supervisors for a given period divided by the total maintenance cost of labor for the same period. These data are obtained from accounting records. In this case, it is 15 percent.
4. Material cost as a percentage of total maintenance cost can also be derived from accounting records. This is the total annual maintenance material costs—48 percent in [figure 6–10].

The point value for expenditures is 70.

Work-Planning Evaluation

The four values on this chart [figure 6–11] are developed as follows:

1. Percentage of maintenance overtime is obtained from accounting records by dividing the overtime hours by total maintenance hours. In this example, it is 12.5 percent.
2. Determining the percentage of labor effectiveness requires some judgment. One way to develop this value is to estimate the work pace, from work sampling, of maintenance personnel (expressed as a percent) and multiply this value by the crew activity level that was developed previously for the service capability chart. If it is difficult to develop this estimate of work pace, 70 percent multiplied by the activity level will provide a reasonably accurate value of labor effectiveness. This percent is 39 in the example.
3. Percentage of emergency work is obtained by dividing the number of maintenance hours spent on emergency work orders during a selected

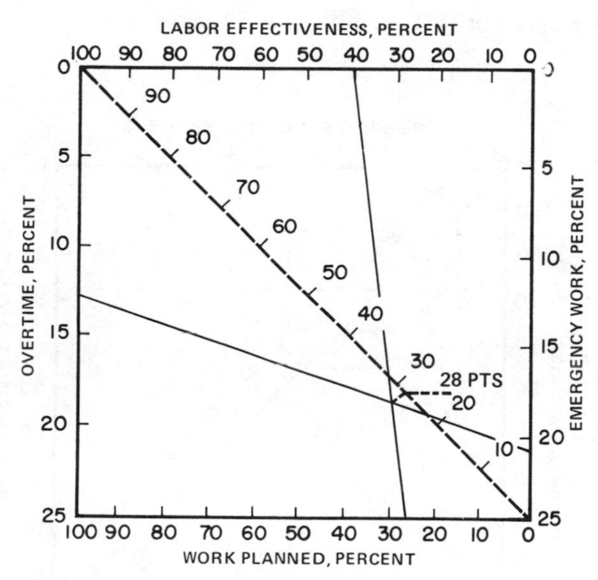

Source: Frank V. Claire, "Evaluating Maintenance Operations," *Plant Engineering,* June 23, 1977, p. 125.

Figure 6–11. Work-Planning Evaluation

period of time by the total maintenance hours spent during that period of time. Example value is 22 percent.

4. Percentage of work planned can be obtained by reviewing a number of maintenance work orders. Each work order should be reviewed with maintenance supervision to identify those jobs that have sketches, or prints, with their material needs and manning requirements fully planned before the job is started. To obtain the desired percentage, divide the number of planned jobs by the total number of jobs by the total number of jobs reviewed. For this example it was 27 percent.

The point value for work planning is 28 points.

Work Scheduling Evaluation

This chart also has a four-value pattern [figure 6–12].

1. Percentage of inplant maintenance can be obtained from accounting records by dividing the cost of inplant maintenance by the total maintenance cost. Total maintenance includes both inplant and contract maintenance work. For this example it is 72 percent.

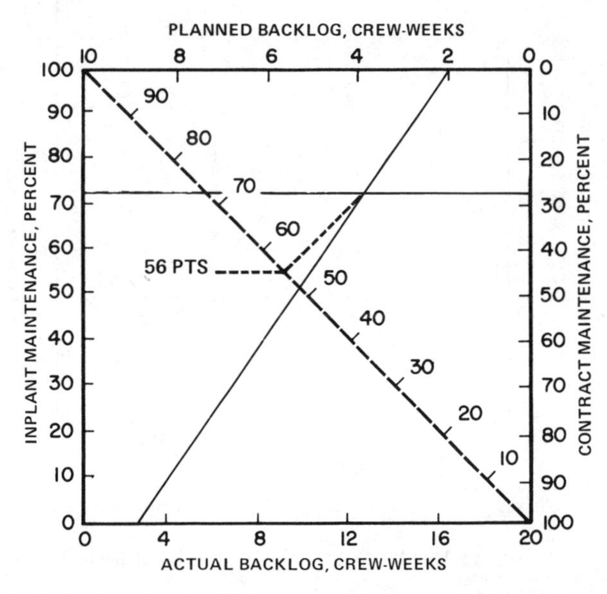

Source: Frank V. Claire, "Evaluating Maintenance Operations," *Plant Engineering*, June 23, 1977, p. 125.

Figure 6–12. Work-Scheduling Evaluation

2. Actual backlog (in crew-weeks) can be reckoned by dividing the hours of maintenance work left on open work orders by the weekly maintenance hours for the crew. If the open work orders have not been estimated, an average estimated number of hours can be developed, based on the work orders on hand. This estimate can be developed through consultation with maintenance supervisors on the completion time for the average work order they process. Multiplying this estimated average by the actual number of work orders on hand yields a figure that can be divided by the total weekly maintenance hours for the crew to obtain the backlog in crew-weeks. The example was a value of 3 crew-weeks.
3. Percentage of contract maintenance is obtained, again, from accounting records. Total contract maintenance cost is divided by the total maintenance cost—28 percent in the example.
4. Desired backlog (in crew-weeks) is the number of crew-weeks of maintenance work backlog that plant engineering feels is necessary for effective crew operation. In this case it is 2 weeks.

The point value for work scheduling is 56.

Composite Service to Operations

The chart in [figure 6–13] utilizes the point values developed on each of the four other charts to reach a composite point rating. The point value from the expenditures evaluation chart (70 points) is plotted on the left axis of this chart under expenditure point value. The value (28 points) from the work planning evaluation chart is plotted along the top axis of this chart under planning point value. The value for the right-hand axis of this chart comes from the service capability evaluation chart (70 points). And the value for the bottom axis of the graph (56 points) comes from the work scheduling evaluation chart. These four values, then, are connected by lines in the same manner as in previous charts. The determination line is again drawn from the intersection perpendicular to the diagonal line to give a composite service point rating of 53.

Note that on the composite service chart there is a point plotted at approximately 78. This point represents the composite service level of an efficient maintenance operation that has good planning and scheduling and uses modern maintenance control techniques.

This method of evaluation provides a relatively quick means of identifying the areas of a maintenance operation that are in need of attention.

Implementing an improvement program of this type has little chance of success unless provision is made for continual, long-term monitoring of the program and its progress. The need for monitoring results from the ten-

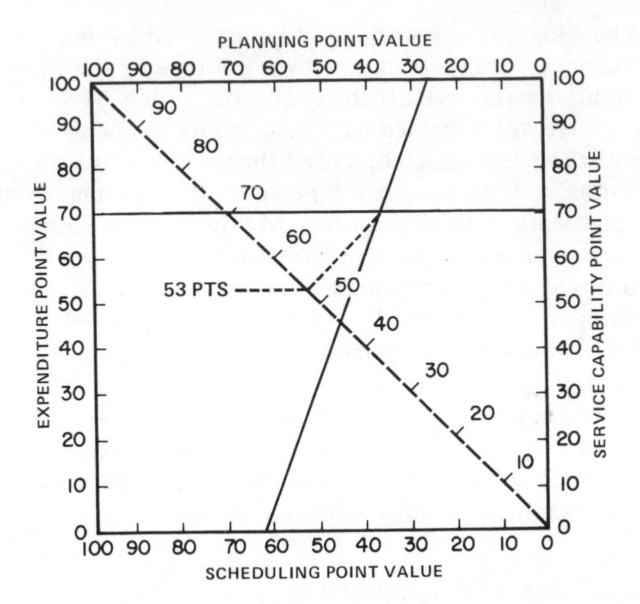

Source: Frank V. Claire, "Evaluating Maintenance Operations," *Plant Engineering,* June 23, 1977, p. 125.

Figure 6–13. Composite Service to Operations

dency in many organizations to give lip service to new methods and then gradually revert to the original methods of operation.

A part of the monitoring procedure should be the preparation of weekly progress reports in the form of graphs representing labor performance, amount of emergency work, size of maintenance backlog, and amount of maintenance work covered by work planning and scheduling.

Notes

1. This section is based on Russell E. Virgils, "Maintenance Work Sampling and Ananlysis," *Techniques of Plant Engineering and Maintenance* 17:173–176. Copyright 1966, Clapp and Poliak, Inc.

2. This section is based on Russell W. Lortz, "Productivity vs. Manpower Requirements," *Techniques of Plant Engineering* 20:177–181. Copyright 1969, Clapp and Poliak, Inc.

3. This section is reprinted from Frank V. Claire, "Evaluating Maintenance Operations," *Plant Engineering,* June 23, 1977, pp. 125–128.

7 Maintenance-Material Control

In most cases, the maintenance-material-control function is, by its nature, somewhat separate from the usual maintenance-management information system (MMIS). After considering the numerous aspects of maintenance-material control, this chapter will discuss integration of this function into the MMIS.

In considering storeroom operations and costs, it is necessary to consider the overall stores operation. This process lends itself to systemization, and most plants already have the stores and purchasing procedures on data-processing equipment.

Material Classification

Before considering material-controls systems and procedures, it is necessary to define the type of material that is normally kept in most stores operations.

Spare Parts

From the standpoint of maintenance, the most important classification of stores is spare parts. Spare parts are categorized as critical if their unavailability would require all or a portion of the plant to stop or slow operations. Critical spare parts make up only a small portion of the stores inventory (3 percent in one large process plant, which stores more than 30,000 items and identifies only 900 of them as critical).

Spare parts might be further categorized as

1. high-cost items, such as spare centrifugal-compressor rotors, sets of alloy liners for compressors, and spare tube bundles;
2. parts for which use is restricted to a single item of equipment in the plant;
3. parts whose absence would cause safety hazards and pollution problems; and
4. parts that have long delivery times.

Normal Stock

Normal stock includes items that are used frequently every day; they account for most of the material used by maintenance—gaskets, valves, pipe fittings, rolled-steel sections, bearings, belts, electrical equipment, and so forth. Generally, there is an order point built into the system to remind stores personnel to reorder this material, so that it is seldom, if ever, unavailable in the plant. The reminding process can be computerized or can be as simple as a number on a ledger card, which reminds the individual who debits the balance to reorder the material. A typical storage-ledger card is shown in figure 7–1.

Equipment and Tools

Tools are defined as items that are stored with the usual maintenance material to be checked out by the maintenance supervisor for a specific work-order task. Equipment is defined as portable machinery, such as cranes, air compressors, welding machines, and tuggers. It is usually stored in the garage area and maintained by the personnel who are responsible for the maintenance of the over-the-road equipment, such as material-handling units and trucks. No matter where tools and equipment are stored, it is necessary for them to be included in the maintenance system, since maintenance planning and scheduling must include those items.

Factors Causing Fluctuations in Materials Inventory

Many factors influence the total amount of material in the storehouse at any one time.

Maintenance-Scheduling Demands

The maintenance planner and scheduler should have a list of the material that is normally stocked and must be able to assume that these materials are always available. For specific equipment, he should also have a list of spare parts that are not normally carried in stock. The most immediate indication of the lack of a minimum amount of materials in stores is an increase in the number of work orders that cannot be worked because of tool, equipment, or supply shortages (stock-outs). When maintenance supervisors begin to experience an unusual number of stock-outs, they make take some action, such as drawing material out of stores before working the order to assure

STOCK NO.	DESCRIPTION		
			AVE. LEAD TIME
			EOQ
			ROP

NO.	VENDOR		VENDOR
1		4	
2		5	
3		6	

BALANCE ON HAND	ORDERED							SHIPPED		RECEIVED			MONTHLY CONSUMPTION				
	Date	Ven.	Order No.	Quantity	Price	Freight Rate	Del'd Cost, Ea.	Date	Days For Dely.	Date	Quant.	Bal. Due.	Mo.	19	19	19	19
													J				
													F				
													M				
													A				
													M				
													J				
													J				
													A				
													S				
													O				
													N				
													D				
													T				

Figure 7–1. Storage-Ledger Card

that the work can be done. This action is an indication of a fault in the system, and attention must be given to the problem immediately. One approach is to reserve material for a specific work order by red-tagging it or physically removing it to holding pens in the stores area or processing units. Such systems have several disadvantages: workmen do not necessarily treat material as reserved merely because it is tagged or in a pen; and, usually, no one returns material to stock if a work order is canceled.

Production-Downtime Considerations

Since the process plant is very sensitive to unforeseen production downtime, it is necessary to consider the effect of downtime when determining the amounts and types of spare parts to stock. Again, the dynamics of the process plant must be emphasized. Decisions made at one point in the economic cycle change over a short period of time. Thus, certain operating equipment might be critical at one time but not so critical at another time, and stores could allow some of the spare parts for that equipment to become depleted.

Quantity Discounts

Suppliers may offer quantity discounts for purchase of more material than is absolutely needed in the plant. Serious attention should be given to the economic question of whether it is best to buy less at a higher price or buy for discount and suffer the cost of storing and having capital funds tied up in material that is not needed in the immediate future.

Decentralized Stores

When a plant has a number of satellite storage areas, the likelihood that there is an overlap in the materials stored in each place is great. Many plants have decentralized stores areas to decrease the amount of traveling time of the maintenance craftsmen and are willing to pay for duplicating material in order to minimize travel and idle time of craftsmen.

Duplication and Material Identification

A classic example is a plant that orders two pumps from two different manufacturers and receives spare parts with each, among them bearings. Investigation reveals that the bearings from both pump suppliers were produced by the same bearing manufacturer and are, in fact, identical. Therefore, the plant is storing double the necessary amount of that item. This example indicates the hazards of lack of accurate identification of spare parts. Increasingly, plants are requiring manufacturers to furnish identifying parts lists for their products so that needless accumulation of the same items will not take place.

Plants that desire to minimize the number of spare parts might consider restricting their purchases of generic items, such as pumps, compressors, and instruments, to two or three specific manufacturers, with the idea that additional purchases can use spare parts already in stock. Plants may also restrict their purchases of pipe, for example, to certain sizes so that storage of flanges, valves, fittings, gaskets, and the like, may be minimized. The plant can then request all their users of pipe to restrict new designs to the pipe sizes normally stocked by the plant.

Supplier Location

Plants located remotely from their material suppliers must, of course, stock more material than plants whose suppliers are nearby; however, increasingly better communications, particularly among scheduled trucklines, have greatly improved this problem. Plants that locate in or near high-density industrial areas are better served by suppliers today than ever before. This is particularly true in instrumentation, electronic components, small pumps and compressors, and pipe fittings; a plant can expect half-day service from these suppliers on most of the items it has previously stocked. Bearing and belt suppliers now maintain a complete stock or will carry specialized stock on request. Users of large, expensive components often make arrangements with local suppliers so that they will stock specific items, with the user sharing the cost of maintaining the inventory.

Managerial Decisions

Often, when a firm needs additional operating capital, it will decide to reduce investment in the storehouse to provide a source of funds.

Costs of Storing Materials

Space: The cost of storage space is determined in a variety of ways. The cost of constructing the storehouse may be debited according to the square footage or cubic footage used for the storage of material, and this cost might be depreciated over a period of time. Other plants might consider the replacement cost or the cost of building a new storehouse.

Special Conditions: Many items require a specialized atmosphere—air conditioning, humidity control, heat, or ventilation—in order to prevent their deterioration.

Records. Operation of the stores function involves significant costs of maintaining files and computer programs for all items in stock.

Materials Handling: The cost of the entire system—often from the supplier's factory to the point of usage in the plant—is necessarily a cost of the storage operation. Although many new and innovative systems, such as unit loading, have come into being recently, this cost is still significant.

Obsolescence and the Like: Some portion of stored material will become obsolete, misplaced, lost, or pilfered, or the equipment for which spares were purchased may be scrapped—all of which adds to the cost of storage. Many items have a finite shelf life, and deterioration adds to the other costs.

Taxes and Insurance: The dollar value of the material in the storehouse is taxable. The average value of the storehouse is used to compute both taxes and insurance. Some storehouses carry their own insurance, but this does not relieve the plant of the burden of replacement when those storerooms suffer damage from events such as fire, flood, storm, and tornado.

Does the Plant Need a Centralized Maintenance Storehouse?

There are a number of alternative solutions to the problem of where to store the materials necessary for an adequately managed maintenance program: a centralized stores area; a decentralized stores area; storage of the spares in the vicinity of the equipment for which they are used; decentralized craft-oriented shops areas, such as pipe, paint, steel, and boilermakers shops; or some combination of these. No generalizations can be made about which system will serve a plant best. Each plant must tailor the service to fit its own needs.

Central stores areas avoid the duplication of stocked items that inevitably occurs with the separate storehouses of a decentralized system. They also are better able to make efficient use of space.

Certainly, an inventory-control program is more easily coordinated and managed when all dispensing and receiving of stores occurs in a single, central area. Normally, records are kept in a more orderly and accurate manner when only one record-keeping group exists. When storehouse records are kept on cards rather than computerized, the existence of a central group to maintain these records certainly would yield a more efficient operation than would one or more such clerks situated at each decentralized store area. Centralized storerooms require less personnel for keeping records and receiving and dispersing material. Also, they can better justify maintaining a staff and equipment to deliver the material from the storehouse to the work site.

Of major relevance here is whether or not the plant has a centralized shop area. Where a centralized shop area exists, immediate proximity of the storehouse yields many efficiencies. Where the shops are dispersed among the different areas of the plant, the centralized stores area becomes less of an advantage and even, in some cases, a disadvantage.

A centralized stores area can maintain better security of the material by having at least one clerk available on the off shifts, whereas emergency off-shift access to decentralized areas is usually through the plant guards or area-maintenance supervisors who might have a key. If a representative of the storehouse is not available at all times, there is great likelihood that records will not be maintained when material is withdrawn.

Another consideration is the interface between the purchasing and accounting departments and the stores department. Both of those functions must work closely with the stores department. The accounting function must be informed when material has arrived and must know whether or not that material fulfills the complete obligation of the supplier. This requires an open line of communication, which is best maintained where there is a central stores area, particularly a central receiving area.

When the plant is contemplating the use of material for a large construction project or for a turnaround, there is a problem regarding how to reserve or store that designated material so that it will be available when the maintenance forces require it. This process appears to be more easily accomplished when there is a centralized area than when material must be reserved at different points throughout the plant.

Decentralized storeroom areas do offer some advantages. In general, they allow craftsmen to obtain material with less walking time and less idle time for crew members who might be waiting for them to arrive with the material. Frequently, this aspect of storing material is an overriding one and there are real financial advantages for decentralization. One plant, discovering that approximately 300 items in the storehouse accounted for 85 percent of the withdrawals, placed some of each of these 300 items on a number of trailers strategically located in the plant. The savings in craftsmen idle and travel time more than compensated for the fact that accountability for the material placed on the trailers was often lost.

A special case of decentralized storage develops when spare parts are stored in the vicinity of their intended usage. This is particularly likely to occur when there is a unique item of equipment in a production unit. This situation presents most of the disadvantages of decentralization. The probability that the material will be available when it is needed is somewhat less than if that material were stored in a centralized place, since records most likely will not be kept and stock will therefore not be replenished as it is used.

The Maintenance-Material Storage Organization

In some plants, the storehouse for maintenance materials is completely separate from maintenance; in others, it is wholly under the maintenance organization. The maintenance organizations of most process plants seem to have their own storehouses, probably because, as a result of their bulk nature, the process raw materials require entirely different storage facilities. In the usual hard-goods plant, however, maintenance material is more likely to be in storehouses under the control of production. This arrangement can cause problems of coordination between maintenance and the storehouse.

Some maintenance organizations are operating in a system whereby storage of their materials is under the control of the purchasing department. The philosophy here is that the ordering of material is a primary function of the purchasing department and that storage of that material should be within the same department. Such situations exist when a plant has started from a small, modest operation and has grown, but the organizational structure has not changed. The disadvantage of this plan lies in the fact that the purchasing department often is unaware of relative urgencies. This can result in the purchase order for an item that might be holding up a turnaround or preventing production being considered in the same light as a purchase order intended merely to replenish stock.

Similar problems would exist if maintenance storage were under control of the accounting department.

From the examination of these alternatives, it appears best for maintenance stores to be administratively housed in the maintenance department. In large operations, an individual from the purchasing department is assigned to the stores department to initiate and expedite purchase orders as necessary. He is entirely familiar with all of the operations of the purchasing department and with the vendors.

A Maintenance-Stores Program

The first consideration in designing a maintenance-stores program is its paperwork system. The maintenance-stores system uses a primary document to obtain material and spare parts—the stores requisition, which includes, as a minimum the following information:

1. quantity ordered,
2. parts number,
3. parts descriptions,
4. unit cost,
5. total amount of each item,
6. charge number,
7. signature of issuing individual,
8. approving signature,
9. indication that the material has been deliverd,
10. date, and
11. a sequential number.

The stores-requisition card illustrated in figure 7–2 is a data-processing punch card used to debit the material from the storehouse stock, the records of which are usually on the computer. It is also used to apportion the material cost to the appropriate work-order number. In the absence of a computer, inventory records of material in the storehouse are kept on kardex-type forms on which the arrival and withdrawal of material is recorded manually. These forms require that the posting clerk inform the purchasing department when stock reaches the reorder (order) point. (Such a form is shown in figure 7–1).

Most plants today have their records on data-processing equipment; the inventory record is stored in the equipment and printed out as needed. The program is capable of immediate retrieval, often on visual-display units, so that stores management can know the status of any stock at any time. The charge numbers are the accounting department's classification for the allocation of funds. Thus, the total material charges for any work order or combination of work orders can be obtained from the accounting department. The charge numbers should be designed so that cost centers can be identified readily, because maintenance charges should be retrievable by cost center.

Periodically, usually annually, it is necessary to verify that material that is indicated as in stores on the computer printout actually exists in the storehouse.

What to Stock

When a plant is initiating a storehouse or drastically revising a storehouse system, it is necessary to have a systematic way of determining what material should be stocked.

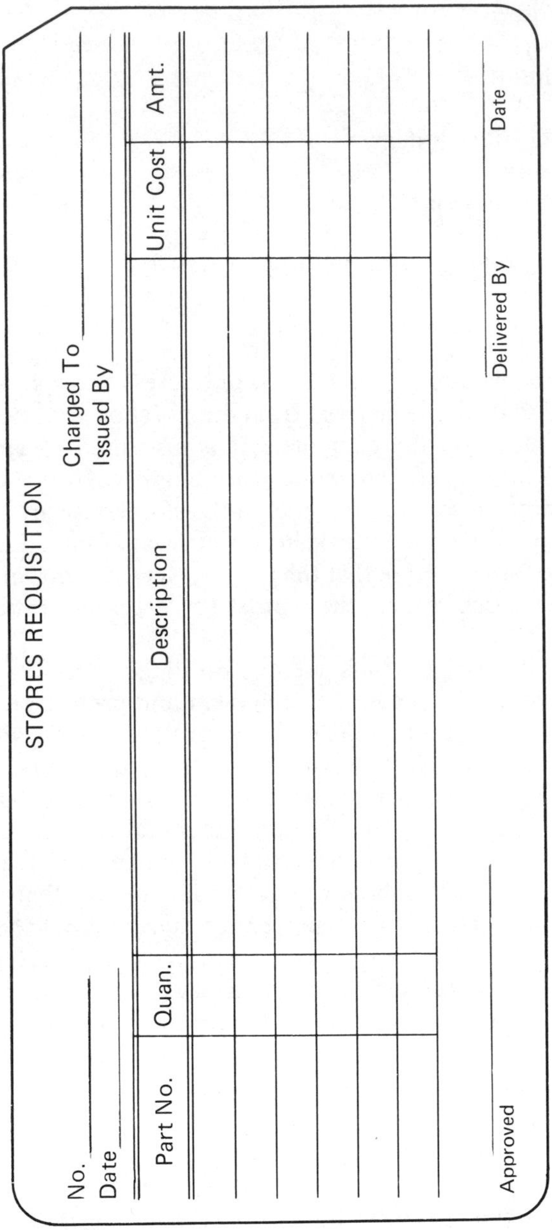

STORES REQUISITION

No. ______ Charged To ______

Date ______ Issued By ______

Part No.	Quan.	Description	Unit Cost	Amt.

Approved ______ Delivered By ______ Date

Figure 7–2. Stores-Requisition Card

Spare parts are usually stocked in accordance with the manufacturer's recommendation and in light of plant experience. If the plant is new and does not have any experience, the manufacturer's list of recommended spare parts should be considered in light of what is available locally and what can be obtained in a short period of time. Thus, although the manufacturer might recommend that bearings, chains, and belts be stocked, when these items are available locally and can be obtained quickly, it is not necessary for the plant to stock them. At first glance, it may seem reasonable to involve operating personnel in the determination of what should be stocked; their main interest is in minimizing downtime, however, and it is likely that their recommendations will be in excess of the funds allocated to maintenance materials and spare parts. If maintenance management is to be responsible for minimizing the storehouse resources and also for keeping the equipment responsive to the needs of operations, then maintenance management itself is most likely to arrive at an optimum program. It is necessary, however, for maintenance management and operations to cooperate in identifying items to be classified as critical, since, normally, critical items are purposely overstocked.

Once it has been determined what items will stock the storehouse, it is usually necessary to obtain upper-level permission to add new stock items.

How Much to Order

Perhaps the most important consideration in determining how much of each item to order or reorder is the average number of units used during a given time. Plots of typical maintenance-material usage are shown in figure 7-3.

Figure 7-3(a) illustrates usage for common items, such as pipe fittings, valves, steam traps and other items that are purchased in large quantities and for which there is relatively constant usage. Safety stock is the quantity stored to protect against stock-outs. Lead time is the time between the identification of a need and the receipt of the material. Time for the purchasing department to react and vendor-reaction time both constitute lead time. During different phases of the economic cycle, vendor-reaction time might change drastically. In manual systems, this variable might be considered on an order-by-order basis. With a computer system, the program must be written so that it factors in the lead-time variable. Rather than considering each item separately, it is suggested that the items be separated into generic families and that the correction constant be reconsidered periodically.

Other considerations include quantity discounts offered by suppliers and the cost of carrying inventory in stock. Most plants compute the average carrying cost as a percentage of the original cost of an item. If the cost

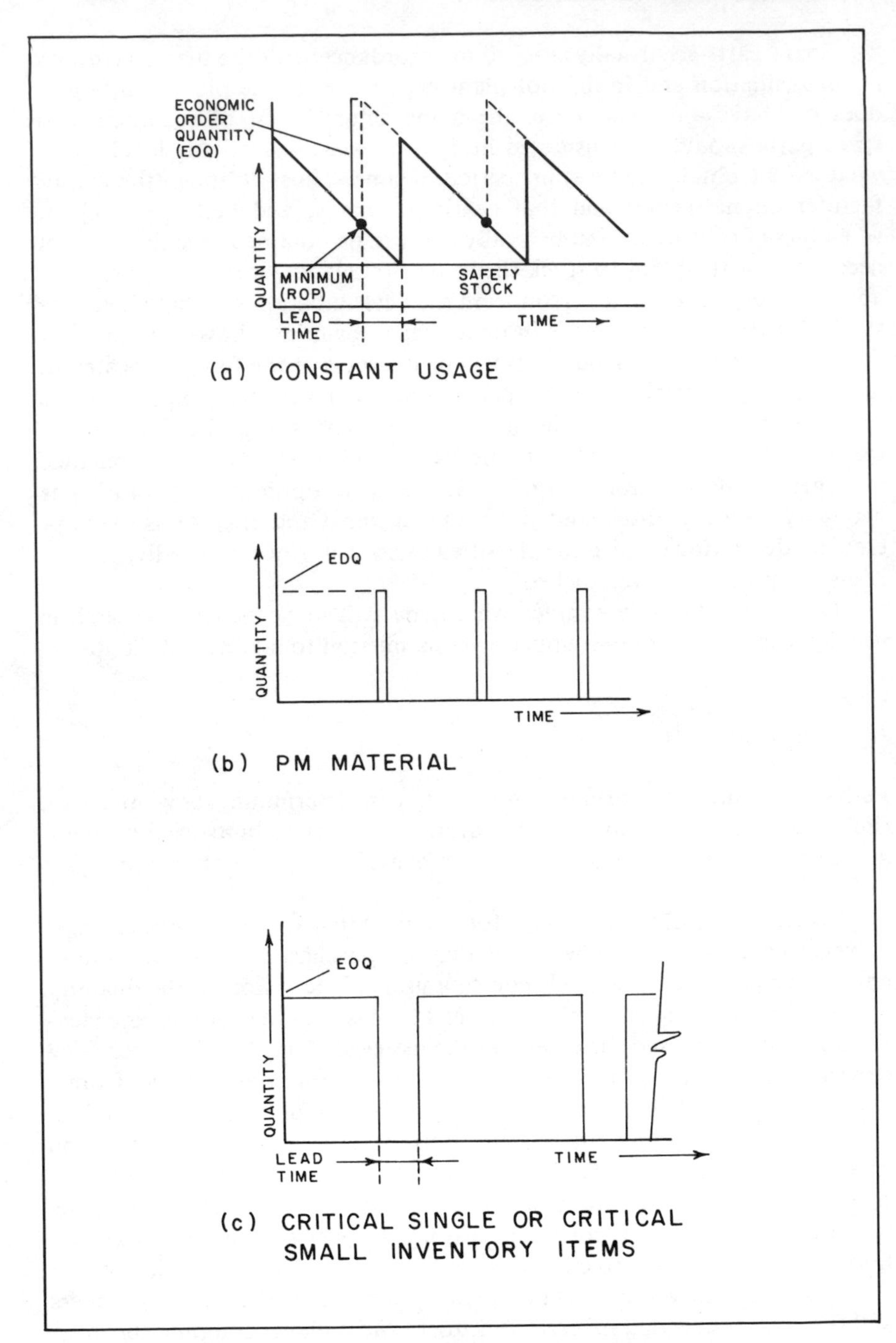

Figure 7–3. Storehouse Item-Usage Patterns

of borrowing money is approximately 6 percent, then a 20 percent carrying cost is appropriate; if borrowing cost is about 17 percent, then a 33 percent carrying cost is suggested.

Generally, some form of the simplified economic order quantity (EOQ) formula is used in determining how much of an item to purchase. The order point, stock-out considerations, and other aspects of more sophisticated inventory-control systems can be added to the expression given here:

$$EOQ = \sqrt{\frac{2AB}{I}}$$

where EOQ = economic order quantity in dollars

A = annual material usage in dollars

B = cost of one purchase order in dollars

I = cost of carrying inventory in percentage as a decimal

To find the EOQ in units, use

$$EOQ = \sqrt{\frac{2AB}{IC}}$$

where C is cost of a stocked unit in dollars.

The annual usage, A, consists of estimates of the number of items required during a one-year period multiplied by the unit cost. Naturally, this figure will vary from one time period to the other, and, in many cases, it will be necessary for maintenance material management to forecast what the usage will be as well as the projected unit price.

The cost of placing one purchase order, B, includes only the variable cost.

Figure 7–3(b) shows usage patterns for preventive-maintenance (PM) material. Under this system, material is ordered shortly before the PM activity and is usually sent directly from the receiving area to the unit. It is then used, thereby incurring minimum storage costs.

Figure 7–3(c) illustrates usage patterns for critical single or critical small inventory items. These are materials that must always be on hand; therefore, the decision has been made to stock them no matter what the cost.

It should be realized that the area under these curves is equivalent to the cost of maintaining inventory.

Such general expenses as indirect-labor salaries and building-maintenance cost are omitted because they do not fluctuate with the number of purchase orders processed. The variable cost attributable to orders includes office supplies, telephone bills, clerical assistance necessary for processing the purchase order, freight, and any time spent on specialized specification.

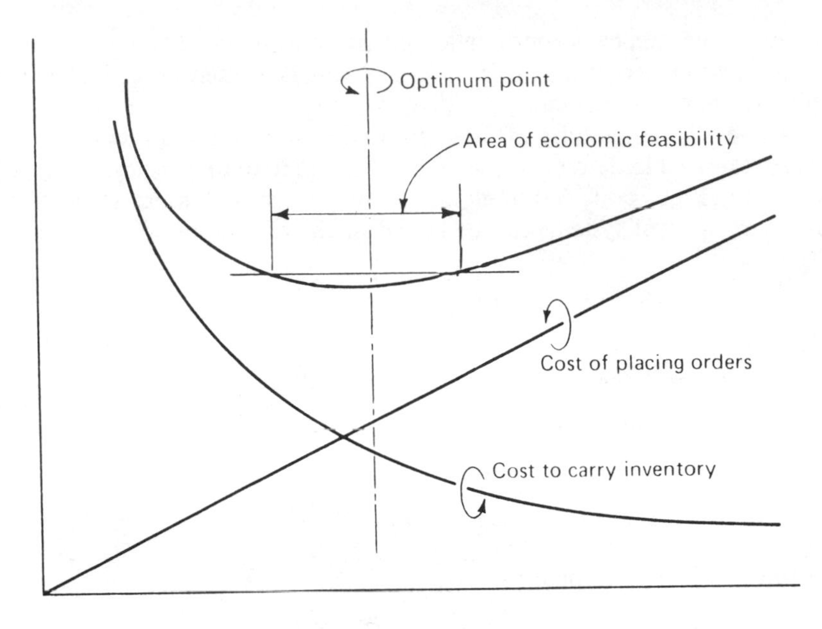

Figure 7-4. Variation of Inventory Costs and Purchasing Costs Relative to Order Quantities

The cost of carrying inventory, *I,* includes taxes, insurance, interest, obsolescence, depreciation, storage space, record keeping, and materials handling.

Figure 7-4 shows a typical curve derived from use of the EOQ formula. It should be noted that the flatter the top curve is in the vicinity of the optimum point, the more alternatives the purchasing department has in adjusting purchase quantities.

When to Order

The order point is the point (number of units) at which the stock of any material is so reduced that, considering the lead time necessary for processing an order and receiving the material, the supplier should be requested to initiate delivery of a fixed quantity of that material. Order points must be reviewed continually and adjusted according to the experienced response time of suppliers.

The inventory in the storehouse may be thought of as being divided into two parts; one part services normal usage and the other part is the safety stock of critical items. The amount of safety stock is the number of units by which the normal order point is advanced ahead of the normally expected requirements for the interim between reorder and receipt of stock.

If 300 six-inch, 90° welding ells are used a year, for example, each costing \$25.00, annual usage would be 300 × \$25.00 = \$7,500.00. If the cost of placing a reorder is \$50 and carrying inventory is 20 percent, then

$$EOQ = \sqrt{[2(7{,}500 \times 50) \div 0.20]}$$

$$= \$1{,}940.00$$

By dividing \$1,940.00 by 25, it is found that one would order approximately 78 ells each time an order is placed; dividing the 300 per year by 78, it is found that an order would be placed approximately every three months.

Open Stock

It is becoming increasingly expensive to monitor and record the transactions that have to do with maintenance management. Therefore, in most plants, there are a number of items—bolts, nuts, gaskets, welding rods, small pipe fittings, and the like—for which no records are kept. The thinking behind this practice is that the unit cost of the items is so small that there is no need to allocate their costs to specific cost centers or work orders. Therefore, the craftsmen merely withdraw these items as they need them from a bin to which they have access. The costs are then apportioned over the entire plant. Often, the plant has an arrangement with a supplier to monitor the bin periodically and refill the supply as the need indicates.

Storehouse Stock Catalogs

The computer has replaced printed storehouse catalogs. Use of a computer for handling storehouse information has several advantages. The computer can print out the current amount of each item in the storehouse. It can accumulate a monthly activity report. It can print out a list of material for which there has been little or no action during some previous period or a year-to-date accumulation of cost and activity within certain categories or for the plant as a whole.

Each item in the storehouse should be assigned a number. Not only is this essential to a computerized system, it ensures accurate identification of material by craftsmen in requesting withdrawal and by purchasing in ordering replacements. All standards should identify material by number, whether or not it is otherwise described. The more complex technology becomes and the more a plant depends on electronic systems, the more necessary it is to refer to material by number rather than attempting to describe it narratively.

A considerable number of man-hours are required to place into the computer the description of all items normally stocked by large process plants. It would require approximately 1½ man-years to computerize 30,000 storehouse items. Nevertheless, most plants agree that, once the descriptions are in storage, the benefits from the ready retrievability and from the ability to analyze stock and traffic in the storehouse more than offset the cost of the effort. In addition, benefits are gained by purchasing and accounting.

When material is received in the plant, it is usually charged to an undistributed inventory account. As the material is drawn out, it is assigned a specific charge number, usually from the work order. The charge number might be an eight- to twelve-digit figure, which could include coding for capital versus expense material, material versus labor and equipment, and the cost center to which the material is being charged. The open-stock material is charged to an overall expense account and is equally divided among cost centers. Most plants use a blanket or open work order with which material can be obtained, either from open stock or from secured storage, for jobs too small to warrant their own work order.

Interface with the Purchasing Department

The responsibility for replenishing storehouse stock as well as for obtaining critically needed spares rests with the purchasing department. Many plants have a priority system by which stores or maintenance indicates the degree to which purchase orders should be expedited. Other plants have a purchasing agent actually working in the storehouse, so that the interests of the stores department are represented at all times. Some plants have one individual in the purchasing department who is responsible for all stores operations, and this individual attends daily and weekly meetings in order to be apprised of the need for material.

Plants in highly industrialized areas can purchase much of their material from industrial-hardware suppliers in the immediate vicinity, and it is not uncommon for a plant to arrange an open purchase order with those suppliers. Maintenance supervisors are then authorized to go directly to the supplier, to obtain material whenever they need it, and to have that material charged to the open purchase order.

Storehouse-Performance Indexes

In evaluating the performance of the stores function, management is most commonly interested in the dollar value of the material in the storehouse. This is not a very useful evaluator, however, particularly since many spares are stored over a period of years. It becomes a problem whether to cost items at the price paid for them at the average price of all items of that type, or at their replacement cost. In addition, during periods of rapid inflation, dollar value will not present a true picture of the movement of material in and out of the storehouse.

Other performance indicators in which management might be interested are the number of items in the storehouse and the quantity of each item, or the total withdrawals—expressed as dollar value, total number of items, and quantity of each item. These values might be compared with a previous base period to determine whether activity in the storehouse is increasing or decreasing. Management is also interested in the spares turnover—that is, how long spares are kept and how often they are used; the number of stock-outs that have occurred; and the time taken to obtain material that was unavailable when requested. Although not within the control of the storehouse itself, management would be interested in knowing the losses that occur when material must be scrapped because it is obsolete or when it has deteriorated in the storehouse. In line with this, management would also like to know the discrepancy between the inventory shown on the computer and the periodic physical inventory. (Other indexes can be found in chapter 6.)

Rework and Returns Records

In periods of rapidly increasing costs, the shops within the plant are called upon to rework more material. This includes, particularly, such items as control valves, pump impellers, shafts, and items that can be made in the machine shops of larger process plants. A consistent, equitable policy for pricing these reworked items must be established. Some plants debit the work order with the entire cost of a new item; others charge the work order only for the cost of reworking the item.

Once material is drawn from the storehouse, it is difficult to get that material returned to its proper place in the storehouse if it is not used. It is the habit of many maintenance supervisors to draw somewhat more material from the storehouse than they need in order to cover contingencies. When this material is not used on the job, it should be returned to the storehouse. One way to facilitate this is to use the stores-requisition form, with a stenciled "stock return" across its face.

Computerized Stores-Information Systems

Although data-processing systems for stores records have much in common, they vary somewhat from plant to plant. The following is typical of systems in large process plants.

Material is received in stores. It is inspected, and the inspector notes on a copy of the order that the material has been received. At the end of each day, these items are keypunched, and the records are updated to reflect the availability of this material. Normal usage occurs until the material reaches a can-order or order point. Previously, the description of the item and five suppliers' names and addresses have been put into the program. The can-order point causes the computer to review all the can-order items furnished by a supplier with which an order is placed so that, if the order point is being approached, the supplier can be asked to furnish those items also. This prevents asking suppliers to reply to requests for bids for only one item.

The computer then prints out a set of forms that includes five copies for the five suppliers. The suppliers are requested to bid on items that have reached the order point as well as on those that have passed the can-order point. If replies have not been received from the suppliers in due course, the computer is programmed to print out follow-up letters. Stenciled across the five initial copies of the request for bids is the notice, "This is a request for bids; this is not an order." Once the purchasing department has selected a supplier, the appropriate individual in the department takes the sixth copy, which has stenciled across its face, "This is an order," and mails it to the successful bidder. The seventh copy goes to the receiving area of the stores department to be used as a notice to the purchasing department, when it has been returned, that the material has been received and inspected. This process is illustrated by the flow sheet in figure 7–5.

Reducing Maintenance Inventories in a Computer-Based System[1]

How can you control the maintenance inventory in a chemical processing plant? Opposing forces are pulling inventory managers in two directions simultaneously: keep the inventory low to avoid the high dollar investment and carrying charges, or keep it high to avoid stock outages that can shut down an entire unit or even a plant for need of a critical spare part.

Controlling an inventory is serious business, and how well the inventory is controlled will significantly affect the total plant profit or loss. . . . Prime

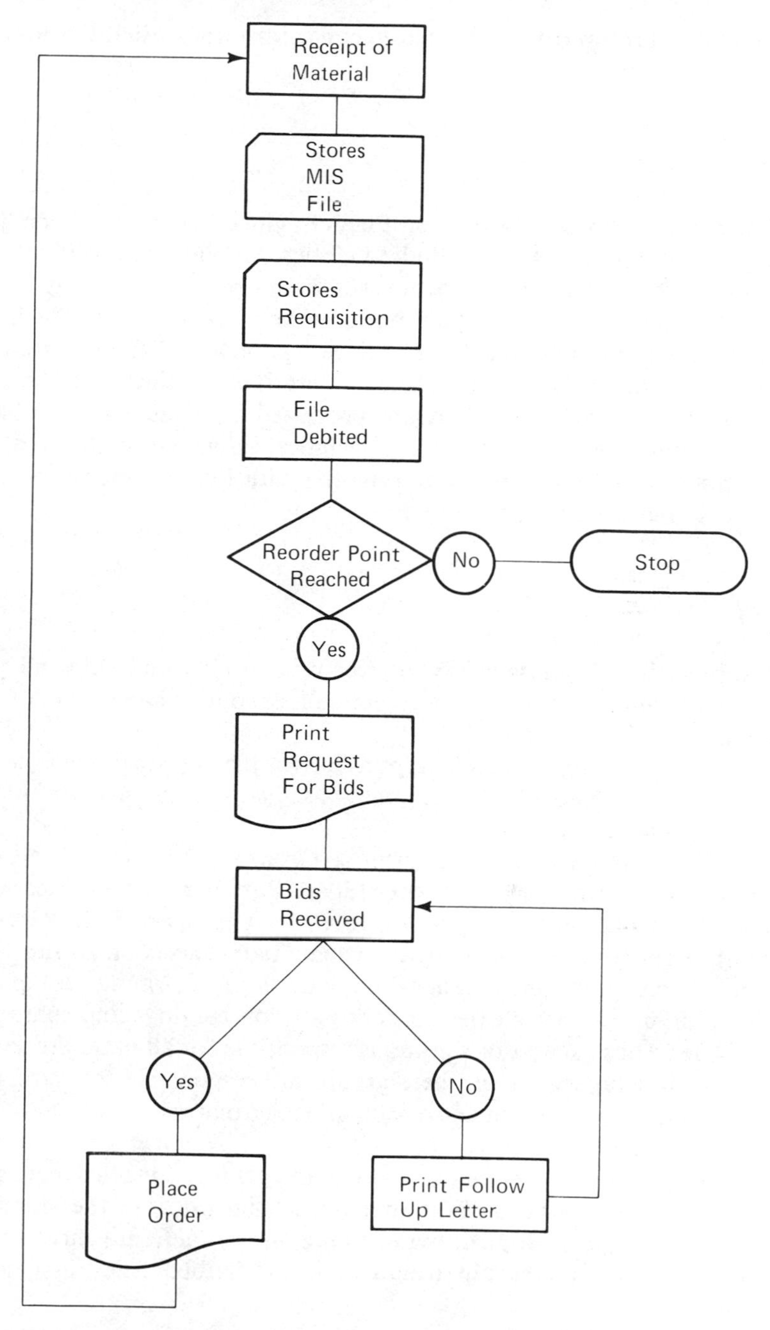

Figure 7–5. Flow Sheet of Maintenance-Material System

responsibility for control of the maintenance inventory usually belongs to the Stores Group.

Organization Stores Group

At [this] point, Stores is a part of Plant Engineering, which is predominantly Maintenance, but also includes Utilities, Special Projects (Construction) and Plant Protection. The Stores supervisor reports to the superintendent of Utilities and Stores, who reports to the plant engineer. The Stores supervisor is a graduate industrial engineer. A technical man is preferred because inventory control is a technical operation. Furthermore, since the inventory records of [this] plant are processed by computer, the Stores supervisor must be able to speak the technical language of the computer programmer, as well as coordinate activities with field engineers, purchasing agents, and vendors.

Activities Stores

The *Receiving and Shipping Group* receives, checks, and ships all plant material into and out of the plant except bulk products (barges, tank trucks and tank cars).

The *Warehousing Section* is responsible for placing stock items in bins, disbursing and delivery of all stock materials. They also furnish forklift service for the entire plant.

The *Spare Parts and Administrative Groups* reporting to the Stores supervisor, consist of the storekeeper (Spare Parts), and the Office supervisor, with clerical groups reporting to each. The Spare Parts Group is responsible for compiling and updating spare parts sheets on all functional equipment, compiling interchangeability sheets on all similar equipment, and keeping up to date all field ordering-station catalogs and spare-parts sheets. The Office supervisor's group is responsible for all other Stores clerical work. The supervisors in these groups are responsible for setting and revising order points and quantities on all stock material.

Facilities. Storage of inventory material at the plant is done in a main storeroom, located on a portion of the first and second floors of the Stores and Shops building and in four outlying warehouses. There are three forklift trucks, three half-ton pickup trucks and one flatbed truck assigned to Stores.

Issue and Delivery of Stores Material: The majority of storeroom items are kept in the main Stores and Shops building. Bulky items and dangerous liquids are kept in field warehouses.

Materials in Stores are issued primarily by telephone orders. Customers may come to Stores only when necessary to compare parts or discuss an order. Stores issue counters were closed a few years back. Currently, the volume of business is approximately 120,000 requisitions written for stock items annually.

There is a telephone answering room [with] three telephone lines. . . . Here, on day shift, two clerks take orders for material over the phone, recording these orders on telephone requisitions.

After the telephone clerk writes the requisition, it is delivered by a high-volume air blower to the appropriate storage area. A stock clerk there files the order and places it at the truck station for delivery to the field requisitioner.

Once a day, all Stores requisitions are accumulated, counted and sent to Data Processing, where the transactions are recorded in the inventory control system.

Stores materials are delivered by either a route truck or one of the three *hot-shot* or special trucks. The route truck is scheduled to circle the plant four times daily, delivering routine bulky items. The three hot-shot trucks handle all other items, . . . Hot-shot service is designed to give delivery within 30 minutes of an initial call. . . . All Stores trucks have radio-communication so an item in transit to the field can be located immediately.

Stores catalogs and spare parts sheets are compiled and maintained in each operating unit in the plant by a spare parts clerk from Stores. . . .

Methods Used to Control and Reduce Inventory

To control an inventory, the system must accomplish the following:

1. Review all stock items periodically:
   a. to adjust order quantities and safety stocks to meet current demands, and
   b. to locate obsolete items for removal from inventory.
2. Control the new items requested for stocking to make certain that only essential items are stocked.
3. Order and receive replacement stock rapidly and accurately.
4. Audit the operations of the Stores organization with a physical inventory of the stock.

Inventory Review: Review of the inventory is done on a cyclic basis on several sections of the inventory each month, so that all items are reviewed once annually. Review is done by the Store supervisor, the Office supervisor or the Spare Parts supervisor. The Order Quantity is calculated by computer, with the recommended OQ on each item printed out. In addition, minimum balance during the year, present Order Point (OP) and Order Quantity, and number of orders placed per item are recorded.

A sample activity report is shown in [figure 7–6]. Order quantities are calculated using the following [version of the EOQ] formula

$$EOQ = \sqrt{(2AS/IC)}$$

where EOQ is the most economical amount of stock to order so as to minimize total system cost. A is the number of items used annually, S is the dollar cost to initiate a purchase order and to receive the material, I is the inventory carrying factor as a decimal percent of unit cost, and C is the unit cost of the item.

New Item Control. Any plant supervisor may request that a new item be set up in Stores stock. To do this, a stock authorization is prepared in full, . . . [and] is reviewed by the originator's supervisor and/or superintendent; the Stores supervisor; . . . and the plant engineer. . . .

Stocked-by-Vendor Items: Stores will not normally stock parts or material valued over $100 per unit if the part is normally stocked by a vendor [nearby]. If the turnover is great enough or the item is critical enough, an exception is made and the item is placed in Stores. . . . A total inventory reduction of 7 percent has resulted from *stocked-by-vendor* items.

Spare Parts Interchangeability: To control an inventory, duplication of parts under different numbers must be avoided. One of the most effective methods for avoiding duplication of spare parts is the use of interchangeability sheets [see figure 7–7].

[These sheets] can be made from only one source, the manufacturer's bill of material, which gives a unique part number for each part. After all part numbers for a piece of equipment are tabulated, they are then compared with existing numbers in inventory to avoid stock duplication. Information from the spare parts interchangeability sheets can also determine whether or not a particular spare part can be obsoleted when a piece of equipment is removed from service. A total inventory reduction of 7.5 percent has been realized to date as a result of interchangeability listing of spare parts.

Shop Fabrication of Spare Parts: The items for Shop Fab are selected jointly by the Spare Parts storekeeper and the Machine Shop foreman, and

INVENTORY ACTIVITY REPORT

Date: 1/15/

MCC NO.	BAL.	UNIT COST	INV VALUE	QUAN ISSUED	DOLLAR ISSUED	DATE LAST TRANS	MIN BAL	NO MIN BAL	OP	OQ	EOQ	ORDER COUNT	REVISED OP	REVISED OQ
50001	10	32.40	324.00	20	648.00	01/07/	2	2	5	10	6	2		
50004	7	9.60	67.20	21	201.60	12/30/	0	3	1	6	12	4		
50005	26	8.71	226.46	48	418.08	01/10/	8	2	8	20	18	2		

Source: Carl Jumper, "What Can Be Done About Maintenance Inventories?" *Hydrocarbon Processing,* January 1970, pp. 129–132.

Note: A sample activity report. Note: the No. Min. Bal. is the number of times the balance has been down to the minimum during the order period. OP is the order point. OQ is the order quantity and EOQ is the most economical order quantity.

Figure 7–6. Inventory-Activity Report

SPARE PARTS SHEET

Spare Parts Sheet For: Brand X Pumps

Date: Jan. 30, 19

Description	Ref. No.	Part No.	Catalog No.	28P3-1	28P3-2		28P26-1		# In Service
Rod Connecting	25	801078	412060	1	1				2
Rod Connecting	25	801061	414735				1		
Rotor	22	900582	414736	1	1				2
Rotor	22	900586	415509				1		1
Screw, Drive Pin Retainer	54	900088	414737	1	1		1		3

Source: Carl Jumper, "What Can Be Done About Maintenance Inventories?" *Hydrocarbon Processing,* January 1970, pp. 129–132.

Figure 7–7. Spare-Parts Sheet

only those items are selected which seem most profitable. . . . Most of this work is done in the Machine Shop, but occasionally one of the other plant shops is involved.

There is a direct savings of several thousand dollars per month of Shop Fab over outside fabrication on the selected items. In addition, there are other benefits: Faster delivery, which means less lead time and less inventory; and use for fill-in work on days and shift, to reduce idle time.

Methods of Reordering

If an inventory system is to operate efficiently, it must have a fast, effective, and flexible system of ordering replacement stock. . . . A good stock-reordering program requires a variety of ordering methods.

Traveling-Card Reordering: Conventional Stores ordering is done by the use of travel cards and by blanket order releases.

Travel cards are kept in the Stores office. Information contained on these cards consists of approved vendors, price quotations, description of the item, stock number and other pertinent information.

When the balance on hand of an item reaches the Order Point, a clerk in the Store office pulls the travel card. At the end of the working day, all travel cards that have been pulled are sent to the Purchasing Department where purchase orders are typed out. After orders are placed, the travel cards are returned to the Stores office with the selected vendor, price and purchase order number marked thereon.

Blanket-Order Releases: Use of blanket orders is one method of cutting costs on common, high-usage items. To order a year's supply of most any item at one time would lower the purchase price of the item and reduce the paperwork necessary to buy, but it would raise the investment and warehousing costs of the item. With a blanket-order system, however, both objectives can be realized, at least to an extent. The buyer negotiates a price based on a year's supply, more or less, of an item. As the stock runs low, a release is issued (or a phone call made) for a small quantity against the outstanding order.

Expensive or slow-moving items should not be ordered by blanket order, but its use is a real advantage with fast-moving, low-cost items supplied by a local vendor.

Automatic Reordering by Computer Printout: The most efficient type [of] ordering for [many] items . . . is automatic reordering by computer. With this system, when . . . the balance on hand [goes] down to or below the order point, an order is automatically generated by the computer. . . . All that remains to be done by the Purchasing Department is to mail the orders.

Naturally, a great deal of work must be done prior to the computer printing of the purchase order. Bids must be sent out, prices negotiated, a vendor selected, order quantities agreed upon, and other details worked out. Once these details are settled between the Purchasing Department and the vendor, all the information can then be fed into the computer. The computer is then ready to begin ordering material. Out of 31,000 stock items, about 90 percent are ordered by computer printout.

Statistical Inventory

Physical counting of stock items . . . is done for two purposes: (1) to correct for past errors, and (2) to provide a basis for auditing the operations of the Stores organization. . . .

[At this plant, physical] Stores inventorying was done by two clerks working approximately full time. Auditing was done by limited random checking as deemed necessary to Accounting.

For statistical inventory, the total inventory is divided into 20 stratum based on extended price (unit price time quantity). Frequency of sampling in each stratum is proportional to the extended price, the higher the extended price, the greater the frequency. The last stratum, "over $1,000," is sampled 100 percent.

The Data Processing Group has a program that randomly selects the cards to be generated. One card is generated for each item to be inventoried. Nine teams (one Stores clerk and one Accounting clerk) do the inventorying. Cards are given out to the nine teams, and after counting, the cards are turned in to Data Processing, keypunched and matched against the balance tape. Items outside pre-set error limits are recounted.

An outside auditor is usually present during the entire inventory, and observes the procedures. He also selects several items at random that have already been counted and personally verifies the counts.

When these cards are turned in to Data Processing, Stores' participation in the inventory is complete. Accounting then evaluates the results of the inventory, based on pre-set statistical confidence limits. This year, items counted represented 27 percent of the dollar value with less than 3 percent of the [31,000] inventory items.

Not only has the statistical inventory method eliminated a dull, laborious counting job, and the resulting manpower, but the accuracy of the auditing has improved greatly. . . .

Are these methods effective? With a 20 percent inventory reduction (in face of rising prices), a Stores reduction of six men, a stockout reduction to less than half the previous level, and improved service to field maintenance, [this company feels it is] moving in the right direction.

A Manual System for Reducing Maintenance Inventories[2]

This second case, which emphasizes the removal of obsolete stock, involves a maintenance department with storeroom facilities designed to support a chemical plant valued in excess of $80 million. The storeroom services 249 hourly maintenance workers and various production personnel. The inventory includes repair materials and all other items except raw materials. Storeroom personnel assemble and deliver repair materials for planned work by individual jobs prior to actual scheduling of the work and routinely deliver repair materials for emergencies to the job sites.

The plant in this example is located about forty miles from the medium-sized city where the closest supplier is located, so deliveries can and frequently do become a problem.

One man—the planning and scheduling and materials-control superintendent—is responsible for the stores operation. A manual record-keeping system covers the repair-materials inventory, including more than 16,000 items. These records are maintained by a materials-control engineer, a clerk, and a typist. The system permits the removal of obsolete stock from the storeroom shelves automatically as soon as obsolescence occurs.

The System

A stock-item cross-reference card (figure 7–8) is prepared for each store item. This card contains the storeroom stock number, the stores-catalog description, min-max information, manufacturer's data, interchangeability information, and a listing of the machines or equipment in which the part is used. When an item of equipment is permanently removed from service, the materials- and inventory-control engineer is notified, and this is his signal to check the equipment data sheet (figure 7–9), which lists, by storeroom stock number, all repair parts for that equipment. He then pulls the proper stock-item cross-reference cards, selects the items that will no longer be used, initiates a request for change in storeroom inventory, and prepares a report of disposal or retirement of property.

Data Collection

To obtain all the information necessary to make the stock-item cross-reference card system work, a manual system is employed that has worked extremely well as a foreman's tool and as a source of information on material obsolescence.

When a new piece of equipment is received at the plant, a previously assigned fixed-asset number—a six- or seven-digit number used for capital-equipment identification—is physically attached to the equipment. (Capital equipment is generally defined as those items that cost over $200 and have a life in excess of a year.) Copies of the purchase order, the vendor prints, and all supporting documents are forwarded to the materials- and inventory-control engineer, who immediately begins to collect information for the equipment data sheet, using the vendor information, engineering drawings, and all other available information, with particular emphasis on the repair-parts listing.

There are five types of data sheets:

1. fan—for fans and blowers;
2. pump—for pumps of all types;
3. speed reducer—for all separate gearboxes and those attached to motors 5 hp and above;

STOCK ITEM CROSS REFERENCE

No. ______ Year Stocked ______

Max ______ Min ______ Date Max – Min Set ______

Description ______

Manufacturer ______ Date ______

Nearest Supplier ______ Date ______

Lead Time Reqd. ______

Interchangeability ______

MACHINES IN WHICH THIS PART IS USED

ASSET NO.		OPER. CLASS	No. REQD	DATE ENTRY	DATE REMVD	ASSET NO.		OPER. CLASS	No. REQD	DATE ENTRY	DATE REMVD

Source: C.S. Powell and T.F. Odom, "Improving Maintenance Materials Management by Controlling Material Obsolescence," *Techniques of Plant Engineering and Maintenance* 24:1–4. Copyright 1973, Clapp and Poliak, Inc.

Figure 7–8. Stock-Item Cross-Reference Form

EQUIPMENT DATA SHEET Sheet____of____

Item ____ Asset No. ____
Mfg. ____ PO No. & Yr. ____
Purch. From ____ C & R Job No. ____
Ser. No. ____ HP ____ Capacity ____ Cost ____
Type ____ RPM ____ Span ____
Size ____ K W ____ Pressure ____ Dimensions ____
Model ____ Style ____ Drwg. No. ____
Ratio ____ Cat. No. ____
Material ____ Rotation ____ Temp. ____ Inst. Bk. # ____
Imp. No. ____ Imp. Size ____ Drive ____ Shp. Wt. ____
Additional Data and Remarks ____

Bearing Date ____ Lubrication Date ____ P.M. Date ____

PARTS LIST

No.	Use	No.	Use	No.	Use

INSPECTION & REPAIRS

Date	Work Done	Date	Work Done

SERVICE LIST

Date	Code	Title

Source: C.S. Powell and T.F. Odom, "Improving Maintenance Materials Management by Controlling Material Obsolescence," *Techniques of Plant Engineering and Maintenance* 24:1–4. Copyright 1973, Clapp and Poliack, Inc.

Figure 7–9. Equipment-Data Sheet

4. motor—for electric motors, gear motors less than 5 hp, and variable-speed drives; and
5. equipment—for all other equipment.

There are four sections on all types of data sheets:

1. The first section is for listing manufacturer's data, purchase data, and the fixed-asset number. The spaces for manufacturer's data are self-explanatory and are different on each of the five types of data sheets because the specifications are different.
2. The second section, the parts list, is for listing stores stock numbers and descriptions of the equipment parts in stock. The part description should be brief enough to list on one line if possible. It should include the part name and the application on the equipment, such as "seal, oil, L.S. shaft," or "bearing, upper." This helps maintenance personnel identify parts needed without referring to a drawing or to another parts list.
3. The third section, inspection and repairs, is used to list the dates and descriptions of inspections, adjustments, and repairs. The description of work done should be as brief as possible but should explain what has been done.
4. The fourth section, the service list, is used to list the date, location, and service of the equipment. The date is the date the equipment was installed in a given location and service. The code designates the location, describing in abbreviated form a production unit or other location.

After a recommended spare-parts listing has been compared with the storeroom catalog, and the parts already in stock have been determined, preparation is made to increase the inventory as necessary. At the same time, the new piece of equipment is added to the appropriate stock-item cross-reference cards. The new repair materials, not formerly stocked, are assigned storeroom stock numbers and are added to the inventory and catalog, and stock-item cross-reference cards are prepared for them.

As the stock-item cross-reference cards are being prepared, a series of numbers designating the priority, or order of importance, of the equipment is inserted in the operating-class column. This numerical designation is based on the type of service, its importance to production, the number of identical items in the plant, and the possibility of using substitute items. This is extremely useful in determining minimums and maximums and in making decisions to stock insurance items. At the same time a master data sheet is prepared. Copies of this master, which includes storeroom stock numbers for the repair parts, are forwarded to the appropriate maintenance personnel. The master copy is filed in the equipment file for future use.

Inspections, adjustments, and repairs are recorded on a repair-data card (figure 7–10), and a copy of the card is sent to the inventory-control

HISTORICAL REPAIR DATA CARD Service Number

M.R. No. ________ Area ________ Cost Center ________ Date ________

Fixed Asset No. ________ Condition ________

Service Title ________

Work Performed Code		Problem Code				Part Code								
								Q			Q			Q
1	ALIGNED	1	BENT	10	MIS-ALIGNED	1	AGITATOR Shaft/Blades		19	HEAD		37	SHAFT	
2	BALANCED	2	BROKEN	11	NOISY—ROUGH	2	BASE/ FOUNDATION		20	HOUSING		38	SHEAVE	
3	CLEANED	3	BURNED	12	PLUGGED	3	BEARING		21	HUB		39	SHIM	
4	MODIFIED	4	CORRODED	13	POOR WORKMANSHIP	4	BELT		22	IMPELLER		40	SHOE	
5	NOT WORKED	5	DULL	14	SEIZED	5	BUSHING		23	INSULATION		41	SLEEVE	
6	REPAIRED	6	IMPROPER LUBRICATION	15	TORN	6	CABLE		24	KEY		42	SOCK	
7	REPLACED	7	IMPROPER OPERATION	16	VIBRATING	7	CASING		25	KNIFE		43	SPRING	
8	OVERHAULED SPARE	8	LEAKING	17	WORN	8	CHAIN		26	OILER		44	SPROCKET	
		9	LOOSE			9	CHUTE		27	PACKING		45	SUB-ASSEMBLY	
						10	CLOTH		28	PIN		46	TUBE	
						11	COUPLING		29	RING		47	VALVE	
						12	COVER		30	ROLLERS		48	VIBRATOR	
						13	FOLLOWER		31	ROTOR		49	WEDGE	
						14	GASKET		32	SCREEN				
						15	GEAR		33	SCREW				
						16	GLAND		34	SEAL-MECH.				
						17	GRID		35	SEAL-OIL				
						18	GUARD		36	SEAT				

WORK DESCRIPTION

____

____

____

____

____

____

New F.A. No. ________

Downtime: ________ **HRS.** ________ **MIN.**

Job Was: ☐ **EMERGENCY** ☐ **PM** ☐ **SCHEDULED**

Date Completed: ________

Maint. Foreman: ________

Source: C.S. Powell and T.F. Odom, "Improving Maintenance Materials Management by Controlling Material Obsolescence," *Techniques of Plant Engineering and Maintenance* 24:1–4. Copyright 1973, Clapp and Poliak, Inc.

Figure 7–10. Historical Repair-Data Card

engineer, who transfers the information to the appropriate master data sheet, an updated copy of which is forwarded to the appropriate foreman.

An item of equipment may be used in several services, and any change in the service is recorded in the service-record section of the data sheet. Eventually, because the service life is ended or the function is no longer required, the equipment is removed from service. When this occurs, the repair-data card, which is completed by the maintenance foreman involved, reflects the fact that the equipment, identified by a fixed-asset number, is no longer in service. This information is the signal to inspect all stock-item cross-reference cards related to the equipment and to begin the process of removing the obsolete items from stock.

The system is an excellent fact-gathering device that permits removal of items from stock as soon as they are no longer needed, and it has reduced stock-outs and obsolete or overstocked items.

The same simple system will work for any organization, regardless of size, but if it is to operate successfully, some basic steps are necessary. A workable stores catalog and numbering system must be developed, and equipment records must be kept up to date. Also, the people who will administer the system must be reasonably familiar with the repair materials that will make up the storeroom stock.

Notes

1. This section is reprinted in part from Carl Jumper, "What Can Be Done About Maintenance Inventories?" *Hydrocarbon Processing,* January 1970, pp. 129–132.

2. This section is based on C.S. Powell and T.F. Odom, "Improving Maintenance Materials Management by Controlling Material Obsolescence," *Techniques of Plant Engineering and Maintenance* 24:1–4. Copyright 1973, Clapp and Poliak, Inc.

8 Maintenance Budgeting and Forecasting

A maintenance budget, like other budgets, is a financial plan that repr sents a projection of expenditures for some future time period. Budgets should be realistic; that is, they should reflect what will actually happen rather than what management would like to happen. The very fact that individuals must work with budgets forces them to plan more than they would normally.

Cost Centers

In most process plants, the maintenance budget is divided among cost centers. A cost center is a geographical portion of the plant, usually represented by a homogeneous production unit. Each cost center is thought of as operating on its own operating budget, a portion of which is for maintenance. The cost-center operation purchases maintenance supplies, labor, and equipment from the central maintenance organization; that is, its budget is debited for the labor and material used to maintain its equipment. The advantage of the cost-center concept is that the operator of each cost center can be reminded constantly of whether his operation is ahead of or behind anticipated maintenance expenditures. One disadvantage is that the operators might save their funds until the end of the time period and then make large demands on the maintenance department so that the normal scheduling procedures are disturbed. Some plants anticipate this problem and, at the end of each month, divert unspent maintenance into a general fund.

Long-Range Planning

The first consideration in long-range maintenance planning is man-hours. In planning, the total number of man-hours currently involved in maintenance must be adjusted according to the anticipated increase or decrease in productivity and in the demands made on maintenance through bringing new equipment on stream or retiring existing equipment. Long-range planning must also take into consideration any managerial decisions to alter the level of maintenance for any or all of the plant.

In many cases, a change in the level of maintenance is ignored in long-range planning. Policy does change, however, and it is reasonable to expect that the appropriate level of maintenance will decline for equipment that is to be phased out in the foreseeable future. Several problems accompany the level-of-maintenance concept. There are separate levels of maintenance for different portions of the plant. Different operators have different concepts of adequate maintenance. The level of maintenance must be quantified so that management can maintain control over maintenance expenditures. In quantifying the level of maintenance, one might consider controlling the level by placing a limit on the man-hours or dollars to be spent on a given unit in specific time periods. As individual units become more or less valuable to management, the level of maintenance should be reviewed and adjusted.

Program Budgeting

Most maintenance budgets are categorized according to labor, material, and equipment. This breakdown may be divided further into more specific subcategories. A different budgeting concept is known as program budgeting. In program budgeting, the same dollars included on a line budget are divided according to where the dollars will be used, that is, the programs for which the dollars will be spent—the lubrication program, the painting program, the program to maintain each specific process unit, the program to maintain the docks, and so forth. When management decides to decrease or increase the funds allocated to a program budget, maintenance management is in a position to inquire where the increase or decrease should be made, whereas the assumption must be made that management wants a uniform change in labor, material, and equipment when a change is ordered in funds allocated to a line budget. This is seldom the case. Management usually has in mind the reduction of the level of maintenance in some areas of the plant and the increase of maintenance in other areas of the plant. The program budget gives all parties an opportunity to understand the results of decisions.

Capital Programs

Most forward-looking maintenance departments have a long-range plan that includes some capital improvements—for example, a five-year program to update power tools, a program to replace the rigging craft with mechanical handling equipment, or an extensive training program. These programs are usually accomplished over a long time period, and an appropriate portion of them should be included in each annual budget.

Usually, maintenance must justify these capital programs by demonstrating a percentage return on investment. Management is familiar with this approach, and program proposals that are presented in this manner are more likely to receive appropriate consideration.

Estimating Return on Investment[1]

Old equipment wears out. New equipment is available that can do the job better and faster, possibly with less manpower. The new equipment requires a substantial capital investment, however, whereas the old equipment has long since been written off the company's books and is working for nothing. How can management be persuaded to give up the old relic and invest in a newer, more modern process?

Proposed capital investments cover all aspects of a firm's business. They are alike in that they are capital projects requiring a large initial investment, which—it is hoped—will be repaid by future profits. How can management select from a variety of potential projects those that will contribute the most to profits? Clearly, management needs a method of estimating the profitability of a proposed project.

In appraisal of capital projects, a commonly used standard is the payback period, the length of time it takes for the initial investment to be recovered; for example, if a new machine costs $5,000 and produces a net annual savings of $1,000, the payback period is five years.

Rate of Return

The payback period is easy to compute and understand, but it can give misleading results, especially when it is applied to projects with widely different costs and lifetimes. For this reason, alert managements have been using the rate-of-return method for evaluating prospective investments.

What is meant by rate of return? Suppose I deposit $100 in a savings account that pays 5 percent per year. (To simplify the discussion, assume that interest is paid only once a year, rather than quarterly, as in most banks.) Suppose I invest on January 1. On December 31, the bank credits me with $5 interest. If I withdraw my funds, I have $105; but suppose I leave my money in the bank for another year. This time, I earn more interest than before because I am earning interest on last year's interest. At the end of the second year, I have $110.25. If I leave my $100 in the bank for five years, at 5 percent interest, I will have $127.63 at the end of the fifth year.

The rate of return earned by a capital project can be defined from analogy with compound bank interest. When we say that a project offers a rate of return of 5 percent, we mean that it yields results that are financially equivalent to the results of depositing the same amount in a bank that pays interest at 5 percent per year (paid each year). A more formal definition of rate of return is that it is the rate of interest earned over the life of the project while we are recovering the investment.

Present Worth. The typical capital-improvement project requires an initial investment followed by a series of annual payments and receipts. Only in rare cases are the payments and receipts neat and regular, like bank interest. How, then, can management determine the rate of return? The secret lies in the concept of present worth. If $100 invested at 5 percent will be worth $127.63 after five years, it is equally true that, at an interest rate of 5 percent, $127.63 in five years is worth $100 today. Put another way, $100 is the present worth of $127.63 in five years if the interest rate is 5 percent. The amount of the investment divided by the amount it will produce after a given length of time is the present-worth factor. In the example given, the quotient is 0.7835; to find the present worth of any amount five years from now (at 5 percent) we multiply it by 0.7835—for example, $1000 five years from now is worth $783.50 today if the interest rate is 5 percent. Mathematically:

$$P = \text{the present worth factor} \times S \tag{8.1}$$

where

$$P = \text{present worth}$$

$$S = \text{future amount}$$

and

$$\text{Present worth factor} = 1/(1 + i)^n \tag{8.2}$$

Here, i is the interest rate and n is the number of years. Formulas (8.1) and (8.2) can be combined as follows:

$$P = \frac{S}{(1 + i)^n} \qquad ((8.3)$$

However, it is usually more convenient to refer to tables, such as table 8–1, which lists present-worth factors for interest rates from 0 to 25 percent for periods of one to twenty-five years.

The simple concept of present worth makes it possible to relate an irreg-

Present-Worth Factors

Year (n)	Interest Rate 0.02	0.04	0.06	0.08	0.10	0.12	0.14	0.16	0.18	0.20	0.22	0.24	0.26
1	0.9804	0.9615	0.9434	0.9259	0.9091	0.8929	0.8772	0.8621	0.8475	0.8333	0.8197	0.8065	0.7937
2	0.9612	0.9246	0.8900	0.8563	0.8264	0.7972	0.7695	0.7432	0.7182	0.6944	0.6719	0.6504	0.6299
3	0.9423	0.8890	0.8396	0.7938	0.7513	0.7118	0.6750	0.6407	0.6086	0.5787	0.5507	0.5245	0.4999
4	0.9238	0.8548	0.7921	0.7350	0.6830	0.6355	0.5921	0.5523	0.5158	0.4823	0.4514	0.4230	0.3968
5	0.9057	0.8219	0.7473	0.6806	0.6209	0.5674	0.5194	0.4761	0.4371	0.4019	0.3700	0.3411	0.3149
6	0.8880	0.7903	0.7050	0.6302	0.5645	0.5066	0.4556	0.4104	0.3704	0.3349	0.3033	0.2751	0.2499
7	0.8706	0.7599	0.6651	0.5835	0.5132	0.4523	0.3996	0.3538	0.3139	0.2791	0.2486	0.2218	0.1983
8	0.8535	0.7307	0.6274	0.5403	0.4665	0.4039	0.3506	0.3050	0.2660	0.2326	0.2038	0.1789	0.1574
9	0.8368	0.7026	0.5919	0.5002	0.4241	0.3606	0.3075	0.2630	0.2255	0.1938	0.1670	0.1443	0.1249
10	0.8204	0.6756	0.5584	0.4632	0.3855	0.3220	0.2697	0.2267	0.1911	0.1615	0.1369	0.1164	0.0992
11	0.8043	0.6496	0.5268	0.4289	0.3505	0.2875	0.2366	0.1954	0.1619	0.1346	0.1122	0.0938	0.0787
12	0.7885	0.6246	0.4970	0.3971	0.3186	0.2567	0.2076	0.1685	0.1372	0.1122	0.0920	0.0757	0.0625
13	0.7730	0.6006	0.4688	0.3677	0.2897	0.2292	0.1821	0.1452	0.1163	0.0935	0.0754	0.0610	0.0496
14	0.7579	0.5775	0.4423	0.3405	0.2633	0.2046	0.1597	0.1252	0.0985	0.0779	0.0618	0.0492	0.0393
15	0.7430	0.5553	0.4173	0.3152	0.2394	0.1827	0.1401	0.1079	0.0835	0.0649	0.0507	0.0397	0.0312
16	0.7285	0.5339	0.3936	0.2919	0.2176	0.1631	0.1229	0.0930	0.0708	0.0541	0.0415	0.0320	0.0248
17	0.7142	0.5134	0.3714	0.2703	0.1978	0.1456	0.1078	0.0802	0.0600	0.0451	0.0340	0.0258	0.0197
18	0.7002	0.4936	0.3503	0.2502	0.1799	0.1300	0.0946	0.0691	0.0508	0.0376	0.0279	0.0208	0.0156
19	0.6864	0.4746	0.3305	0.2317	0.1635	0.1161	0.0829	0.0596	0.0431	0.0313	0.0229	0.0168	0.0124
20	0.6730	0.4564	0.3118	0.2145	0.1486	0.1037	0.0728	0.0514	0.0365	0.0261	0.0187	0.0135	0.0098
21	0.6598	0.4388	0.2942	0.1987	0.1351	0.0926	0.0538	0.0443	0.0309	0.0217	0.0154	0.0109	0.0078
22	0.6468	0.4220	0.2775	0.1839	0.1228	0.0826	0.0560	0.0382	0.0262	0.0181	0.0126	0.0088	0.0062
23	0.6342	0.4057	0.2618	0.1703	0.1117	0.0738	0.0491	0.0329	0.0222	0.0151	0.0103	0.0071	0.0049
24	0.6217	0.3901	0.2470	0.1577	0.1015	0.0659	0.0431	0.0284	0.0188	0.0126	0.0085	0.0057	0.0039
25	0.6095	0.3751	0.2330	0.1460	0.0923	0.0588	0.0378	0.0245	0.0160	0.0105	0.0669	0.0046	0.0031

Source: Joseph Horowitz, "Estimating Return on Investment," *Techniques of Plant Engineering and Maintenance* 20:125–131. Copyright 1969, Clapp and Poliak, Inc.

ular series of cash incomes and expenses to a single common base and permits an accurate determination of the rate of return for any project.

Advantages of the Method. Unlike the payback period and other rule-of-thumb methods, the rate-of-return method yields a single number that is consistent from project to project. Equally important, the results are directly comparable to the interest rates that could be earned from bank deposits, investments in bonds, and so on.

Since the rate-of-return method makes possible comparison of the merits of various kinds of investments, it is also called the investor's method; and, since the procedure is to discount future cash receipts and disbursements back to a present base, it is also known as the discounted cash flow (dcf) method.

Summary of the Procedure. The procedure for finding the rate of return will be illustrated with examples, but first, we look at the steps to be taken:

1. Estimate the initial cost, the useful life, and the salvage value (if any) of the capital investment.
2. Estimate the cash income and the expenditures for each year of the project's life. Note that income taxes are included as a cash cost but that interest is not.
3. Total the cash flow for each year. Cash in is considered plus, cash out is considered minus. This figure is the net cash flow.
4. Assume a trial interest rate.
5. Using the trial rate, discount the net cash flow back to the start of the project by using the appropriate present-worth factors from the table. This gives the discounted cash flow.
6. Add up the discounted cash flows.
7. If the trial interest rate was correct, the sum of annual profits (when discounted) should just equal the initial investment. If these two are not equal, choose another trial interest rate and repeat the calculations.
8. The final interest rate is determined by interpolation between the two closest of the tabulated values.

Example—Equipment Replacement Problem. Consider the problem of replacing a machine. Assume that the machine, which costs $10,000 when new, has been fully depreciated on the company's books. Because of its age, it requires an increasing amount of maintenance.

A new machine can produce the same output with less manpower, and repair costs would be reduced. A trade-in of $1,000 is offered on the present machine toward purchase of the new machine. Purchase price of the new machine is $20,000, including installation, and it is estimated to have a useful life of 5 years. At the end of this time, it could be sold for salvage at an estimated price of $2,000.

Initial purchase cost	$20,000
Less: trade-in of present machine	1,000
Net purchase cost	$19,000
Useful life	5 years
Salvage value	$ 2,000

Table 8–2 shows projected operating costs with the existing machine and with the proposed replacement. Annual costs are shown as increasing, reflecting the increasing age of both machines.

The approximate payback period for the new machine can be found in the cumulative-savings column in the table, which shows that the purchase price of the new machine is recovered during the fourth year. Obviously, if the payback period were longer than the anticipated useful life of the machine (five years), the investment would not be worthwhile.

To find the true rate of return, use of a worksheet (figure 8–1) is convenient. For simplicity, assume that each of the annual costs or savings occurs at the end of the particular year. Since the new machine is purchased at the beginning of the first year, the purchase is assumed to occur at the end of year 0. The estimated salvage value is shown as a negative factor in the capital-investment column.

For the purposes of this example, the annual savings are assumed to include the effect of federal income taxes. (Normally, these would be estimated for each year by the accounting or tax department, using standard methods. When operations result in a net loss for that year, income tax is

Table 8–2
Estimated Savings from Investment in New Equipment

Year	*Operating Costs with Present Machine*	*Operating Costs with New Machine*	*Anticipated Savings with New Machine*	*Cumulative Savings*
1	$10,000	$5,000	$5,000	$ 5,000
2	12,000	6,000	6,000	11,000
3	12,500	6,000	6,500	17,500
4	12,500	6,300	6,200	23,700
5	12,500	8,000	4,500	28,200

Source: Joseph Horowitz, "Estimating Return on Investment," *Techniques of Plant Engineering and Maintenance* 20:125–131. Copyright 1969, Clapp and Poliak, Inc.

CAPITAL PROJECT EVALUATION WORKSHEET

DIVISION	PROJECT NUMBER	DATE

PROJECT: Machine Replacement

LOCATION:

ANTICIPATED RESULTS

RATE OF RETURN 17 % USEFUL LIFE 5 YEARS

PAYOUT PERIOD 3.2 YEARS PREPARED BY ____

INVESTMENT DATA

	COST	SALVAGE OR RECOVERY		COST	SALVAGE OR RECOVERY
A – LAND			E – STARTUP EXPENSE		
B – BUILDINGS			F – WORKING CAPITAL		
C – MACHINERY & EQUIPT.	19,000	2,000	G – OTHER		
D – RESEARCH & DEVEL.					

RATE OF RETURN CALCULATIONS

YEAR (1)	CAPITAL INVESTMENT (2)	INCREASE IN WORKING CAPITAL (3)	INCOME OR SAVINGS GENERATED (4)	EXPENSES (5)	INCOME TAX (6)	NET CASH INFLOW (7)	DISCOUNTED CASH FLOW @ 16 % PRES. WORTH FACTOR (8)	@ 16 % PRESENT WORTH (9)	@ 18 % PRES. WORTH FACTOR (10)	@ 18 % PRESENT WORTH (11)
0	19,000					(19,000)	1.000	(19,000)	1.000	(19,000)
1			5,000			5,000	.8621	4,311	.8475	4,238
2			6,000			6,000	.7432	4,459	.7182	4,309
3			6,500			6,500	.6407	4,165	.6086	3,956
4			6,200			6,200	.5523	3,424	.5158	3,198
5	(2,000)		4,500			6,500	.4761	3,095	.4371	2,841
		TOTAL: YEARS 1 - 5						19,454		18,542
TOT.										

Source: Joseph Horowitz, ''Estimating Return on Investment,'' *Techniques in Plant Engineering and Maintenance* 20:125–131. Copyright 1969, Clapp and Poliak, Inc.

Figure 8–1. Capital-Project Evaluation Worksheet

negative, which means that taxes are reduced by the amount shown. This assumes, of course, that overall operations of the company are profitable.)

To find the net cash inflow (figure 8-1, column 7), add across for each year. Cash paid out is always considered minus, and cash received is plus. For the zero year, there is only one figure—the minus cost of $19,000, which is the net purchase price. The cash inflow includes savings with the new machine and, in year 5, the estimated salvage value of $2,000. Both represent cash in and are plus.

To find the rate of return, we must assume a trial interest rate. For the first trial in figure 8-1, we assume an interest rate of 16 percent. Table 8-1 is used to find the present-worth factors for an interest rate of 16 percent for each year. (Note that, in year 0, the present worth factor is 1.000; this is the same as saying that $1 now is worth exactly $1 now. This is true, of course, regardless of the interest rate.)

Present worth is found by multiplying each of the net cash flows by their corresponding present-worth factor. Algebraic signs are important; if the cash flow is negative, the discounted cash flow must also be negative.

The figures in column 9 give the value of the net cash inflow for each year of project life, discounted back to the start of the project, at the scheduled interest rate. What this means is that the net cash receipts of $6,200 in the fourth year, for example, are worth only $3,424 at the start of the project if the interest rate is 16 percent.

If the sum of the discounted cash flows were exactly zero—that is, if the discounted value of annual savings generated just equaled the initial investment—we would have recorded our entire investment, plus interest at 16 percent. In the present case, however, the discounted cash flows total $19,454, which is greater than the $19,000 initial investment. What this means is that we anticipate earning more than 16 percent on our investment. To find the correct rate of return, we must choose a larger interest rate—perhaps 18 percent—and repeat the calculations, using present-worth factors at 18 percent interest. We then find that the sum of the discounted cash flows ($18,542) is less than our investment of $19,000. The true rate of interest, therefore, lies somewhere between 16 percent and 18 percent. A straight-line interpolation between these two values yields a rate of return of about 17 percent. Since all the figures used in the calculations are estimates, this is accurate enough for practical purposes.

Note that the original cost of the present machine and its present book value did not enter into the calculations; only future costs and incomes affect the rate of return. (There is one exception to this rule; past costs do affect federal income taxes.)

Taxes and Depreciation. The worksheet in figure 8-1 shows a column for income taxes. Some firms prefer to compute the rate of return before taxes so that the results will be more directly comparable to interest rates on bonds, stocks, and so on, which are also computed before taxes. Two projects that are otherwise similar may be subject to different tax rates, however, the effect of which is revealed only if taxes are included as a cost item.

One factor that causes more confusion than any other in capital-project evaluation studies is depreciation. To understand what is meant by depreciation, look at figure 8–1. We have shown the entire purchase price of the new machine at the beginning of the project, which is when the money is spent. We recover the purchase price through the profits in later years.

Accountants must try to estimate the profit-or-loss position of the company each year, however. They cannot wait until a long-term project is completed to find out whether the firm is making or losing money. Showing a large capital item as an expense all in one year would result in a tremendous loss in that year, followed by fictitious profits in the following years. Accordingly, good accounting practice (and federal tax laws) require that the cost of a capital asset, such as a machine or a building, be spread over a period of years approximating its useful life. These annual charges are called depreciation. Note, however, that they are an accounting convention and do not involve any actual cash flows: thus they do not enter into the rate-of-return calculations.

Occasionally, the figures available for annual operating costs shown in column 5 include an allowance for depreciation. If this is the case, the depreciation allowance must be added back to the total in column 7 to obtain the true cash flow. Failure to make this adjustment would result, in effect, in recovering the cost of capital twice.

Using Rate of Return

Of the many elements that must be considered in evaluating a capital project, one of the most significant is risk. The return on a business investment—unlike that from a government bond—depends on future happenings. The numbers that enter into the rate-of-return calculations are estimates—predictions of what the future will bring. There is always the possibility that sales will not live up to forecasts, or that costs will be higher than anticipated. Competitive pressure may force prices down, or new technology may render the product obsolete. Any of these factors may cause the actual results of the investment to differ from the predictions.

Risk is the likelihood that the actual return on the investment will not be anticipated. All other things being equal, a project involving a higher risk should offer a higher rate of return to be considered attractive.

The rate-of-return calculations themselves do not reflect the degree or risk inherent in the project. One way of estimating it is to repeat the calculations, using different assumptions for the various future payments and incomes; for example, the calculations could be repeated for several levels of sales forecasts, ranging from optimistic to pessimistic.

Other factors affecting the decision are the company's policies on investments, the availability of the management or technological skills the investment will require, and the amount of capital that will be tied up. These and many other intangible factors must be evaluated, along with the risk and the rate of return, in selecting suitable projects.

The rate-of-return method is superior to rough-and-ready methods, such as the payback period, in that it takes into consideration the timing of expenditures and it provides a convenient method for expressing a complicated series of cash payments and receipts as a single number. It is only one tool for evaluating proposed projects, however; managerial judgment and experience are still needed.

Variance Reports

Since the control aspects of budgets are all-important, variance reports are necessary to pinpoint areas where expenditures are out of control. Variance reports are lists of projects or categories of expenditures that are greater than or less than some percentage range that is established by management. A common figure used in industry is that all expenditures more than 110 percent or less than 90 percent of estimated expenditures must be explained or justified. In this way, management projects that are outside the range are systematically brought to the attention of the decision maker.

Projecting Maintenance Costs by Factored Budgeting[2]

Maintenance cost control should begin at the foreman level. But, when the foreman is given a ready-made budget and told to "live with it," he has little incentive to exert any special effort to achieve budget objectives. As a result, many budget goals are just not met.

One way a maintenance manager can involve the foreman and encourage him to meet budget goals is through factored budgeting. This approach is simply a means of preparing a realistic program to control maintenance costs by relating equipment repair history to supervisor judgments. With factored budgeting, the maintenance manager can tie together maintenance data and production targets.

Factored budgeting is not complex. It is a means of budget preparation that provides a plan of attack. Every possible area of maintenance cost is logically considered. Primarily, it is a recounting of the number of man-hours that have been needed to maintain specific key equipment; in this way, desired production levels can be sustained. Labor estimates are stated in man-hours for a true picture of the workload. The historic relationship of man-hours to production output establishes the ratio of results (man-hours/ton, for example). Applying these ratios to desired future production levels will ensure an accurate projection of future man-hour requirements. And, if average wage rates are used with projected man-hours, labor costs can be easily determined.

In most plants, the maintenance foreman develops (or works with the planner to develop) the man-hour workload and material requirements. And, he must determine when extraordinary expenditures are necessary.

Labor Requirements

The types of maintenance required and the man-hours involved—and all pertinent data relating to labor activities must be detailed. For example:

Assigned Maintenance	*MH/Year*	*Men*
Machinist	4160	2
Lubrication		
three weekly routes @ 6 hr		
$18 \times 52 \times 1.4^{a} =$	1310	1
two daily routes @ 2 hr		
$4 \times 350 \times 1.4^{a} =$	1960	1
Administrative time (writing reports) =	700	
	3970	2
PM Inspections		
five weekly routes @ 2 hr		
$10 \times 52 \times 1.4 =$	728	1
four daily routes @ 2 hr		
$8 \times 350 \times 1.4 =$	3920	2
Administrative time (writing reports) =	850	
	5498	3
Running Repairs		
Historical allocation (all equipment)	8320	4
Emergency Repairs		
Historical allocation (all equipment)	4160	2

Major Repairs

Allocated by equipment and identified by specific planned/scheduled work. (Normally, planned/scheduled work should represent about 60 to 65 percent of the total work load; therefore, major attention is given to this type of maintenance work when the budget is prepared.)

---

[a]Converts 5-day coverage to 7-day coverage.

Typically, only 40 percent of the equipment may be responsible for as much as 70 percent of the total maintenance cost. When this condition exists, this key equipment must receive special budgeting attention. If the foreman uses historical manpower data, he can establish a definite ratio between the number of man-hours used for the number of tons produced (or whatever measure is being used).

When formulating preliminary equipment repair cost projections, the maintenance foreman must consider:

1. Which items require extraordinary attention (major component change-outs, rebuilds, or equipment replacement)?
2. During which month or quarter should this extraordinary action be planned?
3. What resources are necessary for the maintenance of items that are not identified individually?
4. How much overtime should be budgeted, and where can it be used most effectively?
5. What will be the effect of vacations, sickness, and absence on the annual maintenance plan?
6. Where will contract work be required and when should it be scheduled?
7. How much more will materials cost next year, and how will maintenance costs be affected?

When these questions are answered, the foreman can begin to develop specific aspects of the budget necessary to meet budget goals. Historical trends of relatively stable maintenance activities (lubrication, prevention, preventive maintenance inspections, running repairs, etc.), must be studied closely.

As soon as the foreman has completed preparation of the basic man-hour and cost figure data, these should be checked with his supervisor. If he agrees on the accuracy of the data, a cost distribution plan must be developed. A uniform monthly cost distribution is prohibited because of vacation, sicknesses, holidays, and so on. And, unusual heavy operating expenses (major repairs, for example) should be budgeted several months apart so that their cost impact is reduced.

A key element in the factored budgeting technique is the ability of the foreman to assess the current status of the equipment, and to evaluate special needs for the future, especially for the coming year. The foreman is guided by such historical data as man-hours, man-hours/ton, and, in general, dollar expenditures in previous years. These data also highlight cycle needs-turnarounds, for instance, and indicate whether these major activities will be necessary during the coming year.

The foreman's participation in factored budgeting preparation is vital. He is closest to the overall maintenance picture and is the one most familiar with the equipment. And, when he participates in budget development, he is committed to carrying out his budget.

Forecasting Maintenance Manpower Requirements[3]

When maintenance manpower requirements for the process industries can be accurately predicted, another useful cost-saving technique will be available to management. The purpose of the technique is to formulate a predictive model that will provide accurate maintenance manpower forecasts and to evolve a model which considers those variables. The independent variables which have been hypothesized to affect maintenance manpower, the dependent variable, are:

Level of plant operation

Backlog

Equipment Age

Weather delays

Absenteeism

Number of units concurrently maintained

Impact of predictive maintenance control monitoring

Delays caused by parts procurement

Level of crew experience

Job-related training

Delays caused by job classification restrictions

Shut-down/start-up characteristics

The behavior of those variables is hypothesized. The detailed investigation to verify, reject or add to those hypotheses will, in the main, be left to others.

Traditional Forecasting Techniques

Forecasting techniques are procedures used for predicting future events and conditions. Qualitative forecasting procedures are based on subjective predictions by those who are professionally or academically aware of the field of interest in which the forecast is required. The formal qualitative procedures vary from simple summations of future estimates to the Delphi-type methods developed by the RAND Corporation where a consensus of predictive opinion by a panel of experts is obtained.

Quantitative forecasting methods are statistically based, logically stated and mathematically computed. The procedures require that the historical data be analyzed so that the underlying process that is causing the variable or variables to fluctuate can be determined. Once the underlying process is identified and modeled, extrapolations (even considering all the cautions and danger of extrapolations) can be made for forecasting purposes. The forecast models may be either *time series* models or *causal* models. Previous work on forecasting techniques has almost universally considered only one, or a few, independent variables.

When a set of observations of a variable or variables is sequentially based on time, it is called a *time series*. Analysis of the time series for its underlying generative process may be done through regression analysis, exponential smoothing, or the Box-Jenkins methodology. Multivariate time series exhibit reduced overall variation if opposing univariates are aggregated. Therefore, aggregating variables may subvert valuable information.

Causal models are based on the relationships of two or more variables of a time series. If the variables of the time series are correlated with the dependent variable that is being forecast, the statistical relationship can be modeled and then used for predictive purposes. Regression analysis, with all of its adjuncts, sum of squares, R^2, ANOVA, contribution of each variable, cross-products, and transformations, are used for causal model forecasting. Drawbacks of causal models include the requirement that sufficient historical data be available for each variable in the model and that the independent variables are identified when the forecast is made. The literature is replete with examples of regression analysis, exponential smoothing, and moving averages.

Industrial process plant maintenance includes preventive, emergency, and turnaround manhours. A phase diagram, shown in [figure 8–2(a)], illustrates a theoretical situation where the number of manhours available

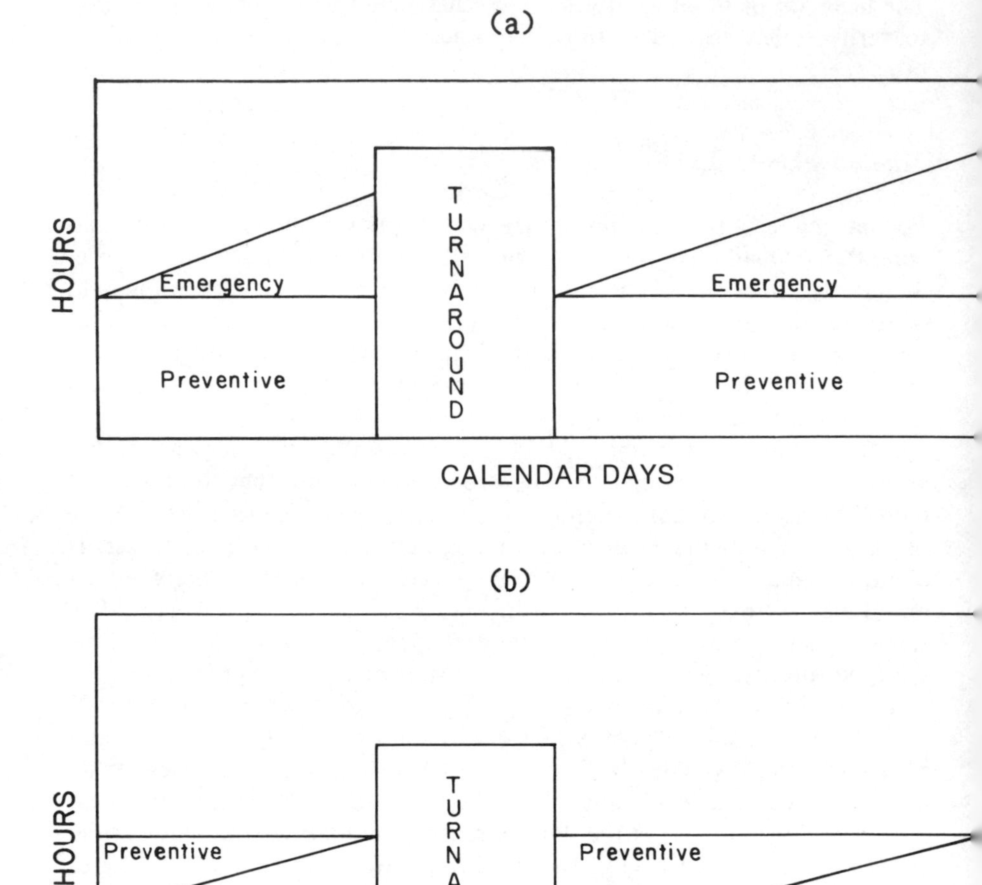

Figure 8–2. Maintenance Man-Hour Requirement Diagrams

to the system is sufficient to maintain a constant rate of preventive maintenance even though the number of hours required for emergency maintenance increases. In [figure 8–2(b)] it is assumed that the work force remains constant over time so that as the number of hours required for emergency maintenance is increased, the number of hours available for preventive maintenance must be decreased. In both cases, it is assumed that maintenance work done during plant shut-down is not short-term preventive or emergency, but is long-term, regularly scheduled turnaround maintenance. Through effective maintenance manpower management, a reliable maintenance manhour requirement forecast model should aid the maintenance manager in his goal of retaining an optimal preventive maintenance level in spite of increased emergency maintenance manhour demands.

Variables that Determine Maintenance Manpower Needs

The *first* step in a system analysis is to ascertain those variables which may influence the desired outcome. Others have considered that influences affecting maintenance work load determination include job cancellations and postponements, intermix of maintenance and nonmaintenance jobs, variety of short-term and long-term jobs, unplanned emergency maintenance, production changes due to changes in sales requirements, materials shortages, and manning problems caused by sickness, absenteeism, or lack of technical expertise. Additional influences on the total maintenance manhour requirement include long term weather trends, the productivity of the maintenance work force, the amount of job-related training of the work force, the acceptable preventive maintenance level, the age of the plant, the degree of technology of the plant (traditional labor intensive steel mill versus a highly automated, capital intensive chemical plant) and the degree of sophistication of the management information system (MIS).

Once the forecasting problems associated with maintenance manpower requirement determination are identified, the *next* step in the analysis procedure is to quantify those problems into a variable format for which data may be collected. Since the final outcomc will bc a list of dependent variables which forecasts a quantity in number of manhours, the initial variable list is compiled so that the data can be collected in or transformed to time-based values. [Table 8-3] illustrates those variables that can be quantified on a unit-time basis.

It must be assumed that all maintenance tasks are well defined through in-plant job classification and remain well defined throughout the historical and forecast time span so that data redundancy can be minimized. To further reduce data redundancy, where an independent variable may exert influence on two or three areas of manhour requirement, the variable is listed in that area (emergency, PM, turnaround) where theoretically it has the greatest impact. Where a variable theoretically may have an equal impact on more than one area of manhour requirement, the variable is listed in each area. It is assumed that either data can be obtained that will

Table 8–3
Initial Variable List

Preventive	*Emergency*	*Turnaround*
Operating Level (%) (+)	Operating Level (%) (+)	Operating Level (%) (+)
Backlog (H) (+)		
	Parts Procurement Delays (H) (+)	
Major Equipment Age (H) (+)	Major Equipment Age (H) (+)	
Weather Delays (H) (+)		
	Crew Experience (H) (−)	Crew Experience (H) (−)
	Union Restriction Delays (H) (+)	
Absenteeism (H) (+)		
Number of Units (N) (+)		Number of Units (N) (+)
Control Monitoring Impact (H) (−)		
		Shutdown-Startup (H) (+)

+ = additive variable; − = subtractive variable; H = hours; N = number; % = percentage. Where an independent variable may exert influence on two or three areas of manhour requirement, the variable is listed in that area where theoretically it has the greatest impact. Where a variable theoretically may have an equal impact on more than one area of manhour requirement, the variable is listed in each area.

show explicitly and quantitatively a specific variable impact on a specific maintenance manhour requirement or that the quantitative impact can be derived from the existing data and/or reliable supporting data bases. For example, data showing the quantitative impact of major equipment age on preventive maintenance manhour requirement may be explicitly available. On the other hand, data may be available showing the aggregated quantitative impact of major equipment age on preventive and emergency maintenance manhour requirement. In this case the analyst would be forced to find the amount of the independent variable impact on each maintenance manpower requirement area.

As shown, the initial variable list is only a guideline for those variables which the analyst may consider in an actual situation. The analysis objective is to identify those variables that prove to be the most significant predictive variables for each maintenance manhour requirement segment.

On table 8–3, the theoretical additive or subtractive impact that each variable may have on the specific maintenance manhour requirement is denoted by the + or − notation shown with each variable.

Preventive Maintenance Variables

Backlog data has been shown to be useful in forecasting maintenance manpower requirements. For control purposes, backlog data should be accumulated on a weekly basis. A satisfactory work priority system must be in force before backlog data collection is useful. To reduce the extent of data collection, and the possibility of data redundancy, worker productivity may be considered as being reflected by the level of backlog. When the weekly use of manpower is stabilized, the backlog data can be used to determine optimum workforce size and composition, where the highest productivity level is attained at the least cost. Similarly, the reduction in manhour requirement due to job cancellation or postponement may be reflected sufficiently in backlog data. For use in a monthly maintenance model, the weekly backlog data would be averaged.

The level at which the plant has been operating during the observation time plays a part in determining how many hours will be required for preventive maintenance. A full-throttle plant operation requires that all preventive maintenance work be done on schedule and to the full extent that has been planned, while a lower plant operating level may require significantly less frequent preventive maintenance manhours. The percentage level of operation may be translated into hours of 100 percent operating level for data consistency.

The age of the major equipment units in a plant will have an effect on the number of manhours required for preventive maintenance. Older equipment may or may not be more easily accessible for preventive maintenance and may require more time-consuming preventive maintenance procedures. Newer equipment may generally be designed for minimum maintenance operation.

Weather may cause loss in manhours where equipment is located in nonroofed areas or where extreme temperatures may make outdoor maintenance work less productive.

Absenteeism among maintenance crews may increase the backlogs and/or increase the number of manhours required to perform the maintenance activities.

As a plant ages, the number of units that are maintained preventively may change. Possibly units that are only maintained during turnaround will be found to be able to be preventively maintained during production time. Design changes during the life of the plant may change the number of units that are preventively maintained.

With increased use of control monitoring equipment, preventive maintenance manhours may be reduced and time between maintenance work may be extended. Predictive maintenance control monitoring may show that more frequent preventive maintenance should be done.

Emergency Maintenance Variables

The operation level at which the plant has run over the time span of the observations will play a part in the number of manhours that will be necessary for emergency maintenance. A high operation level may result in deferred planned preventive maintenance and in an increased number of manhours required for emergency maintenance.

Parts procurement delays may increase the number of hours of emergency maintenance if the crew must spend hours waiting for parts or in trying to find a source for the needed parts.

The age of the equipment that is subject to emergency breakdown will determine the amount of time required for a single emergency repair and will determine the frequency of repairs. Older equipment may be less accessible, more time consuming in parts procurement, and require longer repair times.

The average years of maintenance crew experience will have an effect on the speed with which an item of equipment can be diagnosed and serviced. Hours of job-related training adjusted for learning curve considerations may be included in the experience variable by recognizing that a specified number of training hours is equivalent to a specified number of on-the-job experience hours. Increased training and experience work hours should result in some quantitative degree of workforce ability. A highly trained workforce may result in fewer extended or repeat equipment breakdowns. A useful relationship may also be found between downtime hours and mean crew experience-training in years.

With strict craft union jurisdictional restrictions, manhours may be lost during emergency maintenance while specific craftsmen are located for various aspects of an emergency maintenance task. Less manhours will be lost in plants where general purpose maintenance personnel are allowed to cross craft lines. If during the observation time span, new contract negotiations have brought about a change in craft union restrictions, there may be quantifiable data available to reflect this change.

Turnaround Maintenance Variables

Operation level of the plant over the time span of the observations will play a part in determining the number of hours required for turnaround maintenance. A reduced operating level may allow the time between turnarounds to be lengthened, or it may mean a shorter turnaround at the usually scheduled time.

The crew experience and number of job-related training hours will affect the efficiency of a turnaround and should be considered as a possible aggregate variable where a proportion of the number of training hours adjusted for learning curve considerations, may be added to the experience hours.

The number of units involved in turnaround maintenance may change over the life of the plant as design changes remove some units from turnaround maintenance and place them into the preventive maintenance program, and conversely other units, due to design changes, may be removed from preventive maintenance and placed on turnaround maintenance.

The hours required to shut down or start up a plant will affect the frequency of turnaround and the extent of preventive maintenance. The length of the turnarounds will be determined by the number of units that can not be maintained preventively.

Model Development

If the total maintenance manhour requirement has three major additive components; namely, preventive, emergency, and turnaround maintenance, the overall forecast model should be

$$Y_{t+f} = Y_{PM(t+f)} + Y_{EM(t+f)} + Y_{TM(t+f)}$$

where:

- t is the time increment (month);
- f is the number of time increments ahead;
- Y is the total manhours forecast over a turnaround-to turnaround cycle;
- Y_{PM} is the total manhours forecast for preventive maintenance between turnarounds;
- Y_{EM} is the total manhours forecast for emergency maintenance between turnarounds; and
- Y_{TM} is the total manhurs forecast for turnaround maintenance. The Y_{TM} team will equal zero for monthly forecasts between turnarounds.

The overall forecast model may be used also to predict the number of hours required for preventive and/or emergency maintenance between cycles of turnaround maintenance. The turnaround term represents the number of manhours that would have been spent in preventive and emergency maintenance during production time plus the excess number of manhours required for the turnaround work. When the forecast is made over the entire maintenance cycle which includes the turnaround time, the Y_{TM} term is included.

Preventive Maintenance Model Segment

The dependent variable Y_{PM} must be preadjusted if the productivity records show that more or less manhours should have been available for preventive maintenance over the time span of the observations. How this specific preadjustment is calculated will depend upon the desired level of productivity and its relationship to the number of manhours. Work sampling data may be used to calculate the productivity level. Data for other variables of interest should be transformed to common units of measurement to ease analysis.

For those variables where data is available, multiregression analysis of those variables versus the preventive maintenance manning levels followed by either a step-wise or backward elimination procedure will show which variables are the strongest predictors. If there is high correlation among all or some of the variables, then the correlation can guide the analyst in establishing which variables are the most significant for predicting the amount of manhours required for preventive maintenance. A regression model is then formulated using the most significant variables. In the preferred situation the Y_{PM} regression model will have low slope coefficients since preventive maintenance manhours should be a relatively constant value. Operating level, high backlog, equipment age, number of units, absenteeism, hours lost to weather, will increase the number of hours required for preventive maintenance. Hours gained by control equipment monitoring will vary inversely with the number of hours required for preventive maintenance.

Theoretical data plots of the preventive maintenance variables are shown in [figures 8–3 and 8–4]. As shown in [figure 8–3(a)], preventive maintenance backlog level varies proportionately as preventive maintenance manhour requirement. It should be emphasized that all plots shown in [figures 8–3 through 8–6] represent hypothesized behavior and further research is necessary to define the various relationships. When adequate manpower is available to keep backlog at a constant level, the slope of the regression line will approach zero. As shown in [figure 8–3(b)], preventive maintenance manhour requirement will vary directly as the number of hours of 100 percent operating levels varies. Theoretically, preventive maintenance manhour requirement will vary directly as equipment age, expressed as number

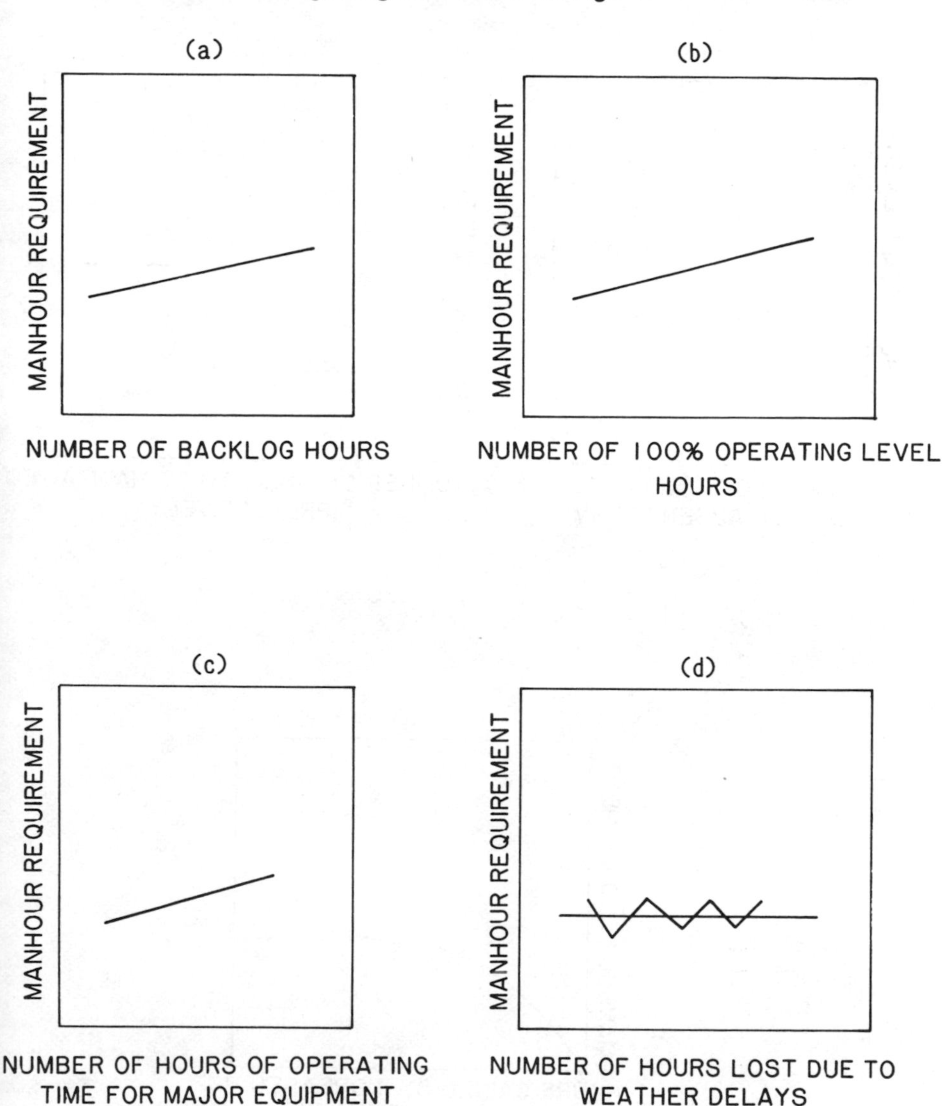

Figure 8-3. Theoretical Data Plots for Preventive-Maintenance Variables

of operating hours and shown in figure 8-3(c). Preventive maintenance manhour requirement will cyclically vary directly as hours lost due to weather. The cyclical variation will be seasonal and over the long term will vary around a constant mean as shown in figure 8-3(d). A single exponential smoothing model might be appropriate.

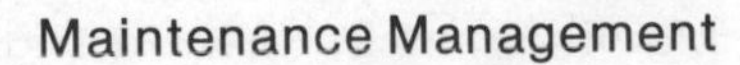

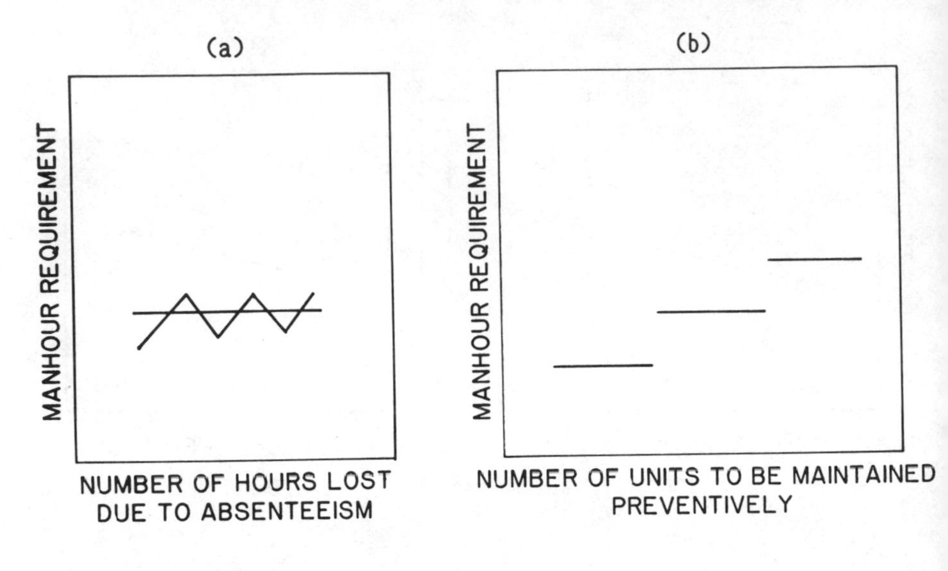

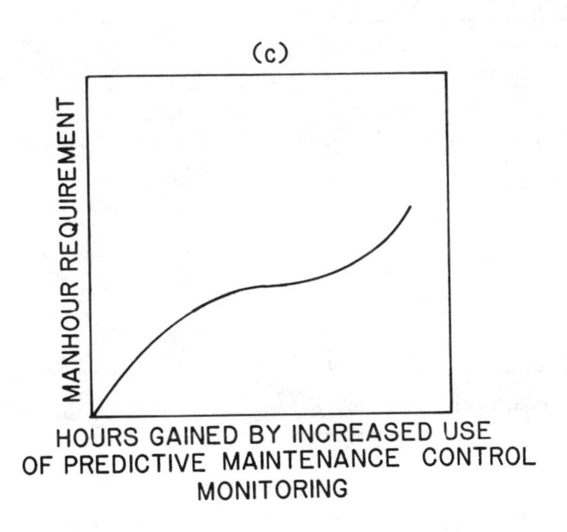

Figure 8-4. Additional Theoretical Data Plots for Preventive-Maintenance Variables

Preventive maintenance manhour requirement will vary cyclically and directly proportionately to hours lost to absenteeism. The cyclical variation will be seasonal and around a constant mean over the long term as shown in figure 8–4(a). The probable model will be a single exponential smoothing model. Preventive maintenance manhour requirement will vary directly as the number of units to maintain changes. The changes will bring about an abrupt change in manhour requirement, one best depicted by a discontinuity causal regression model where the B_0 term is increased or decreased by the change as shown in figure 8–4(b). As shown by the *S*-curve in figure 8–4(c), preventive maintenance manhour requirement will vary either inversely or directly as the immediate impact of predictive maintenance control monitoring is realized. Once the control monitoring is established and backlog is stabilized, the causal regression curve should approach a zero slope or horizontal configuration at a higher level of maintenance if shown to be necessary or at a lower level of maintenance manhour requirement if a less frequent preventive maintenance level is possible. Which of the preventive maintenance variables are finally included in the Y_{PM} model will depend upon the statistical significance of the variable in its relationship to the dependent manning level variable.

The forecasted level for preventive maintenance is then determined with the use of those variables that prove to be significant. The analyst may run simple linear regression analysis on each significant variable and extrapolate the future variable values if the past time series data exhibits low seasonal or cyclical components. Where seasonal or cyclical components are exhibited, more reliable forecast variable values for the Y_{PM} regression model may be gained by exponential smoothing of the variable over the time span of the past observations. Significant seasonal data should be deseasonalized if accepted exponential smoothing procedures are used. The forecast values based on the exponential smoothing simulation are calculated by period ahead extrapolation or by period-to-period extrapolation and updating as new data becomes available. The forecasted variable values are input to the multiregression model which will then yield parameters for the first term of the overall maintenance requirement model.

Emergency Maintenance Model Segment

The dependent variable, manning levels, should be adjusted if productivity over the time span of the observations has not been at the goal level. Too high or too low backlog levels will reflect productivity inconsistently also. The variables for which sufficient data is available should be studied by regression analysis to determine which variables are the most significant for the determination of the number of hours required for emergency maintenance.

Indicator (0 or 1) variables may be considered as qualitative variables for instances where specific numerical data is not available if the general trend of the dependent variable does not reflect a change that may have occurred over the time span of the observations. For example, due to a union sanctioned work slow down, there may have been a drastic increase in number of hours required for emergency maintenance. If the specific numerical data is not available, yet the analyst believes that such an incident could have caused a change in the total emergency manpower requirement, the use of an indicator variable for the multiregression analysis will be appropriate. As data input for this example, the variable may be accorded a binary value of one from the time period of the change and in each time period where the change has continued to exert the possibility of an increase in manhour requirement. The qualitative indicator variable will induce additive or subtractive changes to the values of the regression parameters resulting in a discontinuity if only the B_0 parameter is affected or in slope changes if the B_n parameters are affected.

Among the variables listed under emergency maintenance, the operation level may exhibit a cyclical variation. Major equipment age may provide a regression input that will cause the number of emergency manhours to increase over time. Crew experience and job-related training hours probably will cause emergency manhours to decrease since the worker productivity may increase. Depending on the union-corporate climate, the union restriction delays may or may not be a significant variable for determining the number of emergency manhours. Theoretical data plots showing the effect of these variables on emergency manhour requirement are shown in figure 8–5.

As shown in figure 8–5(a), emergency maintenance manhour requirement may vary directly as the plant operating level, expressed as a number of hours of 100% operating level. Since operating level may exhibit a cyclical variation due to business conditions, an exponential smoothing model might be more appropriate than a simple regression model. Emergency maintenance manhour requirement will vary directly as hours lost due to equipment parts procurement delays, as shown in figure 8–5(b). The directly proportional relationship between emergency maintenance manhour requirement and the age of major equipment, expressed in hours of operating time is shown in figure 8–5(c). In figure 8–5(d), the inversely proportional relationship between emergency maintenance manhour requirement and mean number of hours of crew on-the-job experience and job-related training is shown. In figure 8–5(e) the abrupt changes in emergency maintenance manhour requirement due to changes in union jurisdictional restrictions are shown as discontinuities in the regression curve, which would influence the B_0 parameter.

In the preferred system, the number of emergency manhours will increase as the scheduled plant turnaround approaches. Thus, the importance of the slope constants for total emergency manhours is apparent.

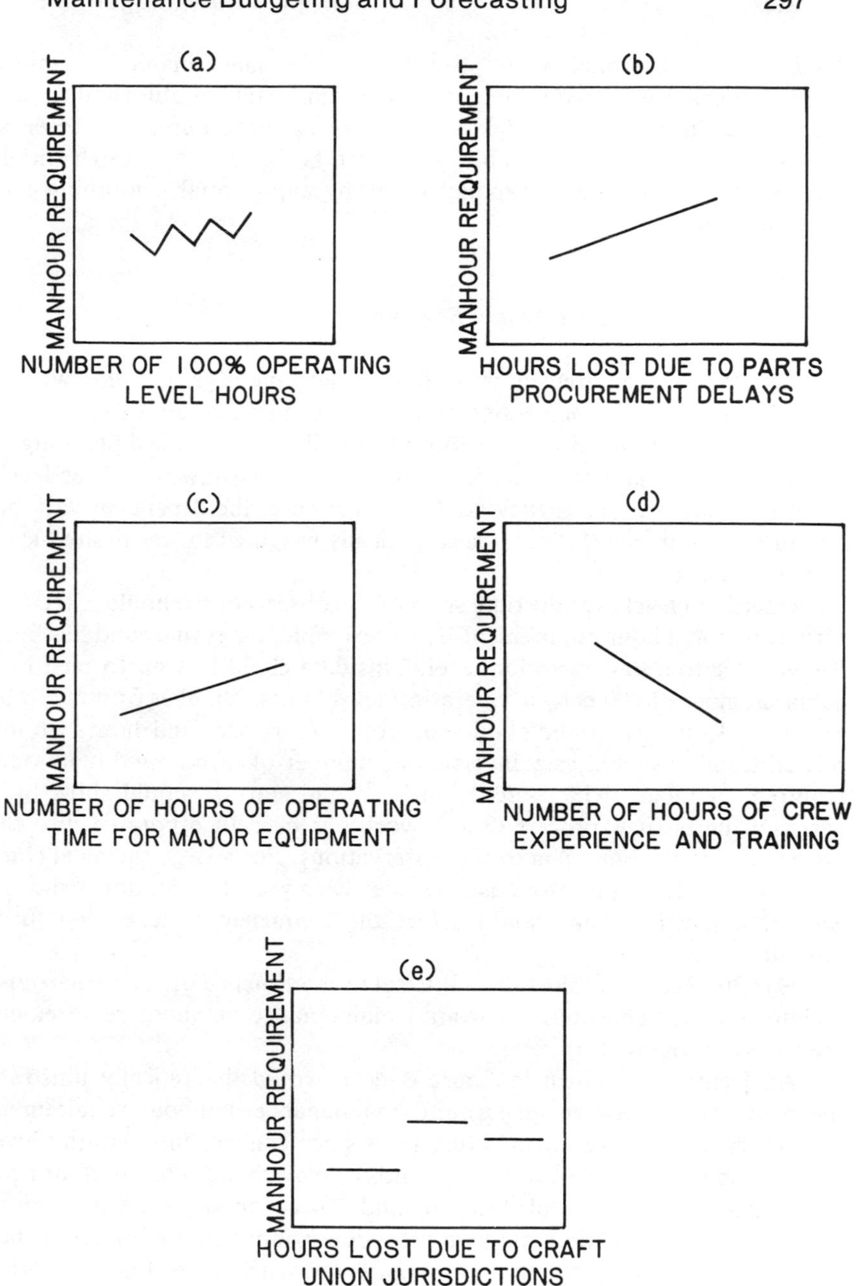

Figure 8–5. Theoretical Data Plots for Emergency-Maintenance Variables

Again, as for determining the preventive maintenance manhours, multiregression analysis is used to determine the significant variables and to provide the multiregression model for the forecast procedure. The forecast values for the significant variables may then be calculated through simple regression analysis and extrapolation or by exponential smoothing and periodic updating.

Turnaround Maintenance Model Segment

Usually manhour requirement at a plant turnaround is greater than what is normally available for preventive and the maximum amount of emergency maintenance. That requirement should be predicted by detailed planning or comparison with past manning levels unless records show that these levels are not aligned with efficiency goals. In that case, the dependent variable (manhour requirement) should be statistically weighted for the desired level of maintenance.

Operation level over the time span of the observations should affect the turnaround manhour requirement if, for example, the turnaround has been delayed due to a low operation level. This data could be transformed into actual number of 100 percent operation level hours. Number of units to be maintained on turnaround, hours of crew experience, and hours of job related training should vary inversely as number of turnaround manhours required. Number of hours for shutdown and startup should show little variation unless new process design work has been incorporated into the system during the time span of the observations. For a large chemical complex with numerous turnarounds scheduled each year, the shutdown-startup ease of a specific plant should affect the frequency of scheduled turnarounds.

Hypothetical data plots that illustrate the influence of the turnaround maintenance variables on turnaround maintenance manhour requirement are shown in figure 8–6.

An *S*-curve, as shown in figure 8–6(a), would theoretically illustrate the relationship between turnaround maintenance manhour requirement and hours of 100 percent operating level since changes in operating level may extend the time between turnarounds or may change the length of time spent on a regularly scheduled turnaround. Turnaround maintenance manhour requirement will decrease inversely proportionately as crew experience and job-related training hours increases as shown in figure 8–6(b). Changes in the number of units of equipment to be included in the turnaround maintenance plan will result in an abrupt change to turnaround maintenance manhour requirement and may be shown by a discontinuity in a regression curve as shown in figure 8–6(c), As shown in figure 8–6(d),

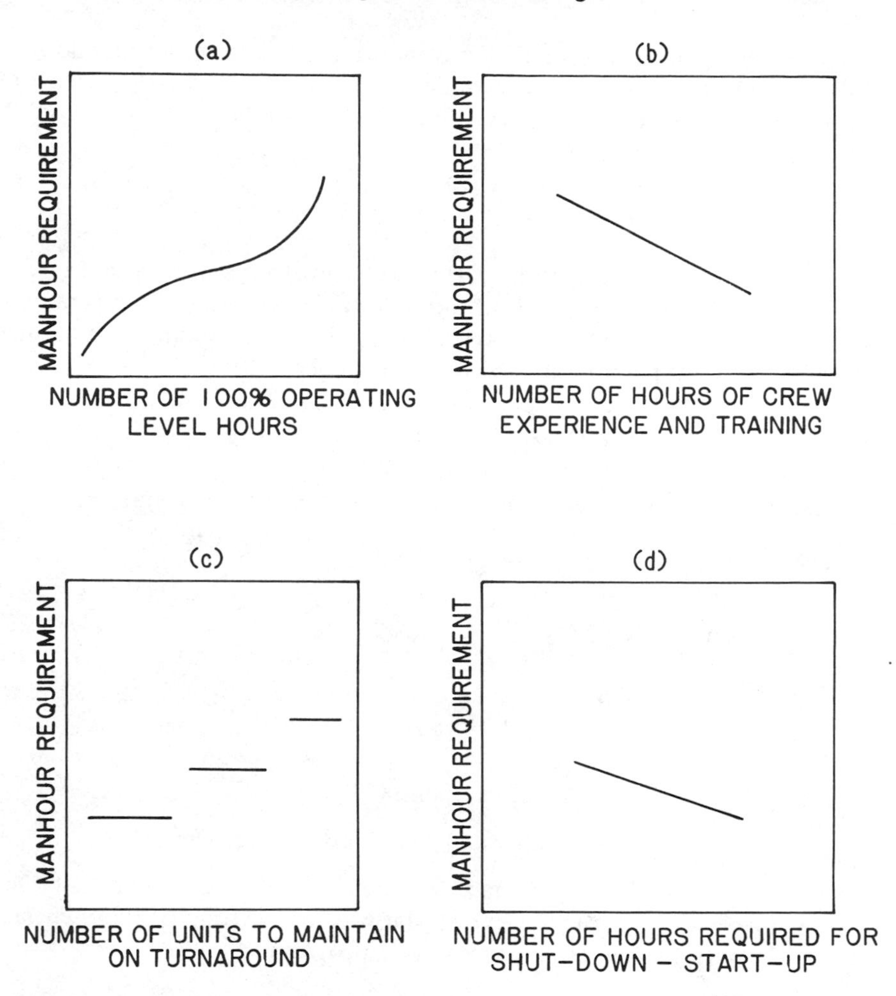

Figure 8-6. Theoretical Data Plots for Turnaround-Maintenance Variables

the turnaround maintenance manhour requirement may vary inversely as the number of hours required to shut down or start up a plant. Thus, a high number of hours required for shutdown or startup will result in less frequently scheduled turnarounds.

Again, multiregression analysis is used to ferret out the significant variables. The turnaround model segment that is then forecast is a variable which is the number of manhours that can be supplied for total maintenance, beyond what is required for emergency and preventive maintenance during production. The turnaround factor will be included in the aggregated forecast model for yearly or turnaround-to-turnaround manpower forecasts. Otherwise the forecast model can be used on a monthly basis to forecast only emergency and preventive manpower needs.

Using the Forecast Model

The specific multiple regression parameters and the specific variables for the preventive, emergency, and turnaround model segments of the forecast model will be determined by the preceding regression analysis procedures on specific industrial plant data. The forecast model

$$Y_{t+f} = Y_{PM(t+f)} + Y_{EM(t+f)} + Y_{TM(t+f)}$$

will be computed from values obtained for the component segments,

$$Y_{PM(t+f)} = B_0^P + B_1^P X_{(t_1+f)}^P + \cdots + B_p^P X_{(tp+f)}^P$$

$$Y_{EM(t+f)} = B_0^E + B_1^E X_{(t_1+f)}^E + \cdots + B_p^E X_{(tp+f)}^E$$

$$Y_{TM(t+f)} = B_0^T + B_1^T X_{(t_1+f)}^T + \cdots + B_p^T X_{(tp+f)}^T$$

where:

t is the time increment;

f is the number of time increments ahead;

X is the variable calculated through either simple regression-extrapolation or through exponential smoothing-forecasting procedures;

$Y_{(t+f)}$ is the total manhours forecast over a turnaround-to-turnaround cycle;

P, E, T denote specific parameters and variables required for each segment of the total forecast model;

$Y_{PM(t+f)}$ is the total manhours forecast for preventive maintenance between turnarounds;

$Y_{EM(t+f)}$ is the total manhours forecast for emergency maintenance between turnarounds; and,

$Y_{TM(t+f)}$ is the total manhours forecast for turnaround maintenance.

p = number of parameters in model, not counting the B_0 constant.

The accuracy of the forecast model will depend upon the quality of the data from which the model segments have been developed. The assumption that preventive and emergency manhours are separate entities is especially important in order to reduce manpower requirement redundancy in the total forecast model. Accuracy of future predictions is also dependent on the method chosen to provide variable forecast quantities. As new data becomes available the model segments should be updated. Tracking signals (sum squared error) should be recorded as the model is updated so that adjustments can be made to the specific model segments as is necessary in order to maintain predictive control. A plot of forecast manhours versus time that is graphically superimposed on actual manhours versus time will provide another view of the modei accuracy.

Summary

The preceding analysis procedure is viewed as a guide for future work with actual data from an industrial process plant where scheduled turnaround maintenance is part of the overall maintenance regimen.

The steps necessary to develop such a model are

1. Actual in-plant data collection
2. Exogenous source data collection
3. Causal regression analysis with plant data
4. Forecast variable value determination from plant data
5. Reliability and sensitivity checks on the overall forecast model

To investigate the general applicability of this analysis procedure, it would be preferable to collect data from several types of process plants and to perform the analysis procedures concurrently so that differences and similarities of actual situations could be realized and investigated.

Actual plant data has not been used in the development of the preceding analysis procedure. Therefore, further work would include actual plant data to insure that the analysis procedure is reasonable. Alternatively, simulated plant data might be used, although it will not provide the analyst with a clear indication of data collection difficulties at the plant. Since the value of the forecast model is based on the availability of accurate, consistent and pertinent data, the use of actual plant data for the procedural testing is preferable.

During actual data collection, the analyst may find other independent variables for which ample data is available and which could be easily included in the causal regression analysis model. The variable list as prepared for this project represents only a hypothesis as to those variables which may influence maintenance manhour requirement.

Once the in-plant data has been collected, exogenous data must be collected for those variables that must be transformed for inclusion in the causal regression analysis procedure. As an example, the in-plant data on hours lost due to weather delays may be recorded as an aggregated value of hours lost due to all types of delays. The weather-induced delays may be determined by analyzing weather trends over the observation time span and statistically weighting the aggregated values of hours lost due to all types of delays.

All data must be sequentially time ordered before the causal regression analysis procedures are performed. From the mathematical coefficient signs of the causal regression model the analyst will obtain an indication of the additive or subtractive contribution that each variable has made to the maintenance manhour requirement. From a t-test statistic result, the analyst will be able to partly determine the significance of each variable to the maintenance manhour requirement. The sum of squares result will help to indicate the fit of the regression model to the data.

Once the overall forecast model has been determined, its forecast accuracy should be verified as additional data is available. It is suggested that a computer program be written to consider the periodic, weekly or monthly variable changes and the resultant forecasts. At least one turnaround-to-turnaround cycle should be included in the time span for which the forecast accuracy is monitored.

Again, this initial stage project has presented an analysis procedure for developing a maintenance manhour requirement forecast model for an industrial process plant. Further work will provide improved and more precise guidelines for data collection and interpretation and more accurate assessment of which of the independent variables may prove to be the most significantly predictive in a specific situation.

Notes

1. This section is based on Joseph Horowitz, "Estimating Return on Investment," *Techniques of Plant Engineering and Maintenance* 20: 125–131. Copyright 1969, Clapp and Poliak, Inc.

2. This section is reprinted from Paul D. Tomlinson, "Projecting Maintenance Costs by Factored Budgeting," *Plant Engineering,* December 26, 1974, pp. 61–62.

3. This section is reprinted from L. Mann and H.H. Bostock, "Forecasting Maintenance Manpower Requirements," *Hydrocarbon Processing,* January 1982, pp. 110–116. Reprinted with permission from *Hydrocarbon Processing,* a copyrighted publication of Gulf Publishing Company.

9 Computerized Maintenance-Management Information Systems

The first step in considering a computerized maintenance-management information system (MMIS) is to define the objectives. The basic requirements must then be defined and the information necessary to solve the problem must be identified. Next, the sources of information must be located and the information to be collected must be determined.

How Necessary Is Computerization?

Intangible considerations are as important as tangible considerations, in determining the need for electronic data-processing equipment. Robert Cash writes:

> Among the intangibles which must be considered is the fact that most users have the computer "on board" and it is being amortized by the usage to which it is being put. In cases such as this, how much extra should we attribute to the cost of the maintenance management information system?
>
> . . . justification for data processing equipment lies in the need to institute devices to better control maintenance and to act as repository for records by which the system might be evaluated and from which other programs such as preventive maintenance and replacement of equipment are evaluated.

Much more attention should be given to the savings that are realized when the process becomes systemized to the point where automatic data-processing equipment has eliminated the tedious, repetitive, manual process in which many errors are made.

Before data-processing equipment is used, the system designer must be careful to emphasize that the computer cannot solve all maintenance-information problems and that an extended debugging period is usually required once the program has been implemented. Those who furnish data for the program must understand the ramifications of careless reporting, and management must be made aware of the extended period before the program begins to carry its own weight.

Recently, minicomputers have become so economically attractive that no maintenance function can ignore their advantages. Paramount among the advantages is the fact that the computer belongs to maintenance and can be used in real-time mode.

What the Computer Can Do

To justify investment in data-processing equipment, it is necessary to list its advantages. The argument most easily understood by management is that the system will result in an overall reduction in maintenance cost. Next, a properly written computer program will substantially reduce the number of human errors that inevitably creep into the system—assuming of course, that the input data into the system are accurate so that the results are useful.

Another advantage lies in the ability of data-processing equipment to rearrange data to serve various ends. Examples of this are the ability of the equipment to calculate backlogs within crafts, total backlogs, and backlogs according to geographical area of the plant and the ability to use data existing in the computer for computing the inspection intervals in preventive-maintenance programs and documenting the history of specific items of equipment.

Routine planning and scheduling activities can be made more creative because the computer can relieve personnel of tedious manual processing, thus allowing them to concentrate on finding better solutions to scheduling and planning problems.

Finally, the computer can store cost data and furnish periodic printouts to indicate where work orders have gone out of control, so that the management of the maintenance function can find assignable causes and thereby render the entire system more effective.

Justifying a Computer for Maintenance Management[1]

Today's maintenance managers must have access to the most modern information systems to help them plan their workforces and control operating costs more efficiently.

Most managers find it increasingly difficult to control rising maintenance costs because of inadequate or outdated procedures. It is now possible for managers to purchase on-line computerized systems that can provide all the needed information about a plant's maintenance operations. The reasonable cost of mini-computers puts them within reach of many small maintenance shops. Space requirements for these units are minimal and they do not need special environmental conditions. And the computers can provide instant, up-to-the-minute data without the need of a costly training program of terminal operators.

However, before considering the purchase of such a system, the company should study its maintenance functions to determine the potential benefits. The study should consider those maintenance areas in the plant that would be incorporated into the system first, as well as those areas that can be included at a later date. The cost to install and operate the system should be included with the initial cost of the computer.

The possible benefits from a computer system along with the overall cost can be determined by a feasibility study; such a study usually takes about 4 weeks and costs $8000 to $10,000. In addition to the benefits derived from having better information on which to make decisions, net savings are possible over a period of years.

A computer information system must be designed to suit the needs and objectives of the company's maintenance operations. All current maintenance practices and procedures should be reviewed, tabulated, and evaluated. After this information has been gathered, ways in which a computer can be applied to maintenance functions should be outlined in detail.

For example, an industrial plant with 45 maintenance craft personnel and a stores inventory of 30,000 parts plans to install a computer system. A feasibility study for this size operation will cost $9000, and installing the system will cost $10,000. Add to these amounts the value of the time plant personnel will spend in developing the system and entering the information: 9 man-months at a cost of $1500 per man-month, or $1500 × 9 = $13,500.

These costs are one-time costs and are included only in the first year's costs. The ongoing costs will be the cost of the hardware and the maintenance of the software. This installation will consist of five CRT display stations, a 300 line-per-minute line printer, an 80 character-per-second printer, and the computer. The computer hardware is available on a lease basis at a monthly cost of $3336, including maintenance of the hardware. The software cost, including maintenance, is $1000 per month. See table 9–1 for cost estimates.

The savings that can be realized through the installation of a computer maintanence system are shown in figure 9–1, "Benefits of a Maintenance Computer System." These savings are based on the experiences of numerous plants that were surveyed over a period of years. An additional benefit is the elimination of a considerable amount of paperwork. Because all additions or changes to the data base are made using online terminals, work orders and job reports are unnecessary and do not become a part of the costly paper chase.

If the median cost saving for each category in figure 9–1 is assumed, the following savings can be calculated:

Labor Productivity—The cost of the 45 man maintenance work-force for the first year is calculated to be $1,170,000. The estimated saving because of better scheduling is 8.5 percent times $1,170,000, or $99,450.

Table 9–1
Implementation Cost Estimate

	Cost per Year (dollars)		
Computer Costs	*First*	*Second*	*Third*
Systems analysis and definition	9,000	—	—
Program installation and tesing	10,000	—	—
Data entry (plant personnel)	13,500	—	—
Hardware (lease, $3,336 per mo)	40,032	40,032	40,032
Software cost and maintenance ($1000 per mo)	12,000	12,000	12,000
Totals	84,532	52,032	52,032

Source: R.D. Mitchell and James Burgess, "Justifying a Computer for Maintenance Management," *Plant Engineering,* August 7, 1980, p. 85.

Improvement because of parts available is 2 percent times $1,170,000, or $23,400. Combining the two figures gives total savings of $122,850 (see [table 9–2]).

Stores Inventory—Plant stores consist of 30,000 parts valued at $700,000. The estimated savings from a reduction in the stock level is 15 percent of $700,000 for a one time saving of $105,000 during the first year; annual savings in carrying costs are estimated to be $10,500 ($105,000 × 10 percent).

Equipment Availability—The annual value of equipment productivity is $5,175,000. Anticipated increase in the availability of this equipment is 1.25 percent through improved maintenance procedures developed by computer programming. A saving of $64,687 ($5,175,000 × 1.25 percent) is estimated.

The examples and potential benefits indicated for this plant are conservative [table 9–3]. Experience has shown that most plants can improve their efficiency in both productive work and inventory reduction by approximately 5 to 20 percent.

Buy, Share, or Rent?[2]

To approach the problem of determining how to get equipment for maintenance, take a step-by-step approach so as to track progress. This also requires that you break down a big problem into a series of smaller, more manageable problems. By identifying the smaller components of the large

BENEFITS OF A MAINTENANCE COMPUTER SYSTEM	Range of Savings, percent	Median, percent
Better Scheduling Planning time per job decreases, resulting in more job plans produced, fewer backlogged jobs, and fewer breakdowns and emergency repairs. Job scheduling is more efficient through availability of reserve stores materials. Only jobs with materials available are scheduled, resulting in fewer schedule changes and reduced time spent by personnel waiting for materials. Job planning is improved by the systems support data, which include planner backlog and status report, quick recall of repetitious job plans, and computerized scheduling and historical job analysis.	5 to 12	8.5
Parts Availability More productive time is assured through parts availability. When parts are not available, craft personnel not only idly await new assignments but also many times spend hours attempting to locate or even fabricate parts.	1 to 3	2
Machine Availability Machine production time increases as the computer contributes to reduced emergency repair through the preventive maintenance program. The ability of the system to automatically schedule preventive maintenance puts this valuable program up front and removes it from neglect.	½ to 2	1.25
Stores Inventory The ability of the system to maintain moment-by-moment inventory levels, automatic ordering, and parts cross referencing results in reduced inventory and fewer stockouts. The automatic cycle counts and updates inventory for efficient parts management.	10 to 20	15

Source: R.D. Mitchell and James Burgess, "Justifying a Computer for the Maintenance Management," *Plant Engineering,* August 7, 1980, p. 83.

Figure 9–1. Benefits of a Maintenance Computer System

problem, you can develop a model and practically write out your plan of action.

A general guide has been developed which can easily be applied to a specific situation to assist in choosing the computer for maintenance opera-

Table 9–2
Savings from Maintenance Functions

	Savings per Year (dollars)		
	First	*Second*	*Third*
Labor productivity			
Improvement through better scheduling	99,450	99,450	99,450
Improvement through parts availability	23,400	23,400	23,400
Stores inventory			
Reduction of stock levels	105,000	—	—
Reduction of carrying cost	10,500	10,500	10,500
Machine availability			
Increase in machine throughput	64,687	64,687	64,687
Totals	303,037	198,037	198,037

Source: R.D. Mitchell and James Burgess, "Justifying a Computer for Maintenance Management," *Plant Engineering,* August 7, 1980, p. 85.

Table 9–3
Net Maintenance Savings

	Savings per Year (dollars)		
	First	*Second*	*Third*
Implementation cost ($9,000 + $10,000 + $13,500)	32,500	—	—
Operating cost ($40,032 + $12,000)	52,032	52,032	32,032
Total costs	84,532	52,032	52,032
Gross savings	303,037	198,037	198,037
Net savings (gross savings - costs)	218,515	146,005	146,005

Source: R.D. Mitchell and James Burgess, "Justifying a Computer for Maintenance Management," *Plant Engineering,* August 7, 1980, p. 85.

tions and management. Because every plant is unique, the guide must be flexible and adaptable. However, a trade-off is involved. Traditionally one attempts to make a plan very practical and easy to use (that is, the old plug-in formula), only to find that it is applicable to a limited number of situations. If the guide is too general and flexible, it is harder to use. Instead of a plug-in formula, think about the appropriateness of the guidelines of a general plan. This trade-off of practicality versus flexibility is most apparent when the computations of costs and savings to choose the computer are discussed.

You are not choosing a computer, but rather you are justifying a computerized management information system (MMIS). The approach is to classify the benefits of the computerized MMIS according to the subroutine which is responsible for the benefit. The subroutines include scheduling, equipment history, downtime, preventive maintenance (PM), and so on. We will have a plug-in formula for savings from each subroutine in the MMIS package. This will give us the advantage of making the guide easy to use and will also give us the advantage of making the guide applicable to many situations, since we can delete the formulas which are not associated with the subroutines in which we are not interested. By attributing savings formulas to specific subroutines, we can readily determine the majority of the variables that do not come into each specific situation. An example would be the case where we already have an inventory control system. Because the savings due to the inventory control subroutine are included in one formula, we can delete that saving from our economic decision calculations.

A major problem in developing the guide is that of apportioning the benefits (savings) to the individual subroutines in the computerized MMIS package. By further breaking down the savings into smaller categories, a better overall estimate of savings is obtained. This is consistent with one of the basic rules of cost (or savings) estimating, which is that of breaking down the proposal into its smallest part and then estimating the cost (or savings) of each.

Finally, we realize that the motivation behind the cost and savings estimates are to economically justify the computer for maintenance. The computer represents a large capital expense, and it is necessary to compute the return on investment. We need to do this because we are competing with production for funds and one of the most important decision criteria is the rate-of-return on an investment.

Although the rate of return criterion is probably the most important, it is not the only criterion. Our forecast of maintenance requirements may exceed the capabilities of our present MMIS. There are also intangible benefits that cannot be reduced accurately to dollar terms. But first, we will develop the step-by-step method of economic justification as noted in [figure 9–2].

Determine Present State

To obtain any degree of practicality in a method, we need to have basic assumptions on which to build. Some of these are: a work order maintenance system, historical records, indexes such as productivity of labor force, overtime, man-hours spent on PM and downtime. The downtime

Step 1	Determine the present state of the maintenance function.
Step 2	Find the lowest cost computer service alternative.
Step 3	Determine the savings from a computerized MMIS.
Step 4	Make the economic decision.
Step 5	Consider all intangible variables and make the final decision.

Source: L. Mann and E.R. Coates, Jr., "Evaluating a Computer for Maintenance Management," *Industrial Engineering,* February 1980, p. 29.

Figure 9–2. Steps to Justify a Computer for Maintenance

needs to be divided into downtime due to normal maintenance, downtime due to emergency maintenance and operations not needing the equipment. In addition, we need a recent work sampling report to determine how much time is spent waiting for instructions, waiting for materials; spent idle and productive, etc. If we don't have a recent work sampling study or other indices, we need to determine them. This is the first step. We need to determine the present state of the maintenance function in order to determine the potential savings that can occur with the computer.

Type of Computer Service

There are basically three ways we can obtain the services of a computer: Buy a minicomputer that will be acceptable for our needs; share time with the company's computer; and rent computer time. Because any of the alternatives will give us comparable savings, we do not need to make multiple alternative economic calculations. We only need to choose the type of computer service (or no service) which gives us the least equivalent annual cost (EAC). Then we will use the alternative that wins out to calculate our rate-of-return. Therefore the second step is to compute the EAC of the three basic alternatives of computer service and choose the lowest one.

Find Lowest Cost Alternate

For the first case, which is buying a minicomputer, the total EAC is the sum of these components:

1. Cost of minicomputer and components such as cathode ray tube (CRT) displays and a FORTRAN compiler.
2. Cost of training people to operate the equipment.
3. Salaries of the people who will operate the equipment (if we hire more people).
4. Cost of installation of MMIS.
5. Cost of maintaining the computer which includes expected repair, paper and power bills.
6. Cost of layoffs (if we find that certain clerks, etc., are no longer needed).

The first and fifth costs can be found from a computer manufacturing company. The cost of training sessions with the maintenance supervisors and personnel, the cost of coding information and the cost of putting the program on the computer can all be obtained from internal estimates. The cost of layoffs can be found by using work measurement techniques to determine how many positions are unnecessary as a result of the computer. The actual costs of layoffs depend on company policy. The most difficult cost item to determine probably is the cost of installing the MMIS. For instance, it should require approximately two man-years to computerize 30,000 storehouse items. The other cost items, which are somewhat easier to compute, are the cost of training sessions and the salaries of new people who will operate the computer (if more people are needed). To compute an equivalent annual cash disbursement, we need an estimate of the life of the equipment. One method would be to use forecasts of maintenance requirements to determine when the capacity of the equipment or obsolescence would be reached. Another method would be to use the IRS guidelines on equipment life. The method which yields the shortest life would be used. Finally, consult your tax specialist to determine the tax advantages that are available when calculating the EAC of that alternative.

The second case is sharing a computer. This may not be feasible if there is not enough time or memory that can be allocated to maintenance. The total EAC for the second case is the sum of the EACs of: sharing a computer, training people to operate it, salaries of people to operate it, installing MMIS and layoffs.

The cost of sharing the computer can be determined from realizing the cost per time unit for using the computer, and estimating the time required by maintenance. Training costs may not exist if a group outside the maintenance department is entirely devoted to the operation of the computer. However, maintenance management still has to be trained to gain access to the maintenance reports. The cost of installation of the system and layoffs would be calculated in the same manner as in the first case.

In the third case, we rent computer time. The total EAC of this alternative is the sum of the EACs of operating time, training people to use the computer, salaries of people working with the computer, installing the MMIS and layoffs.

The cost of renting computer time can be found by estimating the amount of time required and applying the rates charged by the computer time-share company. The remainder of the costs are calculated in the same manner as the other alternatives.

To arrive at the EACs of the alternatives, we need to estimate the amount of time we will use the computerized MMIS. This is necessary because of the one-time costs, such as training and installation, which have to be spread over the life of the system to realize an EAC. Never use a life longer than 15 years because the present system will become obsolete and the capital recovery factor does not appreciably change past 15 years when wc consider minimum attractive rates of return over 12 percent. In [figure 9-3], we find a typical curve of the cost of the three alternatives plotted against the amount of maintenance which is measured by the computer time required to operate the MMIS. Notice that each alternative has its region in which it is the most economical. Now find the boundaries of the region and determine in which region our plant lies.

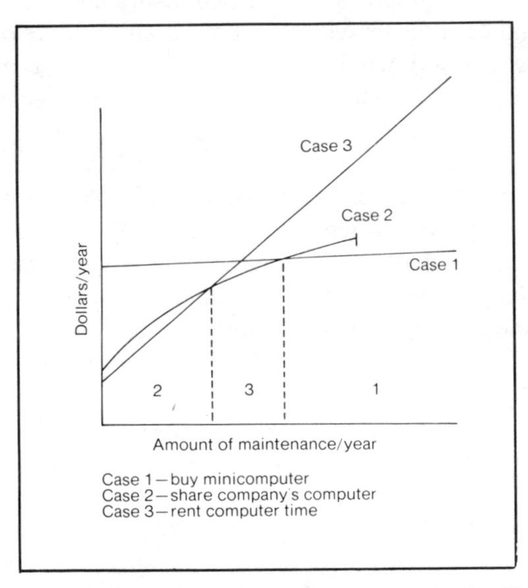

Source: L. Mann and E.R. Coates, Jr., "Evaluating a Computer for Maintenance Management," *Industrial Engineering,* February 1980, p. 30.

Figure 9-3. Comparison of Computer Costs

Finally, we choose the alternative with the lowest equivalent annual cost. Recall that we do not need to consider the savings aspect since there are incremental savings. Only the difference among alternatives are relevant in their comparison.

Determining the Savings

Step three is to determine the savings from a computerized MMIS. We will classify the savings according to the subroutine where the benefit originates. To determine the savings for a particular situation, we determine which subroutines will be used in our MMIS. If the user desires more subroutines, merely add them to the model. Then we calculate the savings corresponding to the subroutines used; these are then added to yield the total savings. Compare that total with the equivalent annual cost of the computer. This is the third step in the justifying of a computer for maintenance.

Before we begin looking into savings, we need to reemphasize an assumption made earlier. We should assume that maintenance management will use the output to monitor, control and improve the maintenance function. In short, we are assuming that the MMIS will be successful in its purpose. When we realize that this assumption is common to all engineering economy decisions, we see the need to estimate the improvement of the maintenance function as a result of installing the MMIS. At first, we may feel uneasy about estimating the improvement due to the MMIS, but all economic decisions involve estimates about the future.

The initial subroutine we will look at is that of financial control. This lists all charges against the work order: direct labor, direct material and equipment. The reports that come from this subroutine are: accounting, labor in man-hours, materials, equipment usage, work order summary, exception (cost) and a cumulative cost center budget status.

One of the easier savings to estimate is the man-hours saved by the computer handling the reports that previously were done manually. Another saving emanates from the labor performance report. By being able to determine trends in cost and productivity from these labor reports, we are able to respond earlier to the problems. Therefore, we estimate the savings that will occur by being able to monitor cost and productivity shifts. In addition, if our goal is to have our maintenance force working at, say 80 percent productive, then we would refer to a recent work sampling study to determine the number of man-hours that can be saved as a result of the MMIS. Equipment usage reports will enable us to determine the optimal number of each type of equipment. Exception reports will emphasize work orders which have exceeded some programmed tolerance. This will assist in finding the cause of the exception and will tend to prevent the same happening in the future. The cumulative cost center budget reports keeps management in closer touch with costs and allows an earlier response to out-of-control items. This is more of an intangible benefit and will be discussed later. A formula for the savings from the financial control subroutine is in [figure 9–4].

The scheduling subroutine includes preventive maintenance and backlog. These scheduling routines will not only publish the list of work orders that should be worked each day, but will also print the craft schedule. It will consider the priority of the jobs as well as the resources required to perform the work orders. The saving resulting from this subroutine includes the man-hours saved with computerized preparation of schedules and backlog reports. Additionally, the scheduling subroutine frees the planner-scheduler from the routine scheduling and allows more free time for creative scheduling. Further, the human mistakes that were once present when formulating routine scheduling are reduced.

We determine how much time he has been spending preparing the basic schedules. Then we determine the improvements he makes from creative scheduling. Next, we determine how much more improvement in the schedule he can make with the additional time he will have. A formula for the savings from the scheduling subroutine is shown in [figure 9 – 4].

The work standards subroutine places work standards in the computer memory. This allows us to accumulate standards for most maintenance jobs. When a new job comes us, we can often create the new standard from elements of other job standards. The work standards subroutine is required for the scheduling subroutine. The savings from this subroutine include the paper work and man-hours involved with keeping up with a manual work standard system. Additionally, there is the time saved when we are creating standards for new jobs because the standards are easier to retrieve from the computer.

The materials control subroutine, if a materials program already exists, frequently needs to be adapted to the MMIS. This subroutine tells purchasing when to reorder and prints out completed purchase orders. Also, when a work order is received, this subroutine will assure that the material is available before it allows the work order to be scheduled. If the material must be ordered, then the work order is delayed in the schedule until the material is received. With this system, we can see a considerable amount of paperwork being saved for both the purchasing department and the maintenance planner. We also have another saving because we have eliminated the situation where maintenance personnel can't start a scheduled job because the material hasn't been ordered. We can find this saving by looking at a recent work sampling study to determine the man-hours that are attributed to this situation. The formula for the savings from the materials control subroutine is given in [figure 9–4].

Financial control subroutine

Savings = man-hours of paperwork saved × wages
+ savings due to management action to variance reports
+ (productive goal − productivity) × number of maintenance craftsmen × 1 man-year × average wage
+ savings from equipment reports

Scheduling subroutine

Savings = man-hours of paperwork saved × wages
+ improvements from more time for creative scheduling
+ improvements from lack of human mistakes

Work standards subroutine

Savings = man-hours of paperwork saved × wages
+ man-hours of craftsmen waiting for instruction × fraction of waiting time due to not having existing standard available × average wage
+ man-hours saved in making new standards from elements in computer files

Materials control subroutine

Savings = man-hours of paperwork saved × wages
+ man-hours of craftsmen waiting for material × average wage

Downtime subroutine

Savings = hours of downtime due to emergency maintenance × (fraction estimated to be eliminated due to better management due to downtime reports) × cost of downtime

Equipment-history and inspection-interval subroutine

Savings = man-hours of paperwork saved × wages
+ man-hours of PM work order generation saved × wages

Source: L. Mann and E.R. Coates, Jr., "Evaluating a Computer for Maintenance Management," *Industrial Engineering,* February 1980, p. 30.

Figure 9–4. Savings Formulas

The downtime subroutine results in a downtime report which indicates the total downtime and how much is due to normal maintenance and to emergency maintenance. The downtime resulting from emergency maintenance may be separated into idle time and maintenance time. The downtime report will help maintenance management spot trouble areas that cause downtime, and it will help remove the idle time that occurs before maintenance people start emergency work. Therefore, we need an estimate of the reduced downtime that will occur as a result of management monitoring the causes of different categories of downtime.

To estimate this saving, we need to set a goal as to how much we intend to reduce the idle time portion of the emergency maintenance downtime. We also need to determine how much the emergency maintenance downtime will reduce when our methods group becomes aware of the causes of that downtime.

The final subroutine to consider is the equipment-history and inspection interval subroutine. This subroutine maintains a history of the maintenance that has been performed on every numbered item of equipment. With these reports, we can determine when to replace or determine the correct inspection interval. This subroutine will automatically generate preventive maintenance work orders based on the inspection interval.

There are savings in paperwork involved in keeping equipment history files. We save the maintenance scheduler man-hours by having the computer generate the PM work orders. Also we can determine more accurately when equipment needs to be replaced and we avoid overmaintaining the equipment. These last two savings are more difficult to determine because they require an advanced knowledge of the optimum amount of PM or the length of equipment service. Therefore, these benefits will have to be considered intangible since it would be difficult to estimate accurately their dollar amounts. The formula for the savings from the equipment-history and inspection-interval subroutine is shown in figure 9–4.

We have now determined the savings that will occur from each subroutine and a summary of the process is shown by the savings formula in figure 9–4.

Make the Economic Decision

Step four is that of making the economic decision. We have already considered the equivalent annual cost of three cases of using a computer and we chose the one with the smallest. Now we are comparing the computer alternative of the status quo. We must include the savings from the computerized MMIS because it represents a true difference between the alternatives.

Because we already have the yearly cost by finding the EAC at the minimum attractive rate of return, we need only determine whether the saving is greater than the EAC. If so, we can conclude that we would get a return on our investment that is greater than the minimum attractive rate of return.

If the computerized MMIS is competing against other projects, it would be a good idea to start with the component costs and savings and calculate the unknown interest rate. This is the interest rate which makes the costs and savings equal. When we find that the return on investment is equal to or greater than the minimum attractive rate of return, we make the preliminary decision to go ahead with the proposal. When we find the rate of return of the proposal unsatisfactory, we make the preliminary decision to reject the proposal. This is a preliminary decison because we have considered only tangible factors and have ignored the intangibles.

Final Decision

Step five includes the consideration of intangible factors and the overall decision. Every good economic decision maker gives consideration to the intangible factors before making his final decision. Intangible factors are those that can not be quantified but are important to the decision maker. The computerized MMIS has many benefits to which we cannot apply an estimated dollar value accurately. Some of the benefits include:

Routine preplanning and scheduling are handled by the computer so that skilled personnel are allowed to do more challenging work elsewhere.

Materials and tools are preordered for delivery to the job site without involving the time and effort of supervisors and craftsmen.

Historical records can be quickly recalled for reference.

Historical records accumulated in the data files can be periodically analyzed to help improve performance in the future.

Other benefits are

Reports make it easier to forecast requirements and make more accurate budgets.

Backlog reports tell us when we have the correct labor force size.

Reports bring trouble areas into the spotlight, enabling management to act on them faster.

Historical records enable us to validate standards and install an incentive wage plan.

Reports allow us to evaluate maintenance supervisors as well as maintenance personnel.

Records enable us to quickly determine the economic benefits of improved methods and tooling.

We avoid overmaintenance by reviewing the equipment history file.

We can more accurately determine when to replace equipment based on trends in the equipment history files.

Because the computer has the ability to look at records from many points of view, management has a new opportunity to take a systems view of the maintenance situation. This frees management from the day to day hassles and routines and gives him an overall outlook which leads to better management.

Finally, there is a saving in the absence of human errors in the routine schedules and other calculations that are not caught by anyone but are still costly.

We need to consider these benefits before we decide about that borderline case where the rate of return is near the minimum attractive rate of return. Also, these benefits should be brought out when comparing the computerized MMIS with other proposals. When we do this, we make a better overall decision because we consider all factors rather than just tangible factors.

Figure 9–5 defines the variables in figure 9–6, our justification model for an MMIS.

Summary of Maintenance Tasks Performed by the Computer

The two previous formats define in detail some tasks usually performed by the computer. Usually, it is not necessary to use all of those tasks(subroutines), but it is advisable to understand what each one ecnompasses. Ideally, maintenance management would indicate the type of information needed for decision making, and the computer program would be written around those needs. The program should serve the user, not the user serve the program.

The financial-control subroutine lists all the charges against the work order: direct labor, direct material, and equipment. Normally, overhead costs are not included.

The scheduling subroutine includes preventive maintenance and backlog. In order to schedule on the computer, it is necessary to have a priority system that delineates at least four levels of priority for the work

AW	—Average yearly wage of people concerned
CBUGS	—Cost of putting program on computer and debugging the program
CCC	—Yearly cost of company's computer (case 2)
CCH	—Cost of computer hardware (for case 1)
CCI	—Cost of coding information such as action and fault codes, equipment numbers, craft numbers, area codes, inventory codes, etc.
CDT	—Cost of downtime
CLAY	—Cost of layoff
CSS	—Savings due to creative scheduling; computer frees scheduler to allow more creative scheduling
CTCP	—Cost of training computer personnel
CTS	—Cost of training sessions
DTPM	—Savings in downtime due to PM program
DTS	—savings in downtime due to management action initiated by down report
EEB	—Expected electricity bills
EHFA	—Time for maintaining an equipment history file after installing system
EHFB	—Time for maintaining an equipment history file before installing system
EPB	—Expected paper bills
EQS	—Equipment expenses saved by equipment utilization program
ERB	—Expected repair bills
EUSE	—Expected use of company's computer by maintenance department
EXRS	—Savings due to management action initiated by a work order exception report
HMS	—Savings due to eliminaton of human mistakes
MHS	—Man-hours of report generation saved by subroutine
MHW	—Man-hours of craftsmen wasted per year because material is not available
NP	—Number of people required for specific situation
NPO	—Number of purchase orders for all items in one year
NR	—Number of receipts of orders in one year
PMMH	—Man-hours required to generate PM work orders
PS	—Productivity shift; upward shift in productivity due to management action initiated by a labor report
PUSE	—Present use of company's computer (case 2)
SBR	—Man-hours of schedule and backorder report generation saved
THPW	—Total hours of paper work per year performed by maintenance management personnel which is eliminated by the computer
TPO	—Time required to look up vendor information and fill out a purchase order
TR	—Time required for paperwork with receipt of inventory system
TSR	—Time share rate (for case 3)
TWSA	—Time required to maintain work order files after installing system
TWSB	—Time required to maintain work order files before installing system
WFI	—Man-hours of craftsmen waiting for instruction × fraction of waiting time due to not having existing standard available
WNO	—Number of times per year we retrieve work standards
WTRA	—Time to retrieve work standard after installing system
WTRB	—Time to retrieve work standard before installing system
XINV	—Capital investment in inventory
XINVS	—Fraction of XINV which is eliminated by material control subroutine

Source: L. Mann and E.R. Coates, Jr., "Evaluating a Computer for Maintenance Management," *Industrial Engineering,* February 1980, p. 30.

Figure 9–5. Variable Definitions

(Note that all cost components should be converted to EAC)

Cost of case 1 computer service
= CCH + (CTCP × NP) + ERB + EEB + EPB
+ CTS + CCI + CBUGS + (THPW/2000) (CLAY-AW)

Cost of case 2 computer service
= CCC (EUSE/(PUSE + EUSE)) + CTS
+ CCI + CBUGS + (THPW/2000) (CLAY-AW)

Cost of case 3 computer service
= TSR × EUSE + CTS + CCI + CBUGS
+ (THPW/2000) (CLAY-AW)

Cost of computer service = MIN (cost of case 1, cost of case 2, cost of case 3)

Financial control subroutine
Savings = MHS + PS × NP × AW × 1 man-year + EQS + EXRS

Scheduling subroutine
Savings = SBR × AW + CSS + HMS

Work standards subroutine
Savings = (TWSB-TWSA) × AW + WFI × AW + (WTRB-WTRA) WNO

Materials control subroutine
Savings = TPO × AW × NPO + TR × AW × NR + XINV × XINVS + MHW × AW

Downtime subroutine
Savings = CDT × DTS

Equipment-history and inspection-interval subroutine
Savings = DTPM × CDT + (EHFB-EHFA) × AW + PMMH × AW

Total savings = SUM (all relevant savings formulas based on the subroutines one plans to use)

Economic decision = Total savings − cost of computer service

If we get a positive number, our rate of return is greater than the minimum attractive rate of return. If we get a negative number, the system does not pay for itself.

Source: L. Mann and E.R. Coates, Jr., "Evaluating a Computer for Maintenance Management," *Industrial Engineering*, February 1980, p. 30.

Figure 9–6. Justification Model

orders. One approach is to assign a date needed to each work order so that the priority will change as the date needed approaches. These scheduling routines not only will publish the list of work orders that should be worked each day but will also print the schedule from the craft standpoint. Preventive maintenance is a level of priority. If the date-needed approach is used, a work order might be of a preventive maintenance nature until the date needed is approached, and then might assume a higher priority so that the work will be completed in the specified time.

The backlog subroutine accumulates the work indexed from the standpoint of individual crafts, total work force, geographical areas of the plant, and priority levels. Some periodic review device should be built into this subroutine so that work orders that remain in the backlog for a certain period of time are reviewed by maintenance management to ensure that work orders for which work has already been done do not stay in the backlog and work orders that are no longer desired are purged from the backlog.

The work-standards subroutine stores job standards in the memory in the computer so that they may be retrieved when desired. The standards could include the method for performing the work order, the time to perform the work order, and the tools and spare parts necessary for the job. It is recommended that standards be stored only for work orders of a recurring nature. Any attempt to place the standards for all possible work orders into the computer is economically unjustifiable. The computer can be programmed to print copies of the work order.

The materials-control subroutine, in many instances, has already been put on the computer through a homgeneous inventory-control program. Frequently, this program has been integrated with the purchasing and warehouse functions with little or no thought given to maintenance, in which case it will be necessary to integrate into the MMIS the portions of that program that have to do with the availability of the spare parts and material necessary for the maintenance operation. In other cases, the operation of the storehouse is under the direction of the maintenance department, and materials control will be an integral subroutine to the overall MMIS. This subroutine has the ability to print lists of the materials and spare parts necessary for specific work orders when these materials have been listed on the standard.

If the storehouse stock catalog is to be printed by the computer, the materials-control subroutine must have the ability to print, when necessary, a list of the quantity of parts in the storehouse and to remind purchasing when that quantity is approaching the order point.

The downtime subroutine will print a list of the number of hours for which each item of equipment is out of service, from both the periodic standpoint and the cumulative standpoint. Care must be taken in this subroutine to indicate whether the downtime is a result of the equipment being repaired or waiting for repairs or a result of operating personnel considering that the equipment is not needed at that time.

The equipment-history and inspection-interval subroutine—historical records for identifiable items of equipment—is necessary to design preventive maintenance programs, to evaluate replacement of equipment decisions, and to pinpoint areas where extra efforts are needed to reduce the cost of maintenance. For this subroutine, it necessary to identify and code common problems for such equipment as pumps, compressors, shell and tube exchangers, and instrumentation. These are identified at the completion of each work order and the information is stored in the computer for periodic printout. Periodically, this subroutine also prints a list of necessary specific inspection jobs and, in some cases, the computer can be programmed to print out the work order itself.

Preparation for Initiating the System

Every attempt should be made to design the work order to accumulate as much of the necessary input information as possible. This usually entails a redesign of the work order, and orientation sessions should be held to explain the use of the data and the importance of accurate reporting to the personnel. Nothing is so demotivating as to require people to report information when they have no feel for the use to which the information will be put. Problems or lack of problems with the data-input system will indicate whether or not the initial meetings have been successful. If they have not, it might be necessary to hold additional sessions to reemphasize the necessity for accurate reporting.

Certain information that does not normally come to mind when designing a work order form is necessary for a computerized system, including the following:

1. The description of the work to be done must be specified in more detail than is necessary in the normal work-order system. It must be borne in mind that information must be introduced in the computer in quantitative terms and that the usual qualitative explanations of work orders will not suffice.
2. The identifiable items of equipment in the plant must be numbered for computerization.
3. The work done must be identified by a finite number of categories, and those categories must be coded for introduction into the program. Categories might include a description of problems found with the equipment, the action taken based on those particular problems, reference to individual subassemblies, and so forth.

It is probable that, during the initial phases of the program, the work order will have to be redesigned in a number of steps. Individuals initiating work orders should be aware of this possibility so that they will be tolerant of the successive changes that will probably have to be made during the shakedown portion of the system development.

Operating Characteristics of the System

Not all of the subroutines described here are necessarily required by every plant. A decision about desirability and degree of involvement must be made for each subroutine. It is assumed that the plant introducing the system described here has had some experience in manual maintenance-management information systems and is familiar with the terminology and problems of those systems. Input for the system described here is derived from the work order, the time card (which is usually completed by the foreman at the end of the day), and the material-requisition cards by which material is drawn from the storehouse.

Work orders for preventive maintenance can be preprinted. Preventive-maintenance jobs can be introduced into the master schedule along with other routine jobs or can be separately identified as preventive maintenance. The latter course might be the more desirable when a separate work force is assigned to preventive maintenance.

Financial Control

Financial control includes the labor, material, and equipment cost reports and the storage of data for future analysis.

Every man-hour spent on maintenance must be accounted for. Maintenance tasks too small to warrant a separate work order should be accrued on an open or blank work-order number. Labor costs should be entered in both dollar and man-hour terms. All costs should be attributable to a cost center and, when relevant, to an identifiable item of equipment.

Figure 9–7 illustrates one type of report, in which labor and material costs are reported by cost center and work order number. It can be issued as often as desired.

Other reports include repair reports by processing unit and by work-order number; preventive-maintenance reports by processing unit; and material and spare-parts cost reports by cost center and by work-order number.

The total material and labor costs are accumulated for each work order. It is necessary to inform the system when a work order is closed out so that the total cost will be accumulated and compared against the estimate

CURRENT WORK ORDERS, ____________, 19__

Cost Center	Work Order Number	Description	Equip. No.	Estimated Matl.,$	Estimated Labor, MH	Spent to Date Mat.,$	Spent to Date Labor, MH
526	35992	Tighten Seal	P1263	25	2	25	1
526	35996	Replace Steam Trap		20	2	20	2
662	35841	Tie in Air Line		10	8	10	6
683	35850	Replace Liner	C1023	150	14	150	0
701	35872	Remove Tube Bundle	452	0	8	0	2

Figure 9–7. Material- and Labor-Cost Report

for printout of variance reports. Variance reports, in this context, are a separate printout of work orders that exceed some programmed tolerance.

Cost of preventive-maintenance materials-such as grease, welding rod, and the like—can be accumulated along with miscellaneous time on an open or blank work order number.

As in any management information system, it is necessary to disseminate different combinations of the same data to separate levels of management for the purpose of decision making. The specific analysis to which the data will be subjected determines the detail in which the information will be collected and stored. The higher the level of the management, the more categorical will be the report, which will be used primarily to identify problem areas and trends. Therefore, the system should be developed with the categories to be analyzed in mind. Normally, much of this categorization has already been arranged by the accounting department for the accumulation of data needed by that function. Frequently, accounting's category identification is not satisfactory for the maintenance-management information system, and dual systems would be confusing. Therefore, in some cases, it is necessary to reestablish a common language that both the maintenance function and the accounting function can understand and live with.

In addition, the codes must be simplified so that different individuals will not classify similar activities in a dissimilar manner. The cataloging of these codes must be widely disseminated and, in some cases, must be explained to the users in an orientation meeting.

Historical Cost

Maintenance-cost records are not to be confused with equipment history. Few plants maintain accurate records. With the MMIS, it becomes convenient to print out a monthly labor and material report that shows monthly and cumulative annual costs for any cost center or identifiable item of equipment.

The cumulative yearly report can be filed and referred to for historical analysis. It should be searched constantly for trends that could indicate increasing maintenance cost. The cumulative report also might be used for budgetary purposes to discover areas that require future maintenance expenditures. The primary advantage of the monthly printout is to locate areas, either by expenditure or by trend, that indicate that they will require amounts of resources other than those that were budgeted.

The cumulative report usually lists each process unit and cost center by its identifying number and breaks down expenditures into labor, material, and equipment cost. It can be made more sophisticated by printing pro-

MONTHLY TREND REPORT

Cost Center	Equip. No.	Costs This Mo. Labor	Costs This Mo. Matl.	Avg. Costs This Mo. Last Yr. Labor	Avg. Costs This Mo. Last Yr. Matl.	% Decrease (–) or Increase (+)

Figure 9–8. Material- and Labor-Cost Trend Report

jected as well as cumulative expenditures and last year's expenditures for the same period. The program can also include starring of items that are outside the limits of tolerance so that they may be investigated. One suggested format for the trend report is shown in figure 9–8.

A compilation of several monthly trend reports could identify longer trends. In order to appreciate long-term changes in maintenance costs, it is necessary either to plot the monthly increases or decreases or to print out a year-end report.

Whenever a trouble area is located by the trend reports, the follow-up investigation should include comparison of accumulated estimated man-hours versus accumulated required man-hours. There should be discussion with the field forces to ascertain the reasons for any large differences between the two. Frequently, this process results in a revision of the standard or a realization that the equipment is being employed to a greater or less extent than was planned when the maintenance budget was formulated.

Figure 9–9 illustrates the financial-control information flow.

Scheduling

Maintenance work can be divided into three groups for purposes of scheduling: emergency work orders preventive-maintenance work orders, and normal work orders.

The scheduling subroutine should consider the priority of the jobs as well as the resources required to perform the work orders. The object of this subroutine is to obtain a report of all work orders that should be scheduled

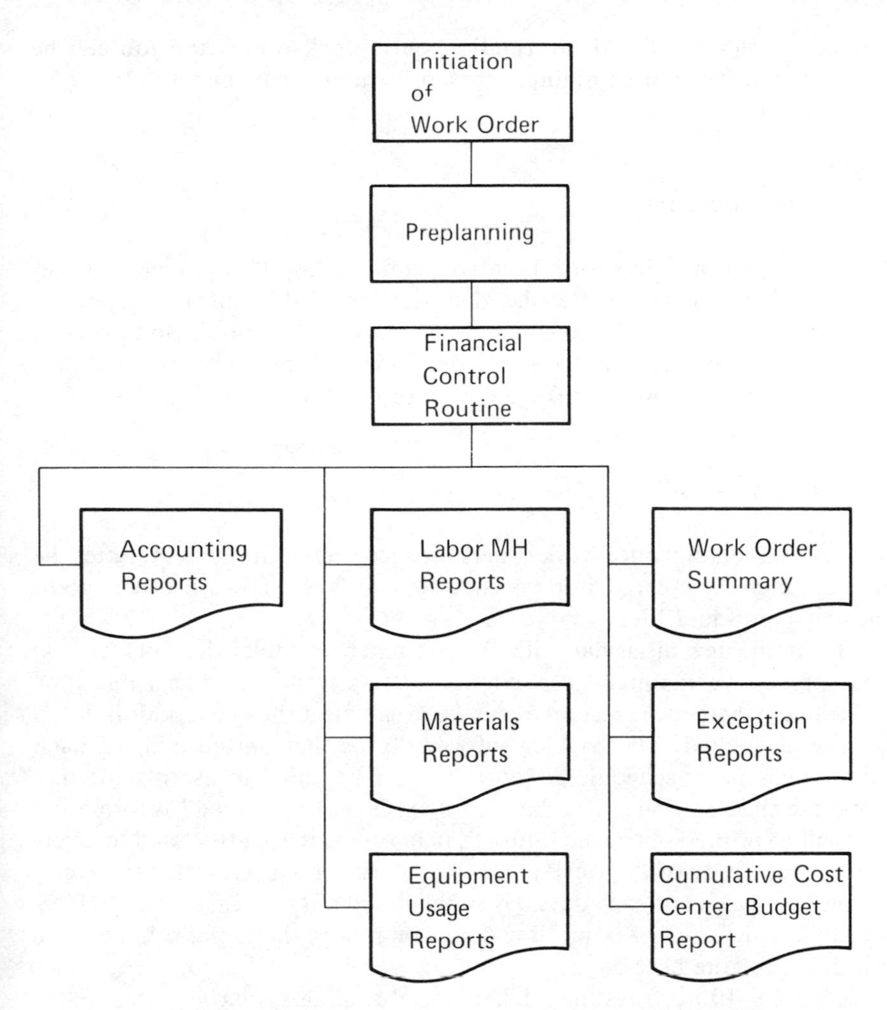

Figure 9–9. Information Flow: Financial Control

on a particular day. Since the backlog is determined by what can or cannot be scheduled, consideration of the backlog and backlog reports are included in this subroutine. Special attention will be given to standards later.

Emergency Work

Emergency-work orders are usually written after the work is accomplished. If the job does not involve identifiable items of equipment and if it is not desired to retain detailed information on a work order, the job can be charged to an open or continuing work-order number from which the work is not accountable.

Preventive Maintenance

In initiating work orders for preventive maintenance, the equipment or unit must be identified first. After the identification of the unit, the inspection interval is determined. Following determination of the inspection interval, the job method and standards are developed. Time standards are then estimated, and the entire package is sent to the new input-data file.

Normal Maintenance

The normal-maintenance work orders are received from the field. After the jobs are either preplanned or assigned standards, these work orders go to the new input-data file.

From the new input-data file, the computer schedules the work orders, both preventive maintenance and normal maintenance. The subroutine determines whether the resources available can meet the needs and, if not, a backlog is created. The backlog information is then sorted and fed back into the computer schedule periodically, usually daily, to ascertain if that work can then be done. The data-storage file accumulates all information that will be printed out in the future to manage the work effort and to determine productivity of the work force. The work is then scheduled to be performed; at the end of each day, it is either closed out or is incomplete. If it is incomplete, it is fed back into the daily computer-scheduling routine to be scheduled for the next day.

Figure 9–10 portrays the scheduling-subroutine system.

Work Standards

Work standards can be placed into the computer memory. Emergency maintenance, because of its nature, is usually not standardized, that is, written on a standard, but some plants have what is known as an emergency or action manual. This manual gives the step-by-step procedure for dealing with anticipated emergency problems.

If a standard does exist, it can be retrieved as printout from the computer memory or withdrawn from the files to accompany the preventive- or normal-maintenance work order, which can then go to the scheduler.

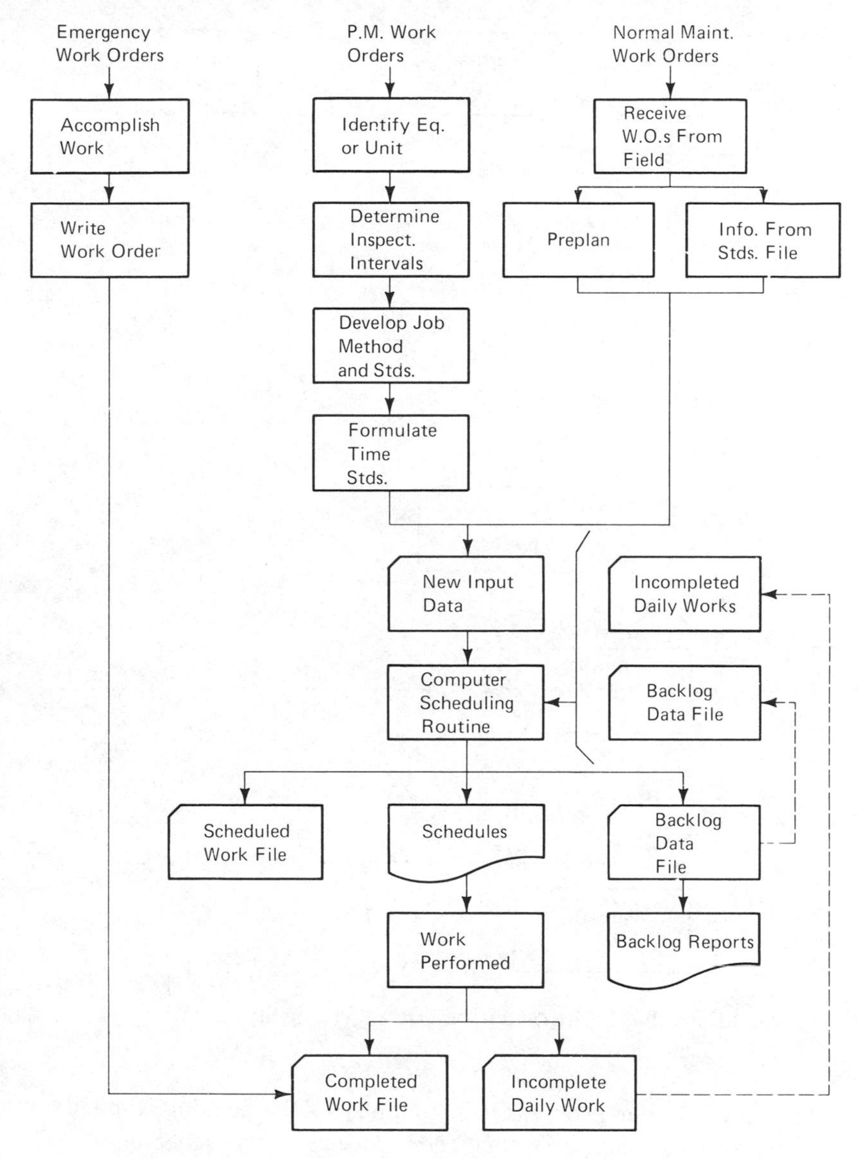

Figure 9-10. Information Flow: Scheduling and Backlog

If a standard does not exist, a decision must be made whether or not to create one. If the decision is made to write a standard, then the appropriate individual or group within the plant receives the work order with a request for a standard. Following the creation of the standard, it is attached to the

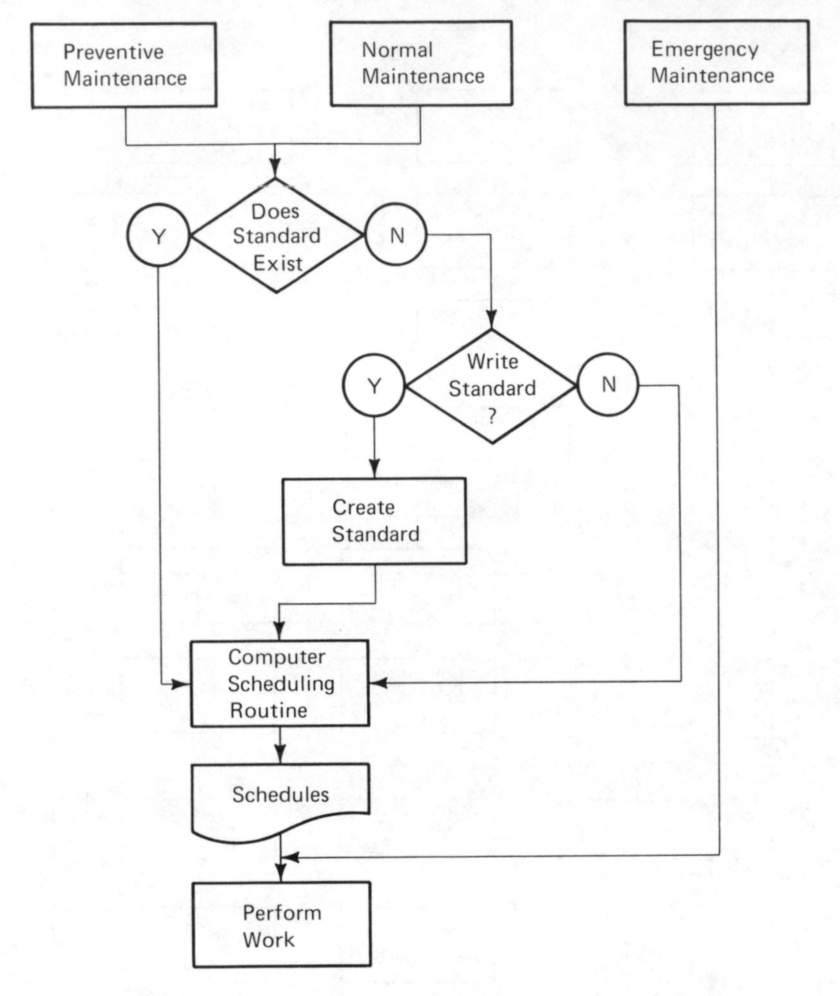

Figure 9–11. Information Flow: Work Standards

work order, which then proceeds to scheduling. If the decision is made not to write a standard, then the work order proceeds to the scheduler.

After the scheduler has obtained all the information, the work orders are scheduled and the work is performed. Figure 9–11 shows the information flow for the work-standards subroutine.

Materials Control

The consideration of materials control in this section assumes that there is an inventory-control program elsewhere in the computer. Only aspects of materials control that directly affect the work-order system are included here.

The inputs into this subroutine take three forms. The first form is nonprelisted-material withdrawal. This usually takes place when the original work order failed to list the material or when the scope of the work order changed to include material not originally anticipated. Such withdrawals are charged to the appropriate work order and are recorded by data-information storage for future reference.

A second input into this subroutine includes material that must be reordered. In these cases, the normal procedures are handled by the purchasing department, the material is received, and this information then goes into data storage. The third type of input includes material that is listed on the work order during preplanning. This list of material is then compared with the list of material in the storehouse, using the data-information storage file. If the material is in the storehouse, the work order can proceed. If the material is not in the storehouse, then the purchasing department must take steps to obtain the material. Exceptions to this include a situation in which the plant has an open purchase order with local suppliers and the maintenance foreman has the right either to order or to purchase the material over the counter.

As can be seen from figure 9–12, the heart of the system is the data-information storage, wherein the initiators of the work orders can ascertain whether or not the material is on hand. If the material is not on hand, then procedures are initiated to obtain that material and, at the same time, to inform the maintenance scheduler when the material has been received so that the work order can be scheduled.

Downtime

The downtime subroutine accumulates data so that analyses can be made on downtime for identifiable items of equipment and areas of the plant. The work order includes an equipment-identification code and a classification of whether the work is emergency, normal, or preventive. If downtime is normal or preventive maintenance, it has been scheduled, and the work order would record the duration of the downtime. The entire time the equipment is out of service is classified as maintenance time since, by definition, the equipment would not be removed from service until maintenance was ready for it.

The downtime that results from emergency maintenance may be separated into idle time and maintenance time. Idle time is the time between the point at which the equipment is removed from service and the point at which maintenance starts. Maintenance time is the time during which the equipment is actually being repaired. These distinctions must be made on

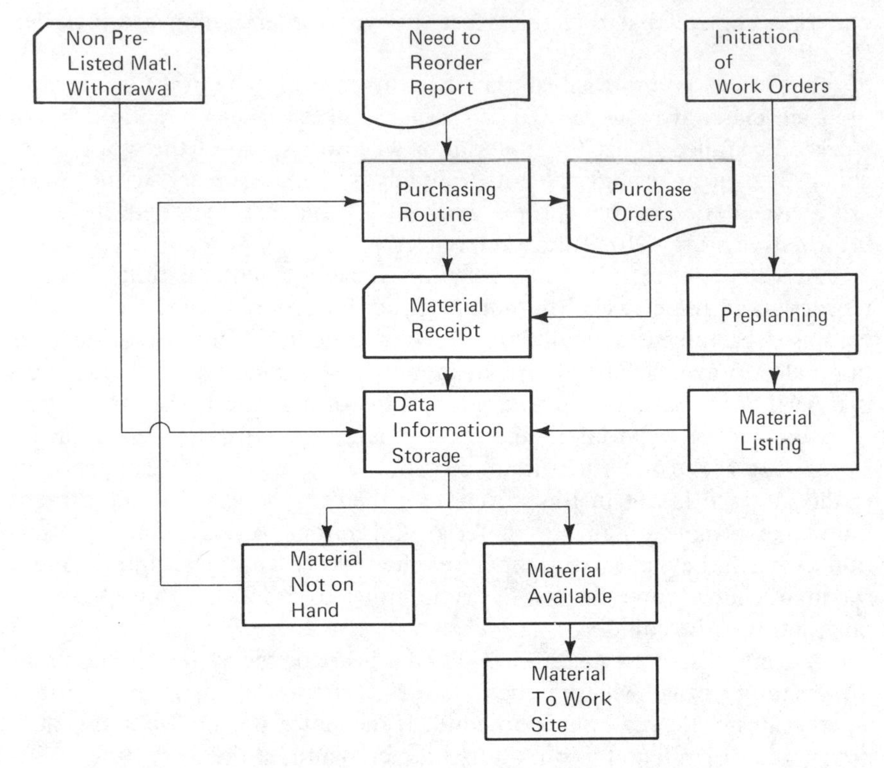

Figure 9–12. Information Flow: Materials Control

the work order, and that information must go to data storage (see figure 9–13).

Periodically, data storage can print out downtime reports, in which downtime is reported by items of equipment that are out of service, by generic types of equipment, or by geographical categories.

Equipment History and Inspection Intervals

Equipment history is necessary for decision making regarding the level of maintenance to be given to that equipment and the point at which it must be replaced.

Traditionally, equipment history has been kept on cards or in separate files, with a folder devoted to each identifiable item of equipment. The work order is then inserted in the folder. When the card is used, the historian records relevant information from each work order on the equipment cards.

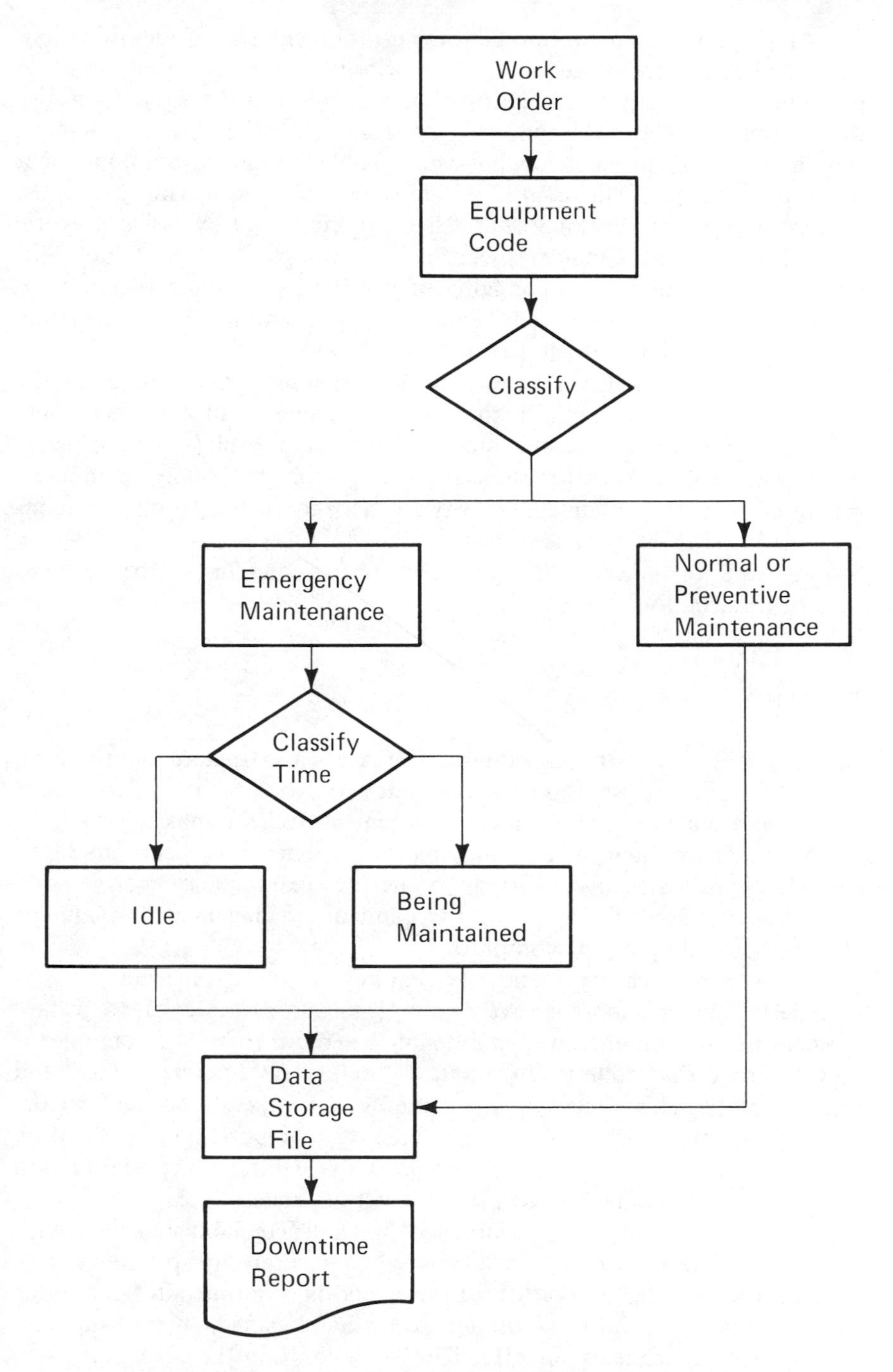

Figure 9–13. Information Flow: Downtime

After the work order is initiated, the priority, equipment-identification, fault, and action codes are added. The priority codes have been discussed previously. The equipment identification is merely a number that identifies the equipment. The fault code is somewhat more difficult to design. It should be a listing of the items that usually make up the reasons equipment needs maintenance. This could include such faults as bearing problems, alignment problems, leaking seals, shell thinning, and excessive pressure drop. The action code indicates what is done as a result of the faults. This might include realignment, changing of bearings, cathodic protection applied, and other steps that are taken as corrective action. This information then goes to the data-storage file.

Inspection intervals can be established, extended, or contracted on the basis of periodic equipment-history reports. These reports can also help develop or adjust methods and standards that could result in a more realistic level of maintenance for the equipment involved. Finally, computer-generated preventive-maintenance work orders can result from storing inspection intervals into the data-storage file.

Figure 9–14 indicates the equipment-history and inspection-intervals flow of information.

The MMIS

Actually, the MMIS is an extension of the basic maintenance-control system. Its purpose is to use the computer-control system to furnish the reports necessary for maintenance-management decision making.

Maintenance-management information systems have been employed successfully as effective working tools by maintenance supervisors, engineers, cost-control personnel, and company management in plants of varying size and operating complexity.[3]

The system does not necessarily require a computer specialist. It has been designed as a comprehensive yet simple means for more-effective management of medium and large maintenance departments. The computer is only a device that collects information, audits and organizes data, and prints reports. The effectiveness of the system depends entirely on the maintenance manager's ability to make use of the reports and information in planning and controlling department activities. Unless appropriate action is taken by these managers, no progress will be made.

Information from input documents—work orders, labor tickets, equipment and inventory records—can be used to generate up to fifteen basic management reports. The content of these reports is outlined in figure 9–15. They provide all the information needed by maintenance, engineering, and cost-control departments for effective maintenance in the plant. Any one

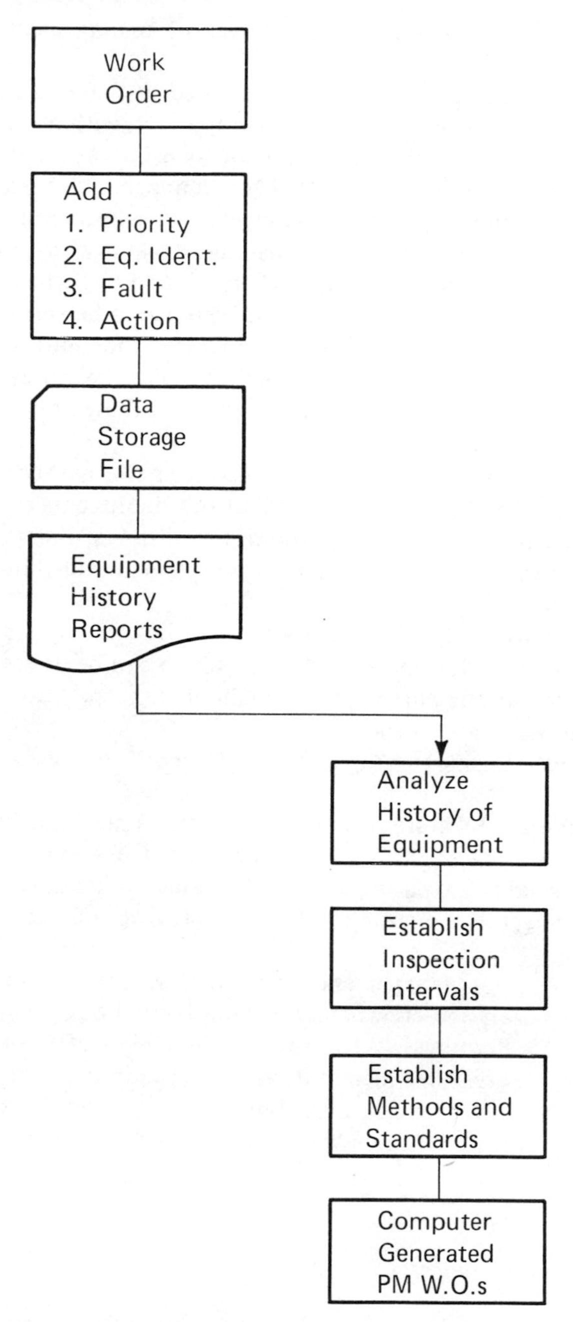

Figure 9–14. Information Flow: Equipment History and Inspection Intervals

report can be used or not, as the degree of management sophistication demands.

Maintenance managers and planners need to know the current project backlog and performance record for comparison with a standard, budget, or estimate. This comparison can be made by using the information given in Reports A, B, and C (figure 9–15). Most managers also want to know (as does the accounting department) what jobs have been completed, the type of work involved, any delays and their causes, and which jobs are causing schedule or cost problems. Reports D, E, F, and G perform this function.

Most planners and managers would also like to be able to obtain, easily and quickly, specific printouts of work required for shutdown or special activities; preventive-maintenance schedules and backlogs; weekly craft schedules; and monthly cost-center charges. For these printouts, the sources are Reports H, I, J, and K.

For the best results in planning, management, and control, inventory information also should be supplied to the maintenance and purchasing departments, including purchase-requisition listings, open orders, priority needs, and inventory status. This information is provided by Reports L, M, N, and O.

The one word that perhaps best describes this system of reports is *visibility*. Timely information and data, which can be used to organize and discipline the various phases of any maintenance operation, are produced. Potential benefits are as follows:

Planning—Backlogs are current. Everyone involved knows exactly what has to be done.

Scheduling—Jobs are performed and crews are used in the most efficient manner. Given current schedules and work orders, foremen can establish logical work patterns. Use of the lead craft designates to one individual the overall responsibility for successful completion of a given work order.

Performance—Management is provided with the opportunity for critical review of craft and crew activities on a factual basis. Cost per standard hour can be determined. Areas of low and high efficiency can be pinpointed, as can areas in which craft and crew activity are not good enough.

Control—Cost of work, both labor and material, is readily apparent and can be applied to the appropriate area. Hidden costs are virtually eliminated.

Working with Modules

A broad-scale maintenance system, like a building, must have a sound structure if it is to survive. To provide a framework capable of supporting

To be effective, a maintenance management information system requires data from a variety of sources. Here's a breakdown of the data contained in a suggested series of reports for such a system. With these forms, plant engineering and maintenance personnel can keep up-to-date on all maintenance activities in the plant.

Report A (Weekly) — **Backlog by Craft** Distr. ★△
1. Contains information needed by planners and foreman to define future schedules and manpower requirements.
2. Includes all incomplete and pending work orders, except preventive maintenance.

Report B — **Backlog Summary Report** Distr. ★
1. Provides an analysis of backlog of work on hand for the next 1- to 52-week period.
2. Provides planners with a basis for scheduling dates for new work orders.

Report C (Weekly) — **Craft Summary Report (Performance)** Distr. ★△
1. Provides information needed to evaluate craft performance.
2. Contains information about completed work orders and craft activities in the past week. By comparing the preceding week's reports to the present period, trends and abnormalities can be detected.

Report D (Weekly) — **Completed Work Order Report** Distr. ★
1. Contains information about all work orders completed during the week.
2. Provides the details necessary to evaluate planning efficiency and performance on specific jobs.

Report E (Weekly) — **Class of Work Summary Report** Distr. ★
1. Indicates the type of maintenance that is being performed.
2. Serves as a planning and evaluation tool.

Report F — **Maintenance Delays by Reason** Distr. ★△
1. Gives information on the cause of lost time and shows the hours lost, per category, related to total hours worked.
2. Forms the basis for evaluating maintenance activity and coordination effectiveness.

Report G — **Exception Report** Distr. ★△
1. Pinpoints all work orders that have, for example, exceeded predefined limits such as:
   - Job behind schedule by specific amount.
   - Job over or under predetermined performance percentage levels.
   - Job over or under predetermined estimated cost levels.

Source: J.J. Wilkinson and J.J. Lowe, "A Computerized Maintenance Information System that Works," *Plant Engineering,* March 18, 1971, p. 69.

Figure 9–15. What to Report

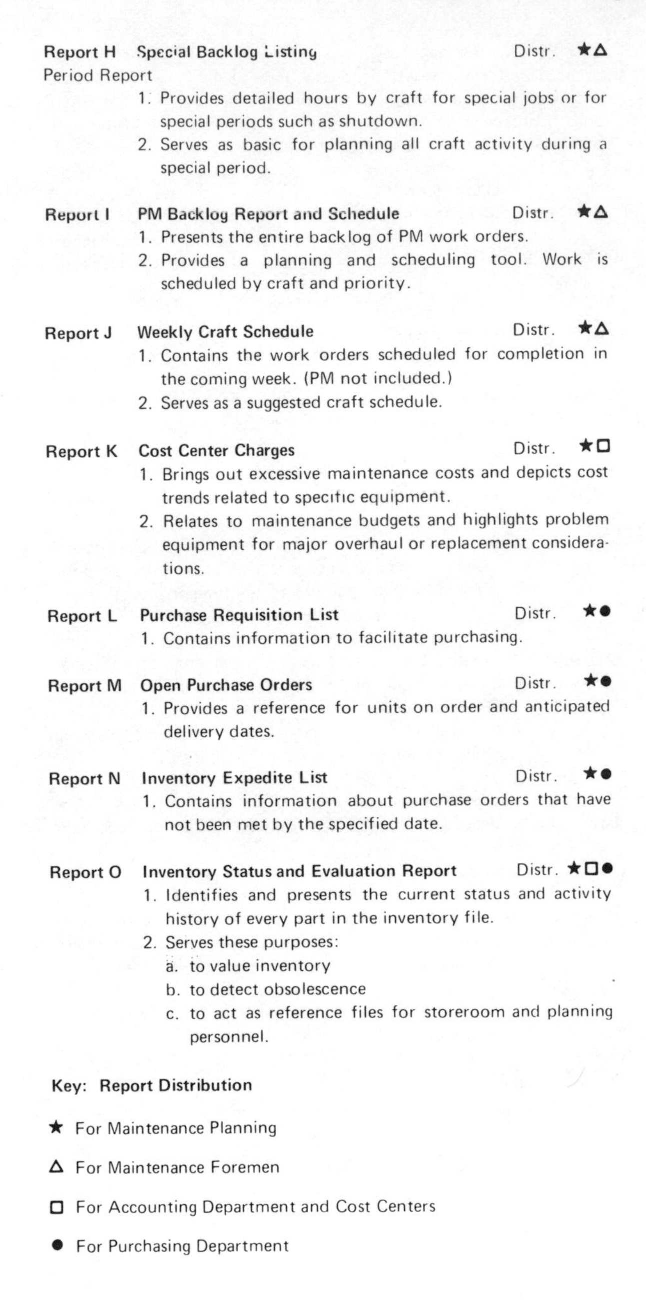

Report H **Special Backlog Listing** Distr. ★△
Period Report
1. Provides detailed hours by craft for special jobs or for special periods such as shutdown.
2. Serves as basic for planning all craft activity during a special period.

Report I **PM Backlog Report and Schedule** Distr. ★△
1. Presents the entire backlog of PM work orders.
2. Provides a planning and scheduling tool. Work is scheduled by craft and priority.

Report J **Weekly Craft Schedule** Distr. ★△
1. Contains the work orders scheduled for completion in the coming week. (PM not included.)
2. Serves as a suggested craft schedule.

Report K **Cost Center Charges** Distr. ★□
1. Brings out excessive maintenance costs and depicts cost trends related to specific equipment.
2. Relates to maintenance budgets and highlights problem equipment for major overhaul or replacement considerations.

Report L **Purchase Requisition List** Distr. ★●
1. Contains information to facilitate purchasing.

Report M **Open Purchase Orders** Distr. ★●
1. Provides a reference for units on order and anticipated delivery dates.

Report N **Inventory Expedite List** Distr. ★●
1. Contains information about purchase orders that have not been met by the specified date.

Report O **Inventory Status and Evaluation Report** Distr. ★□●
1. Identifies and presents the current status and activity history of every part in the inventory file.
2. Serves these purposes:
   a. to value inventory
   b. to detect obsolescence
   c. to act as reference files for storeroom and planning personnel.

Key: Report Distribution

★ For Maintenance Planning

△ For Maintenance Foremen

□ For Accounting Department and Cost Centers

● For Purchasing Department

Figure 9–15 continued

such an activity in both medium- and large-size organizations, the MMIS has been designed on a modular basis. Each module is a building block for the next level of the complete system. The modular arrangement of the system is depicted in figure 9–16, which shows the key functions of each segment.

Mod I can be installed first and later expanded to include Mods II and III. Thus, the system can be expanded as and when further management information is needed, or as the scope and complexity of maintenance activities increase.

Mod III, which completes the MMIS program, provides a fully integrated system. In this ultimate module, maintenance information is divided into three sections: labor, equipment, and inventory. These categories, in turn, are subdivided into six basic files: work order, preventive maintenance, personnel, equipment, operation analysis, and inventory information.

The files (figure 9–17) form the data base or backbone of the MMIS and are used to store the large amounts of historical and current maintenance information needed. This information is periodically updated and screened.

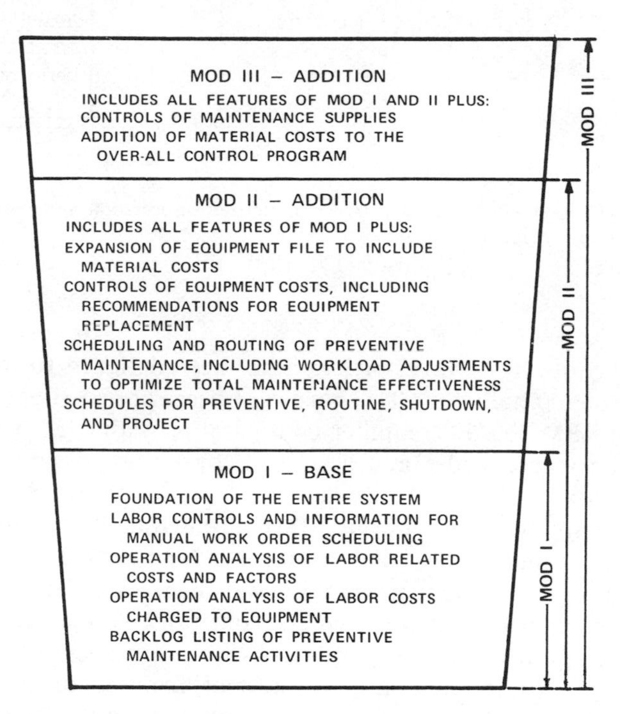

Source: J.J. Wilkinson and J.J. Lowe, "A Computerized Information System that Works," *Plant Engineering,* May 13, 1971, p. 95.

Figure 9–16. Key Functions

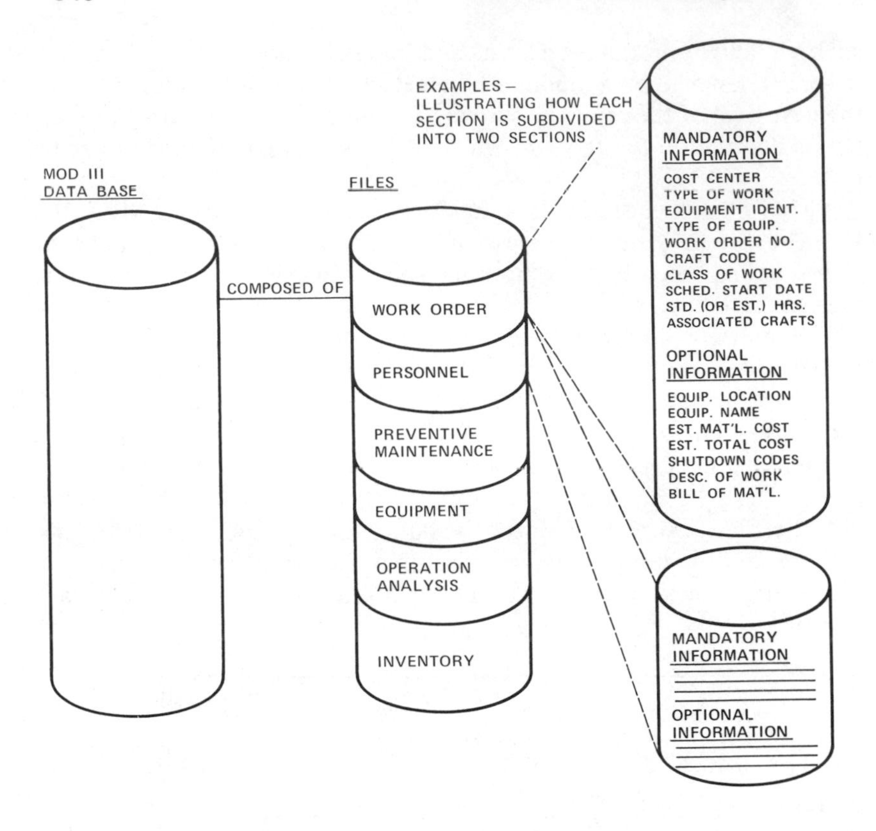

Source: J.J. Wilkinson and J.J. Lowe, "A Computerized Information System that Works," *Plant Engineering,"* May 13, 1971, p. 95.

Figure 9–17. Basic Files

System Structure. The MMIS program has been designed for a relatively small computer installation. The hardware necessary to support the full system is an IBM 360/30 computer configuration with 65K memory, two discs, two tapes, card reader/punch, and a 132-position printer, using OS or DOS.

The assumption underlying any MMIS installation is that there exist or will exist (1) a maintenance-planning function; (2) a work-order system; (3) a method of reporting maintenance labor against work orders; (4) a coding system for equipment and cost centers; and (5) a scheme for reporting material receipts, disbursements, and use by work order (applies to Mod III systems only). While these are all normal features of any well-run maintenance department, up-to-date information and "discipline" are needed to assure optimum results from the MMIS.

A complete MMIS flow chart appears in figures 9–18, and 9–19. The work request and work order, in phase I, are the initiating documents and provide the driving force for the entire system. Actual execution of the maintenance work (and the development of related materials data for Mod III) takes place in phase II. The computerized aspects of the system, including the data-base format and all output reports, make up phase III.

When the maintenance-work orders and time tickets of a Mod III system are filled in, the master data base is used to provide pertinent information on preventive maintenance, equipment, personnel, inventory, and related points for the needed planning, controls, and reports. Thus, when the maintenance-work requests are submitted, the system immediately begins planning and scheduling all of the maintenance work.

Assume, for example, that a job is scheduled to start in two weeks but that the necessary material has not arrived. An inventory-expedite item will then be generated from the inventory file, and the work order will be held until the material is in stock. When the overdue material is received, the work order will be activated automatically and scheduled into the backlog of maintenance work ready to be started.

If, for any reason, an input transaction (for example, a work order to change a valve) cannot be matched against the data files (no such valve available, incorrect equipment number, and so forth), it is reported as an invalid transaction on an error listing. In this way, it is flagged and will be investigated by the maintenance coordinator.

File Structure. The MMIS program is flexible, and provisions have been made to accept many pieces of information. The amount of data actually set up in the computerized maintenance data base is optional, however. Each of the six files is subdivided into two sections (figure 9–17), mandatory and optional.

A skeleton system can be installed by supplying data only for the mandatory sections of each file. Then, as the maintenance function matures and increased sophistication is desired, the optional information can be added to expand the system to its full capabilities. Naturally, the more complete the information supplied to the files, the more comprehensive will be the output reports.

Several coding techniques have been developed for use with the file data. A craftsman may have eight different reasons for an unavoidable delay, for instance. On his labor ticket, the cause is recorded as a single number from 1 to 8. The hours of delay recorded against each of these codes on the maintenance time ticket (figure 9–20) are accumulated in the operations analysis file and form the basis for evaluation Report F, "Maintenance Delays by Reason" (figure 9–18).

A two-digit priority code facilitates work planning and sequencing.

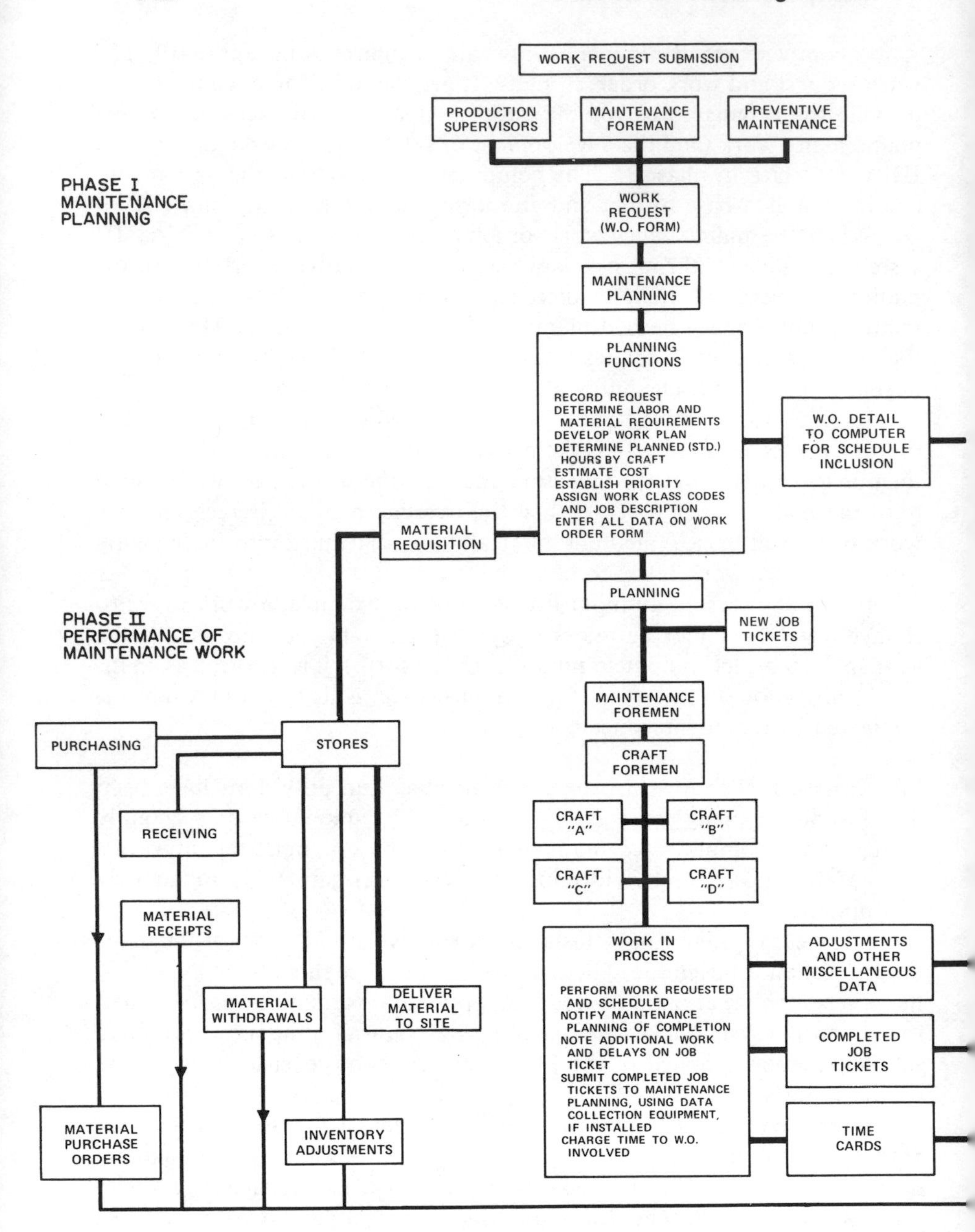

J.J. Wilkinson and J.J. Lowe, "A Computerized Information System that Works," *Plant Engineering,* May 13, 1971, p. 95.

Figure 9-18. Phases I and II—MMIS

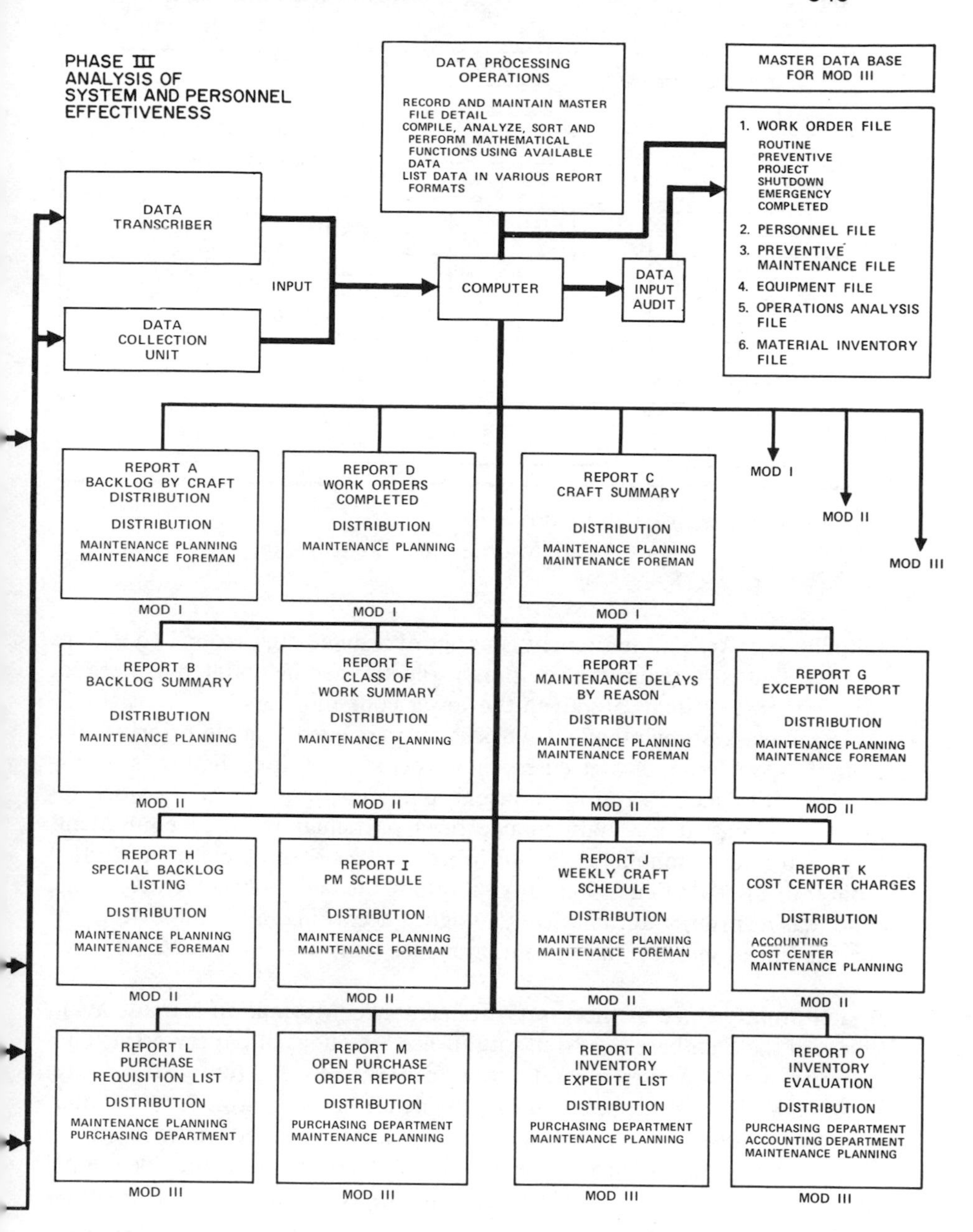

J.J. Wilkinson and J.J. Lowe, "A Computerized Information System that Works," *Plant Engineering,* May 13, 1971, p. 95.

Figure 9–19. Phase III—MMIS

Maintenance Time Ticket

Employee No. Craft Code Date Month Day Year

Work Order Number	Elapsed Hours	Hrs. Worked Category–Check One				Turn 1, 2 or 3
		Straight Time	Reg. O.T.	Sunday Prem.	Holiday	

DELAYS WAITING FOR:

1. Material ☐
2. Other Crafts ☐
3. Tools ☐
4. Equip. Shutdown ☐
5. Instructions ☐
6. Add Men ☐
7. Blueprints ☐
8. Other ☐

Elapsed Delay Time

Employee's Signature ______ Foreman's Signature ______

Source: *Plant Engineering,* May 17, 1971, p. 69.

Figure 9–20. Maintenance Time Ticket

Initially, each type of equipment is assigned a single digit from 1 to 9. This priority digit is stored in the equipment file and can be updated as needed. (More important items are given the lower code numbers.)

Each class of maintenance work is also assigned a single priority digit from 0 to 9. (Zero indicates emergency work; 9 identifies the lowest priority—activities such as routine housekeeping functions.) When a work request is submitted for maintenance on a particular piece of equipment, these two priority numbers are multiplied by the system, yielding a product from 0 to 81. (See figure 9–21.) As a result, the computer-generated work order has its priority automatically assigned according to the importance of the equipment and the class of maintenance work to be carried out.

System Input. Once the necessary maintenance information is made available and the data base has been established, the normal paperwork is adequate to maintain current and accurate files. Periodic (daily or weekly) input data for the computer can be provided in several ways. Some companies, for production recording, already have installed data-collection devices at key points throughout the plant. Here, maintenance personnel can immediately transmit information to the computer center from their work areas.

Other companies have simply retained a conventional maintenance-paperwork procedure for work orders and material requisitions. Mark-sensing or manual entries are made and a keypunch operator enters this information onto data cards, which are fed into the computer. This flow of

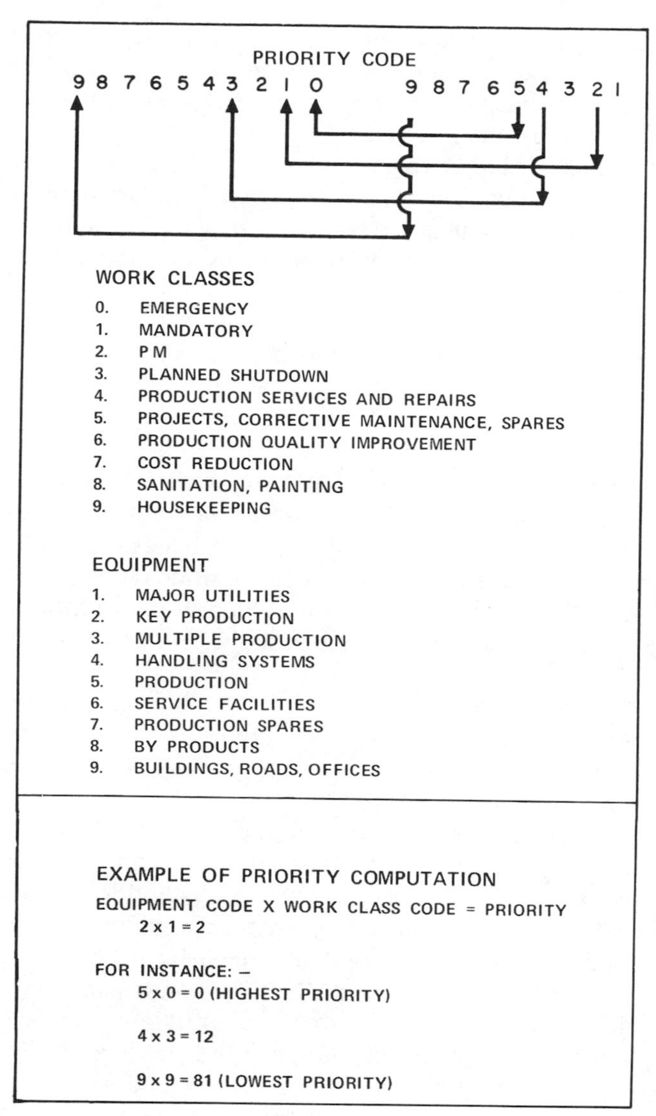

Source: *Plant Engineering,* May 13, 1971, p. 99.

Figure 9–21. Priority-Coding System

information is shown in phases I and II of the system flow chart (figure 9–18).

Data-input audit, an extremely important feature of the paperwork procedure, is shown diagrammatically in phase III of the flow chart. Input records are checked as they enter the files. If an invalid record is found, it is rejected by the program to ensure, as much as possible, that the files will contain only valid information. Unfortunately, erroneous input data, such as an inaccurate inventory piece count or man-hour tabulation, cannot be easily monitored by any program.

Factors that need to be considered in setting up a data-collection system include size and layout of plant; number of maintenance personnel; number of data-input transactions per day; need for current output data; and, of course, the timing of computer runs versus the accumulation of data records.

System Output. The fifteen basic reports produced by MMIS Mod III are pertinent for various levels of management. They provide the information essential to making decisions and initiating improvements. Report titles and levels of distribution are listed (figure 9–15) and are shown as computer output in phase III (figure 9–19). Figure 9–22 represents the basic data files, the scope of each module, and the relationship of the various reports to the three modules. It demonstrates how Mod III expands to broaden the use of the equipment and operation analyses files—while retaining all the advantages of Mod I—and how Mod III retains the features of Mods I and II and adds the inventory files, with all materials, stores, and purchasing reports for all maintenance-related activities.

With few exceptions, computer reports are available from the system on demand. This feature assures that only information of immediate interest will be generated, thus minimizing computer time and the proliferation of unneeded or untimely reports.

Performance reports are prepared and categorized by craft, class of work, cost center, or equipment. In most cases, control reports showing performance are compiled by exception. This method ensures that the most routine analysis activities are delegated to the computer and that only unusual situations are referred to management for action.

The fifteen basic reports of the MMIS are constructed to give management sufficient information to control the entire maintenance operation. System modifications are possible, however, and need not be limited to any particular type or number of reports. Once the data base has been established, it can be manipulated in limitless combinations to generate fewer reports or more—if there is a willingness to pay for the additional programming time.

Installing the System

Planning an effective MMIS on paper is one thing; putting it to work is quite another. A four-phase installation procedure is suggested because of

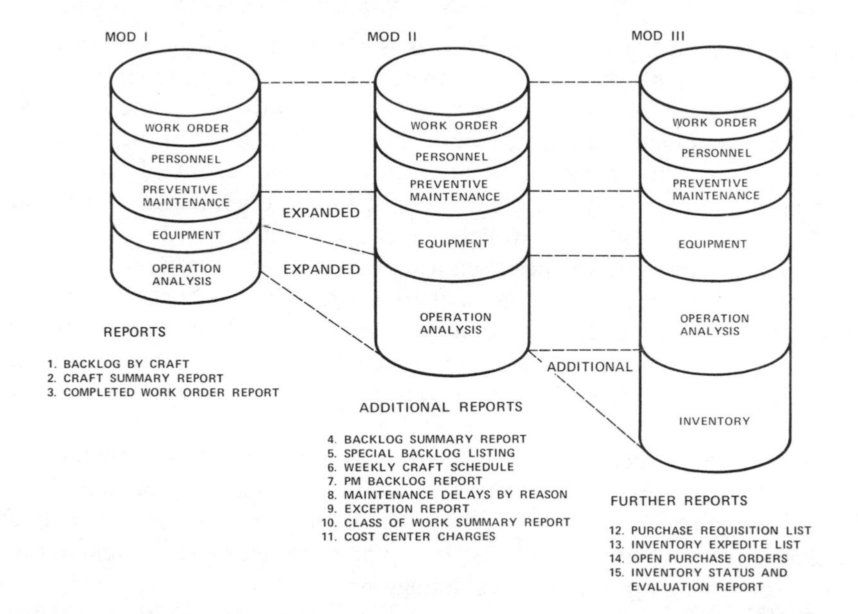

Source: *Plant Engineering,* May 13, 1971, p. 99.

Figure 9–22. Module Reports

the impact a management-and-control system can have on the maintenance department and its personnel. The followng approach assures an orderly transition from manual to computerized procedures:

1. introduction and training of data processing and maintenance personnel;
2. preparation of maintenance department;
3. preparation of data-processing department; and
4. installation and testing of the system.

The training sessions should include maintenance supervisors who will be involved directly with the system, the maintenance coordinator (who will be responsible for all coordination between departments), the programmer, and the systems analyst. The systems analyst must participate in training sessions on data-processing aspects of the system. Subjects covered during the week-long sessions should include maintenance and data-processing concepts, coding methods, paperwork-input procedures, output reports, and the timetable for system implementation.

The most important task of the maintenance department is to gather all the information that is to form the data base. The format of the work orders must be decided. Personnel files must be checked to gather mandatory information (name, clock number, and craft) and optional information for data-base files. Preventive-maintenance routines must be set up or finalized. Equipment numbers, locations, and specifications must be updated, and, in some cases, key spare parts must be identified. Decisions must be made regarding the depth desired in the operation-analysis and management-controls output (which affect the input).

When this information has been accumulated, it is coded, recorded on the proper data-base forms, and sent to the data-processing department. Meanwhile, the various coding techniques and new procedures are being explained to the foremen and craftsmen who will be helping to provide input to the system with data from work orders, time cards, and material requisitions.

In the data-processing department, this information is keypunched onto cards and is used to construct the six files that make up the data base for the computer. If the software is being developed in house rather than purchased from outside, this phase also involves systems analysis and a trial period for setting up, testing, and debugging programs.

During this final phase of installation and testing, unexpected problems are often encountered. In one installation, for example, the files were initially constructed with what was thought to be a good coding system for cost-center charges. Halfway through the final phase, however, the need for a new coding system became apparent, and a table had to be constructed so that maintenance personnel could convert it to the new code and, finally, build up a new file.

Frequently, the biggest obstacle is getting people to accept the system and use it properly. In the beginning, foremen and craftsmen may show little interest or confidence in the system and its reports. Several weeks of experience and strong management support may be required before everyone becomes accustomed to the new way of recording information, understands the reports, and begins using the reports in a meaningful manner.

In most cases, it takes about four months to implement the four-step installation of the MMIS. The bulk of this time, which varies widely from plant to plant, is spent in accumulating and coding the data needed for the six files of the data base. When this step is completed, usually only one month is required to put the program on the computer, test the validity of the output results, and start with meaningful reports flowing to shop supervisors on a regular basis.

Benefits of the MMIS

The prime achievement of an MMIS is the combination of reduced costs with better service, which is accomplished in several ways:

Routine preplanning and scheduling are handled by the computer. In this way, skilled personnel are allowed to do more challenging work elsewhere.

Materials and tools are preordered for delivery to the job site without involving the time and effort of supervisors and craftsmen.

Maintenance work is closely coordinated with operational equipment schedules

Planned maintenance largely eliminates inefficient and cost emergency activities.

Historical records can be quickly recalled for reference. With such information at hand, repetitive jobs can be better planned and estimated (if valid job standards are not already available).

Repair-history reports–by making available complete maintenance records on every major piece of equipment—can pinpoint antiquated or misused equipment for economic replacement studies.

Poor maintenance- or engineering-stores inventories can be made more effective by expanding recorded information to include such items as vendor lead time, delivery performance, safety stock, reorder points, maximum inventory levels, and economical order quantities. As a result of these additions, the right parts or materials are more likely to be in the right place at the right time.

Historical records accumulated in the data files can be analyzed periodically to help improve performance in the future.

Notes

1. This section is reprinted from "Justifying a Computer for Maintenance Management," *Plant Engineering,* August 7, 1980, pp. 83–85.
2. This section is reprinted from L. Mann and E.R. Coates, Jr., "Evaluating a Computer for Maintenance Management," *Industrial Engineering,* February 1980, pp. 28–32.
3. See J.J. Wilkinson and J.J. Lowe, "A Computerized Maintenance Information System that Works," *Plant Engineering,* March 18, 1971, p. 68.

10 Maintenance Training

Maintenance training is part of any productivity-improvement program. Although most plants pay some attention to training, it is all too often done on a periodic basis and without a well-planned program. The range between the minimum proficiency that management will tolerate and the highly trained skill it sometimes demands is tremendous. It has been said that the need for training becomes apparent to management only when the problem becomes overwhelming. Maintenance management has a habit of ignoring training and assuming that new and inexperienced workers will become proficient in their jobs through some nebulous on-the-job training effort. In most cases, the craftsman learns both his job and his notion of what constitutes a fair day's work from his fellow employees, who may or may not have received formal training.

Not only craftsmen, but all participants in the maintenance program—planners, schedulers, supervisors, and so forth, should be trained in their respective tasks. Unfortunately, coordination of the training effort frequently falls to a member of the industrial-relations department, for whom it is one of several responsibilities and, as such, does not receive the attention it would if it were the sole responsibility of a single individual.

Large plants are most likely to have a function entirely devoted to training activities. This function includes fulfilling the training needs made known by various levels of supervision within the plant as well as continuous independent assessment of the work force for discovery of needs that can be met by training.

Identifying Training Needs

Before considering a training program, it is necessary to identify training needs. The training department is accountable for the resources allocation to it in much the same way that other departments are accountable. Therefore, there should be some productivity measurements to assist in determining the cost-effectiveness of the resources allocated to training.

Management could use available published industrial standards against which to compare, for example, the time taken by a group of welders to perform representative welds. If it is found that the performance of the welding group is appreciably below standard, then it would be reasonable to consider a training program. The cost of the training program would be weighed against the benefits received by the plant from the increased abilities of the welding group. This would then be evaluated to determine the return on the investment by the company. In this manner, welders and other craftsmen can be evaluated, and the benefits of the training programs can be quantified.

There are a number of sources in the plant from which management might receive indications of a need for training programs. The first and primary of these is the maintenance foreman, who is responsible for the productivity and the quality of work performed by people who work with him. Supervisors should be made aware that their problems might be solved by training programs. The foremen should be consulted about specific needs within their crews.

Operations personnel and maintenance planners become aware of the need for training when equipment still does not operate satisfactorily after work orders have been completed. The number of rework work orders might indicate a need for a training program.

Particular attention must be given to training needs during times of rapid turnover of maintenance personnel. The more rapid the turnover of personnel, the more necessary are training programs. If a number of trained, experienced craftsmen are scheduled to retire in the foreseeable future, then the maintenance manager must ensure that he has the required number of craftsmen to fill vacancies when they occur. This frequently requires an extended recruitment and training program, and these needs must be forecast early enough to conduct the program.

If a plant is acquiring new and different equipment with which the maintenance forces are not familiar, then it is necessary to anticipate training needs. Similarly, when a plant is changing operating conditions (higher pressures, higher temperatures, and so forth), it would need to consider a training program.

As an example, some years ago most plants had an active craft called riggers. With the advent of mechanical equipment, cranes, and other lifting equipment, the rigger craft has diminished greatly. In its place are such crafts as crane operators, equipment operations, and the like, which needed a training program. In a similar manner, plants that began to acquire aluminum vessels and piping needed a training program so that their welders could weld on aluminum as well as on ferrous metals.

Technical and supervisory personnel also require training programs—particularly those in maintenance management who are involved in planning and scheduling. All personnel involved with the management-information system by which the plant is operated must be trained. Indeed, one would be hard put to name any segment of the plant that could not benefit by training programs. Of course, few plants can include everyone in training programs, but the needs must be assigned priorities, and a program must be worked out to meet these needs.

Locations of Training Programs

On-the-job training programs are conducted primarily in the maintenance shops and in the field, where, along with experienced craftsmen, the apprentice works on the equipment itself. Frequently, on-the-job training programs are accompanied by some amount of classroom work. The classroom can be a meeting room in the maintenance area or some other appropriate meeting place in the plant. Some plants conduct a continuing program, so that it is helpful for them to have a number of professionally designed classrooms in the plant. In large oil refineries, for instance, it is not at all unusual for an operator-training program to be always in progress, and the classrooms that have been built for the process trainees can also be used to train maintenance craftsmen. Demonstrations can be conducted in an appropriate area of the plant.

Many industries rely on vocational-technical schools in the immediate vicinity for a supply of trained or partially trained employees. Usually the plant expects to add to the basic training these individuals obtained in the vocational-technical school.

Another location of training should be the homes of the employees. There are a number of correspondence schools through which a craftsman can better himself in his job. State departments of education provide vocational training on an extension basis, whereby an instructor conducts courses of interest in the area for one or two days a week. Many new and innovative ways of increasing the technical knowledge of maintenance employees are now coming into being, and the future promises that more progress will be made in this area.

Still another location of training might be the union hall. Many unions operate educational programs for their employees, and management should not overlook this possibility when the need for training manifests itself.

Types of Training Programs

On-the-job training programs are perhaps the most popular. In this type of program, the trainee is paired with an experienced employee, a portion of whose job it is to train the new individual. Care must be exercised in choosing the experienced employee, since some employees might not be inclined to participate in this type of program. Positive action must be taken to ensure that the trainee is subjected to all levels of the task he will be expected to accomplish. The program must be monitored to ensure that appropriate transfer of information is taking place. Most on-the-job programs are conducted on actual equipment in the plant, although some simulated situations are used, particularly in cases where failure would be very costly and would impair safety.

When training programs include a general area that would be necessary to many crafts, such as mathematics, a programmed-learning course might be satisfactory. Experience has shown that this type of program would only serve when the craftsman is motivated to learn. It could be said that any training program would only be effective when the craftsmen are motivated to learn, but a programmed-learning sequence requires even more motivation than does thc typc of program in which a portion of the instructor's task is to motivate the individual students.

Large-equipment manufacturers often offer maintenance-mechanical courses in their plants for customers or prospective customers, or the manufacturer might send an instructor to conduct a maintenance-training program on the site where the equipment will be used. This is often offered as an inducement to purchase new equipment. Frequently, equipment representatives will conduct evening courses or short courses at local hotels or motels for training craftsmen in the operations and repair of the equipment they have sold or hope to sell.

Vocational-technical schools usually offer courses that are useful to local industry. In most cases, these courses will offer basic training and the plant must supplement it by on-the-job training programs with the specific equipment. Vocational-technical schools also offer continuing-education programs away from the school; an instructor will travel to the plant vicinity if there is sufficient demand. In such cases, the instructor will use the facilities of the plant if all of the students are from a specific plant, or he might use the facilities of the local department of education.

Correspondence courses continue to be a significant source of training, and a plant will usually bear the expenses of such a program when a student can prove that he has completed it successfully. Examinations for correspondence courses often are administered by local school districts.

University courses in industrial engineering and business administration might improve the performance of maintenance planners, schedulers, coordinators, purchasing agents, and the like. Seminars and short courses probably offer the best opportunity for maintenance-management personnel to improve their skills. A number of universities and consultants offer seminars and short courses at appropriate locations throughout the country; these programs give maintenance-management personnel an opportunity not only to improve their individual performance but also to ascertain the opportunities that are available to improve the entire maintenance system.

Performance-Based Training[1]

Although programmed training techniques continue to find acceptance in industry today, many programs have been modified and their scope has been enlarged. As complexity of machinery has increased, so has the need for qualified maintenance people to keep it operating. An outgrowth of this change has been recognition that training of craftsmen beyond the basics in their field is needed.

Knowledge of troubleshooting and how to repair mechanical devices is needed to maintain much of today's equipment, thereby reducing the downtime. Management has had to face longer time periods and higher costs to train people. A training system that could provide maximum improvement of skilled job performance with a minimum investment of training time and expense was needed.

Performance-based training (PBT) systems were judged to be the answer to those needs. PBT systems are considerably more efficient and less costly than most traditional training programs, particularly when technical skills training is involved [see figure 10–1]. And the systems embody many of the principles of programmed instruction familiar to industry.

What is PBT? PBT is best defined by listing its principles and the procedures with which it is carried out:

PERFORMANCE-BASED TRAINING	MOST CONVENTIONAL TRAINING
1. Performance requirements are detailed.	1. Curriculum requirements are general.
2. Objectives derive from task analysis.	2. Course objectives usually are not specific.
3. Achievement of performance objectives is measured.	3. Number of hours of training is measured.
4. Measurement of effectiveness is built in.	4. Measurement is often lacking or subjective.
5. Individual differences in learning ability are fully utilized.	5. Individual differences are not utilized.
6. Trainees progress at their own pace.	6. Trainees progress at group pace.
7. Procedure-decisions are based on trainee input.	7. Trainees have minimal voice on procedure decisions.
8. Major instructional burden is carried by PBT materials.	8. Most instruction is given by instructor.
9. Learning effectiveness is documented.	9. Documentation of effectiveness usually is not built in.

Based on material developed by A. E. Oriel & Associates, Ltd., Geyserville, CA.

Source: W.H. Weiss, *Plant Engineering,* May 14, 1981, p. 91.

Figure 10–1. Comparison of Performance-Based Training with Conventional Training

Trainees move from one phase of training to another only when they can perform the required tasks, never because a specified number of hours of study have been completed.

The performance requirements of the training grow directly out of the demands of the job, never out of general curriculum requirements. Training is designed to produce specific, measurable performance.

Training efficiency is maximized when individual differences in learning ability are recognized. PBT allows each trainee to progress at his or her own pace.

Measurable job performance is held constant, and individual trainee learning time is allowed to vary. All trainees are required to achieve a similar high level of performance, but because of difference in learning ability, some achieve it faster than others.

Modular training units are used at the convenience of the trainees: they are not held to a structured, time-assigned curriculum. The modules contain instructions and theory previously given by the instructor to enable each trainee to acquire skill and knowledge needed in a self-paced manner.

Performance test and measures of training effectiveness are built into the system. When a trainee feels confident he or she has learned a task, the instructor administers a test. The trainee may need to repeat a learning module if he or she cannot prove that the required skill has been learned.

PBT versus Conventional Training

One important difference between conventional training and PBT is the role the instructor plays. In a conventional training program the instructor does most of the teaching in a group classroom. With PBT, the instructor rarely teaches a group in the classroom. Instead, he provides individual guidance, develops additional task modules as required for specific crafts, and systematically evaluates trainee performance using the materials and methods provided by the PBT system. Trainees learn in a craft-learning center—a fully equipped facility with tools, measuring instruments, simulators, plant equipment, and any other material that helps them become familiar with the requirements of the craft.

Requirements of a Craft-Learning Center

A craft-learning center should be a separate area within the plant where trainees are taught specific procedures and tasks before working on actual plant equipment. Although the center will vary in size according to plant size and number of employees, it should occupy at least 4000 sq.ft. to be maximally efficient.

The center should be divided into four main sections: the vestibule, the performance systems manager's office, a study area for individual study, and a resource library. Because training is individualized and self-paced, there is no need for a classroom.

If heavy equipment is not accessible for trainees to work on in the plant, components of the equipment should be moved into the learning center. The vestibule should be designed so that equipment, simulators, tools and, whenever possible, actual plant equipment can be used by the trainees for hands-on learning. The resource library houses the instructional modules and the training materials.

A unique facet of apprentice training is testing. Trainees should be encouraged to take a pretest on a subject they feel they know. There are two reasons a company would want to pretest trainees. First, if a trainee can achieve a passing score on knowledge of the subject or a hands-on task, there is no need for the trainee to spend time going through a training module. If the trainee already knows the contents of a learning module, it would be a waste of training dollars and the trainee's time for him or her to formally study that subject or task. Second, if a trainee fails the pretest, both the company and the trainee have identified what he or she does not know. Both have determined that the trainee should go through the learning module. After the trainee does this and feels confident that a test can be passed, he or she can be retested. The difference between the score on the pretest and post-test establishes the trainee's increase in learning.

Industries Using PBT

In the last 15 years, many industrial, governmental, and military organizations in the United States have converted all or part of their manpower development programs to PBT systems. Lockheed-Georgia Co. recently began using a PBT machine-operator training system. The system has reduced the time required to produce competent operators from 2 years (approximately 3680 hours) to an average of 500 hours of training through the use of selfpaced, task-oriented, audio-visual modules. Using these modules, trainees with deficient skills have achieved a high level of proficiency in about one-third the time required with a conventional training program.

Chrysler Corp. is using modified PBT systems to train machine operators in its tank plant. The Ford Motor Co. has changed the shop portion of its apprentice training program to task-oriented, achievement-based PBT systems; the company has been able to make reductions in training time comparable to those achieved by Lockheed.

The Goodyear Tire & Rubber Company's experience with PBT has been rewarding. Trainees using purchased training systems and corporate-developed performance guides have been able to perform plant tasks the first time they tried with an acceptable quality of performance in the same time or less than an experienced would require. And they have been able to do the tasks without the help of an instructor.

Task analyses or task lists were developed for nine separate or combined crafts [see figure 10–2]. Then, more than 100 published training programs were evaluated and compared to the performance objectives of each trade. Programs were reviewed in an effort to find suitable materials available on the market, thus reducing the number of modules that would have to be developed by Goodyear's training department staff [see figure 10–3].

Effectiveness of PBT

Research over several years has revealed that in approximately 90 percent of the cases in which PBT systems have been properly developed and administered, training time has been reduced an average of 40 percent and instructor time an average of 60 percent. More important, better job performance over that of workers trained with conventional and traditional methods has resulted.

Although each craftsman in a plant performs many types of jobs or tasks relevant to his particular skill, many tasks are common to more than one craft. Here are some common subjects and some electrical subjects that might be taught in a training center:

COMMON	ELECTRICAL
Blueprint reading	Basic electricity
Basic mathematics	Electrical measuring instruments
Measurement	Batteries and DC circuits
Engineering drawings	Electrical protective devices
Hand tools	Making a hot tap
Portable power tools	Bending and installing conduit
Metal materials	Grounding electrical apparatus
Nonmetal materials	Installing lighting circuits
Lockout procedures	Installing electronic components
Energy conservation	Installing motors
First aid	Maintaining motors
Safety	Wiring a control panel
	Electrical troubleshooting

Source: W.H. Weiss, *Plant Engineering,* May 14, 1981, p. 91.

Figure 10–2. Task Lists

Plant engineers considering a PBT program may wish to contact some of the companies listed here, or others, whose training materials may be of help to them in developing such a program.

COMPANY	SUBJECTS COVERED
Dantrain Daniel Building Greenville, SC 29602	Mechanical tasks
E. I. du Pont de Nemours Educational & Applied Tech. Div. Wilmington, DE 19898	Safety Mechanical and electrical tasks
Educational Methods, Inc. 500 N. Dearborn St. Chicago, IL 60610	Basic math Print reading Linear measurement
Edu Pac 231 Norfolk St. Walpole, MA 02081	Basic math skills
Handbook for Riggers P.O. Box 2999 Calgary AIB, Canada	Rigging and hoisting
Honeywell, Inc. 1001 E. 55th St. Cleveland, OH 44103	Instrumentation
McGraw-Hill Book Co. Order Fulfillment Department 13955 Manchester Rd. Manchester, MO 63011	Textbooks on all technical subjects
Power Tool Institute, Inc. 605 E. Algonquin Rd. Arlington Heights, IL 60005	Power and hand tools
TPC Training Systems Technical Publishing 1301 S. Grove Ave. Barrington, IL 60010	All craft skills
Vickers Sperry Rand Corp. Troy, MI 48084	Hydraulics
Westinghouse Learning Corp. Westinghouse Building Gateway Center Pittsburgh, PA 15222	Electric welding AC and DC circuits
Xerox Learning Systems 30 Buxton Farms Rd. Stamford, CT 06904	Programmed instruction Print reading Pipefitting identification
Yarway Corp. Blue Bell, PA 19422	Steam trap use and repair

Source: W.H. Weiss, *Plant Engineering,* May 14, 1981, p. 92.

Figure 10–3. Sources of Instructional Material

Who Is Responsible for Maintenance Training?

It might be said that everyone in the plant is responsible for training. The craftsman himself is responsible for learning all aspects of his job. The foreman is responsible for determining what the training needs are and for meeting those needs if he has the resources. At all levels of management, management personnel are responsible for an adequate state of training in their subordinates.

More specifically, large plants usually have one or more individuals responsible for the training program. This training function must locate sources of trainers and must arrange the physical facilities for training to take place. It must be aware of all seminars and programs that might be interesting or useful to plant personnel and must disseminate that information to those in the plant who might find them profitable.

Smaller plants that are unable to devote a full-time individual to training usually assign the training function to someone in the industrial-relations area. In all probability, this individual must make use of resources outside of the plant, whereas training in larger plants uses more internal resources.

Locating Trainers

Although it is certainly expected that training needs will be pinpointed by supervisors, care must be exercised that the plant does not place too much training responsibility on the supervisor, so that he does not have the time to perform the other tasks that management expects of him.

One effective rule is that no one can be promoted until they have trained someone to take their position. This philosophy exists to some extent in most plants. At most levels of management, when replacement individuals are identified, it is possible to subject them to rotational job assignments and thus provide the necessary training for the responsibilities they are to assume. In these cases all levels of management are, in effect, trainers.

Many craft-training programs are staffed by qualified personnel already in the plant. For such courses as piping layout, metallurgy, and concrete design, engineers in the plant can conduct training programs for their fellow technical employees. For courses such as chemistry or arithmetic, foremen or technical people in the plant might well serve as trainers.

From outside the plant, it is possible to contract trainers from such sources as vocational-technical schools, universities, consulting firms, and various fields of specialization. A plant with a poor record in product liability, for instance, might well ask a knowledgeable attorney to speak to the operators who control the quality of the product.

Designing the Training Program

Training programs should result from an identified need in the plant. The need may make itself known through unsatisfactory work performance or through the recognition by management that the plant is entering a new and different phase. Other needs, especially those that have to do with management, make themselves known in more subtle ways.

The initial step in designing a training program is to define the topics that should be covered in the program. These are usually identified in a meeting at which the training function, management, foremen, and craftsmen are represented. Attempts should be made to address only specific needs rather than trying to make the plant needs fit into a training program that is already developed or available from any of the several training-program suppliers. If the plant has job specifications or job descriptions, the designers of the program should consult these documents to ascertain if the training program is consistent with the description of the job.

The second step is to determine who will do the training. In most cases, the training department is aware of the availability of trainers from the various sources already mentioned. Often, the plant has had experience with these instructors and knows their capabilities and specializations. Presumably, the training function has inventoried the employees in the plant to ascertain whether or not there exists expertise in the plant to conduct any or all of the programs.

Step three is the completion of the course outline. The committee should now be enlarged to include the individual who will do the training. The committee then decides the desirable number of hours of instruction and writes an outline to indicate specifically what will be covered during every hour of instruction. Only when an outline is so detailed can everyone on the committee be assured that what they have in mind will really be offered to the class. Generalized discussions do not indicate the depth and coverage of the topics discussed in step one. The outline should be available to the students at the beginning of the first lecture and should include such information as outside reading or problems to be worked and their due dates, films to be shown, and projects or workshops to be done in class.

Step four is determination of all the educational resources and material needed to conduct the class. This includes textbooks, notes, films, homework-problem sets, examinations (if any), slides, projectors, and any other necessary demonstration or model material.

Step five establishes where the class will be held. When adequate facilities do not exist in the plant, it will be necessary for the training function to seek them elsewhere.

The final step is program assessment. Most plants do not like to subject trainees to an examination or test, in view of the trainee's fears that the results of these examinations or tests might affect their advancement and salary structure. In cases where management does desire to use the results in determining advancement, this should be made clear to the trainees. This will give a trainee an opportunity to withdraw from the program if he desires.

Management uses a number of devices to evaluate the results of the training program. Classroom quizzes might determine whether the information is being received and understood by the student. When programs such as welding training, meter and instrument training, and the like, are being conducted, the practical-application test appears to be a solution for monitoring the program. Where possible, the use of such monitoring devices should be built into the program. When the programs are sufficiently long and the number of students is small, it is possible for the instructor to maintain a periodic rating system for the students. This process appears to offer a good compromise, in that the periodic rating of the students could serve for their personnel files as well as for a feedback device to assure management that the program is performing the task for which it was designed.

Some of the topics included in such a grading system could be quantity and quality of work, trainee initiative and cooperation, tasks required of the group that apparently are beyond the capabilities of the group to perform, and work habits of the students.

The follow-up portion of the program would include checking to ascertain whether training did, indeed, improve individual performance. Another meeting of the committee should be held at the end to critique the program. Notes should be placed in files so that errors would not be repeated the next time a similar training program is offered. Comments from the students should be considered and, if deemed relevant, changes should be made to incorporate criticisms.

Training Maintenance Supervisors[2]

Training maintenance supervisors is costly, and if the training is inadequate or ineffective, it will cost the company much more. If training is bypassed altogether, the company ends up paying many times what the training cost would have been.

Most maintenance supervisors are promoted from the ranks of maintenance personnel, and the majority receive little or no formal supervisory training to equip them to handle the new job. By functioning less effectively than they would have had they received supervisory training, they can cause thousands of dollars' losses in production inefficiencies, materials waste, machine downtime and delays.

Undertaking a maintenance supervisor training program involves several steps. Maintenance engineers should review their efforts against the following proved steps to maximize the results:

Establish Objectives

Step one is defining the program's purpose and aim, as clearly as possible, in writing and in terms of measurable results.

Typical objectives that have guided companies in training maintenance supervisors include:

To prepare craftsmen for promotion to supervisory positions.

To increase effectiveness of existing maintenance supervisors.

To overcome an existing or foreseeable problem, such as excessive absenteeism, manpower shortage, introduction of new equipment, or excessive downtime.

To impart knowledge and develop skills in the handling of new responsibilities, including interviewing applicants for maintenance positions, evaluating subordinates' performances, handling first-step grievances, or training personnel.

The more specific the objectives are, the more likely it is that training will achieve the desired results. The objectives can also be the standard against which the training program's effectiveness is measured.

Determine Program Content

This step should identify what the maintenance supervisors need to know, feel, and do to achieve desired results. The program content may have five segments.

The first segment is an orientation to prepare maintenance supervisors for their roles as members of management, and it is essential for all newly promoted supervisors. The principal objective of the orientation is to change their attitudes and approaches to maintenance work from those of the worker to those of the supervisor. They must learn what management expects of them in their new role. They need to become familiar with the personal adjustments to be made in moving from being a member of the crew to being the person in charge.

The orientation "tunes in" new supervisors to the requirements of their job and forms the foundation of their learning. Without orientation, individuals tend to remain worker-minded instead of becoming management-minded.

Topics to include in a typical orientation program are

The company—its products, organization, key personnel, philosophy, and history

The maintenance department—its structure, functions, relationships to other departments, philosophy, and values

Types of equipment to be maintained—operating, standby, utilities, and material handling

Company and department terminology—a glossary or dictionary of words, symbols, and initials commonly used in the company and in the department.

The second segment of the program content includes information or knowledge the supervisors must have to succeed. This segment covers the education of maintenance supervisors.

The third segment covers the managerial skills expected of maintenance supervisors. These skills may include the following supervisory abilities:

Interviewing applicants

Communicating

Training and coaching subordinates

Counseling and disciplining subordinates with problems

Handling and resolving complaints and grievances

Evaluating employee performance

Preparing budgets

Estimating time and expenses for maintenance jobs

Scheduling work assignments

Writing reports and keeping records

Motivating subordinates

Analyzing and diagnosing maintenance problems

Writing standard practices and procedures

Improving methods.

General supervisory skills may be acquired by experience or by attending training programs, including publicly offered courses and seminars on specific phases of supervision. Information about the company's equipment may also be obtained by attending programs conducted by manufacturers. These programs are often used when the number of supervisors involved is too small to justify a formal inhouse program. However, if special skills and abilities are desired, the maintenance engineer may have to develop a program specifically designed for the plant's equipment.

The fourth segment contains material for supervisors who need to develop additional maintenance skills. It is not unusual for a skilled craftsman to be promoted to a supervisory position embracing more than the single craft.

For example, a senior electrician was promoted to foreman. In his new position, he had to supervise skilled carpenters, pipefitters, and painters. Although he was a competent electrician, he had little knowledge of the working practices in the other crafts. He had to become more familiar with these crafts to plan, control, and evaluate their maintenance work.

It is common in the average maintenance department for supervisors to manage multicraft units. In addition to learning some of the techniques of each craft, they need to develop knowledge of, and acquire a degree of skill in, various maintenance operations.

The fifth segment deals with problem handling. It should include information on the types of problems maintenance supervisors are likely to encounter in dealing with equipment, process, materials, and personnel. Not only is it necessary for supervisors to be familiar with such situations, but also they must know what actions are expected of them.

A special training program had to be designed in a recently unionized plant. Its objective was to familiarize maintenance supervisors with the provisions of the new labor agreement: how to deal with employee complaints and grievances, and more important, how to work with the maintenance department stewards and reduce and prevent grievances. The program included typical situations that might arise. Management not only identified these situations, but also informed the supervisors how they were to be handled, ensuring consistency and uniformity of supervisory actions and minimizing the possible compounding of problems.

In actual practice, problem handling would be part of the education and skills-training segments of a program. The problems are often used for group discussions, case studies, and workshops. They can also be worked into role-playing or game situations.

Standardize the Program

The third step is the preparation and standardization of the program. This step involves the commitment of the program and its content to writing.

1. **Role and Responsibilities of Maintenance Supervisors**
   A. Program introduction
   B. What is management?
   C. Levels of management
   D. The supervisor's managerial functions
   E. Your managerial responsibilities and amount of authority
   F. Workshops
2. **Planning and Organizing Departmental Activities for Efficiency**
   A. Types of maintenance planning
   B. Methods of planning and organizing maintenance work
   C. Essential considerations in planning and organizing
   D. Planning tools of the maintenance supervisor
   E. Planning workshops
3. **Understanding Maintenance Employees, Their Behavior, and How To Motivate Them**
   A. Law of human behavior
   B. Fundamentals of human behavior as applied to maintenance personnel—review of behavioral theories
   C. Exercise—What do your employees expect from their jobs?
   D. Techniques for motivating maintenance employees
   E. Motivational workshops
4. **Communication as a Tool of the Maintenance Supervisor**
   A. Communication exercises
   B. Fundamentals of supervisory communications
   C. Leveling communication roadblocks
   D. Essentials of effective supervisory communications
   E. Methods of keeping others informed
5. **Establishing Maintenance Performance Standards**
   A. Types of performance standards (desired results/accomplishments)
   B. Purposes of evaluating maintenance personnel and maintenance jobs
   C. Essentials of establishing and installing standards
   D. Why employees fail to achieve performance standards
   E. Exercises in establishing standards and evaluating results
6. **Training Maintenance Employees**
   A. Techniques for effectively training maintenance personnel
   B. Determining what maintenance employees need to learn
   C. Essentials of increasing employee learning
   D. Developing maintenance training aids and materials
   E. Training workshops
7. **Building and Maintaining Maintenance Department Team Spirit**
   A. The role of the maintenance supervisor in employee relations
   B. Types of supervisory-employee relationships
   C. Developing and maintaining morale in the maintenance department
   D. Cultivating employee loyalty and cooperation
   E. Exercise—Attitude vs. behavior

Source: L.J. Smith, "Training Maintenance Supervisors," *Plant Engineering,* June 26, 1980, p. 101.

Figure 10-4. Supervisory Management-Development Program

Figure 10-4 continued

8. Effectively Handling Maintenance Department Problems
   A. Typical maintenance supervisory problems
   B. Attitudes/approaches to problem solving
   C. Considerations for solving problems
   D. Decision-making techniques
   E. Exercises in decision making and problem solving

9. Constructively Disciplining Maintenance Employees
   A. A review of standards of conduct and behavior
   B. Techniques for achieving desired conduct and behavior
   C. Methods of dealing with disciplinary problems
   D. Types of constructive disciplinary actions
   E. Workshops in typical disciplinary problems

10. Resolving Maintenance Employee Complaints and Grievances
   A. Types of complaints and grievances and their causes
   B. Techniques for preventing or minimizing complaints and grievances
   C. A review of the company's complaint and grievance procedure
   D. Methods for resolving first-step grievances
   E. Grievance-handling workshops

The program needs to be structured into sessions to be covered in a predetermined time period. Each session topic is identified with a listing of the main points or subtopics to be included. A program outline for a short course having ten 2 hour sessions, used by a company, is shown in [figure 10-4].

A leader's guide along with instructions should be developed. It should explain how the program should be presented, as well as what audiovisual aids, trainee notes, reference materials, case studies, problems, tests, and assignments could be used by the leader. If the company's main objective is uniformity and consistency in the training program, the program should be as detailed as possible.

Select a Qualified Instructor

The success of the training program rests largely with the individual who will handle the training. In one company, a training program failed because inadequate attention was paid to selecting the instructor. The instructor selected was a good maintenance engineer, but he lacked the teaching skill necessary to impart his knowledge of supervision to others. He used a packaged supervisory training program, but was unable to apply it effectively. The crux of the problem was that he was assigned to the instructor's post without either proper evaluation of his teaching skills or the determination of his desire to teach.

Some companies assign experienced maintenance supervisors as instructors. This approach proves effective only if the individual has been trained and qualified as an instructor. An experienced supervisor, no matter how successful in the job, may lack the required knowledge or skill to instruct. The same result can be expected if a plant engineer, maintenance manager, or plant superintendent does not meet the instructor qualifications. Managerial ability does not necessarily imply training ability.

Many companies have gone outside their organizations to obtain qualified individuals to handle the assignment. These individuals should be qualified not only as instructors but also as experts in various aspects of maintenance and supervision. Their expertise could enhance a company's maintenance and supervisory knowledge and skills. However, qualified outside instructors should be supplemented by an internal individual knowledgeable on matters relating to company policies and practices.

Put the Program into Effect

The first step in executing the program involves selection of a training site or facility. The next decision should consider the program schedule, starting date, length of each session, time of day, and frequency of sessions. Whether the site is inplant or away, a controlled environment, regulated by the instructor, maximizes learning.

Good execution is made possible by following a standardized program. Deviations from the planned program should be documented. These records will help the instructor determine if such changes should be made in future programs.

Evaluate and Recognize

To assess the merit of a training program, the maintenance engineer must evaluate the results in terms of how much was learned and retained by the trainees. The program can also be measured regarding the desired or expected fulfillment of objectives.

Trainees' attitudes, knowledge, and skills could be tested to determine the extent of their learning. The results should be measured against established standards.

Once it has been determined that the trainees have met the minimum desired learning standards, they should be given some form of recognition, such as a certificate of completion, publicity, continuing education units (CEU's), or official status as a qualified maintenance supervisor. Many companies provide recognition at a dinner party with members of management present, at the conclusion of the training program.

Miscellaneous Aspects of Maintenance Training

Before starting any training program involving personnel included in a collective-bargaining agreement, discussion should be held with the union to determine if there are any objections to the program and to enlist union cooperation in the program.

Little attention has been given to selecting the trainees, since the program is usually the result of some detectable deficiency in the trainees themselves. Nevertheless, the trainees should be personally interviewed to ensure that they are aware of the reason for the program and the possible benefits to be derived from it. Management should take this as an opportunity to impress upon the craftsmen that the program is an attempt to help them do their jobs better, and the interviews should result in high motivation on the part of the trainees. Care should be taken that the trainees do not find a negative indication in being asked to participate in a training program. There are few craftsmen or maintenance managers who would not benefit from a training program.

In order to reinforce trainee motivation, consideration should be given to increasing the wages of trainee incrementally as he successfully completes specific steps in the training program.

No discussion of training would be complete without calling attention to the relationship between maintenance standards and training. It was emphasized in the discussion on maintenance standards that one of the purposes of such standards is for training. When a craftsman is given a work order for a task he has not previously performed, the standard would indicate to him the manner in which the job should be done. Repeated applications of such a standard would result in the craftsman performing the job in the correct manner—that is, with training.

Are Training Costs Hidden?

When new maintenance employees are hired and placed directly in the work force, management obviously assumes that they will learn by on-the-job training. If such a trainee is 50 percent effective when he starts the job, then 50 percent of his salary should be attributable to training, not to maintenance costs. Different trainees and craftsmen learn at different rates; at some time in the future, most trainees become 100 percent effective, after which time the entire salary of that trainee should be attributable to maintenance.

A learning curve similar to that shown in figure 10–5 should be drawn for each group of craft trainees so that their percentage effectiveness can be plotted and so that maintenance is not unduly burdened with training costs.

If an abnormally large number of trainees are introduced into the

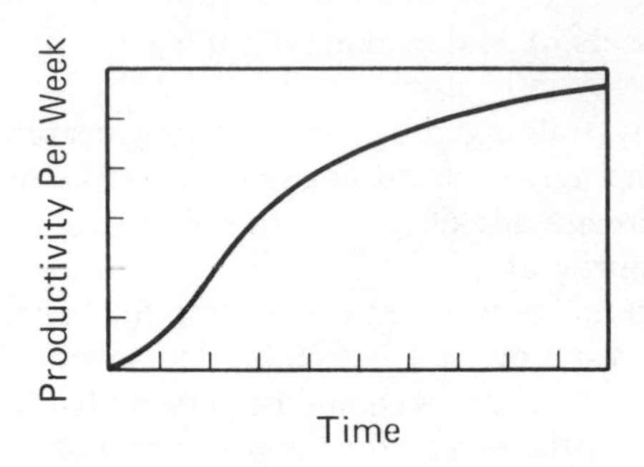

Figure 10–5. The Usual Worker Learning Curve

maintenance program, the effectiveness of that program is artificially lowered by the inclusion of training time in regular maintenance time. This must be considered when a productivity survey is used to monitor the productivity of the work force. It should always be kept in mind that most maintenance departments are continually in the training business.

New Concepts in Multicraft Training[3]

The plant in this example works on the development of new chemical processes, primarily those related to organic polymers, and covers approximately thirteen acres. The maintenance and construction department, which is made up of 78 men in addition to the supervisors, provides services for more the 500 people. The plant is nonunion.

Dr. George Odiorne of the University of Michigan says that any job can be separated into three parts. One part is the normal routine work, the second part is problem solving, and the third is innovation. Normally, maintenance training programs are intended to teach only the first part of a job—the routine work. The aim of the plant in this example is to teach a man to solve technical problems as well as to handle the regular craftwork and, eventually, to give him an opportunity for innovation.

Steps in Advancement

A man who is to help with technical problems and who must be innovative must have a great many more skills than those required of a common laborer or even an ordinary journeyman craftsman. In our example, a man hired as a maintenance helper may have no more and perhaps even less than the skills provided by a high-school education. Over approximately a four-year period, he is trained to the point where his skill level will be very similar to that of a journeyman in a union plant. He is then classified as a development utilityman, first class.

To reach this classification, however, he must have more skills than are usually required for a journeyman in his craft; for example, the skills required by a first-class development utilityman (pipefitting) are

1. pipefitting: (a) steamfitting, (b) plumbing and heating piping, (c) oil and gas piping, (d) air-conditioning piping, (e) chemical piping;
2. plumbing: sanitary;
3. millwright: (a) rigging, (b) equipment installation;
4. carpenter: (a) forms, (b) remodeling, (c) millwork, roofing, builders' hardware, plastering;
5. insulation;
6. masonry: lime, cement, mortar, foundation, brick work, concrete construction, concrete estimating;
7. painting;
8. yard maintenance;
9. lubrication;
10. miscellaneous construction;
11. plant layout: main lines, sewer, steam, water, gas, nitrogen;
12. cutting and burning; and
13. installation of process equipment.

In addition, the man should be able to understand, operate, install, and repair the materials and equipment listed. When a company says it has multiple-craft training, usually it means that a man has acquired skills similar to those listed; that is, a pipe fitter is able to do a little painting, insulating, rigging, and so on. For a man to become a first-class development utilityman, he not only must be proficient in one of the skills listed, he also must have worked, at one time or another, in at least two other skill areas so that he can be moved to another skill area if the work-load demands it.

In most plants, this is where advancement stops, unless the maintenance man becomes a supervisor, but a man should not be kept from advancing further if he can learn other skills and acquire other technical knowledge. So a method of advancement beyond the first-class classification is provided; the next step is to the engineering-technician classification. To become a technician, a man must have one major skill area, but he must also have knowledge of two other skills—at least at average journeyman level. In other words, a pipe fitter would also have electrical or mechanical skills, for example.

In addition to this, the man would have started to learn the engineering concepts in his major area; for example, a pipe fitter would begin to learn something about mechanical engineering or civil engineering or chemical engineering.

From this point, a man can advance to the next classification, which is engineering assistant. Emphasis is still on the multiple-skill concept, but with increasingly greater attention to the technical side.

The next step on the ladder is classification as a senior engineering technician; for example, if a man had majored in electrical construction, electrical engineering would be his major technical area, and he would be able to do the skill work or the problem solving or the creative engineering needed in the electrical field. His salary would then be equivalent to that of a graduate engineer with at least five years' experience.

Thus, a man can work his way up from helper to junior maintenance engineer; but he need not stop even there. The goal of plant management is to increase the man's involvement and his ability to use his brain. He is told that the plant is going to operate on Douglas McGregor's Theory Y, which holds, among other things, that the ability to do creative work is much more widely distributed in the population than most managers suppose. Every man in the plant, from the janitor up to the vice-president of development, has had one week or at least three days' training in this management theory.

The plant is no longer interested in having the man do only normal routine jobs. He is to use his brain to the fullest, to be part of "our" company, and to take part in "our" problem solving—because he is trained.

Techniques Used

How do you train a man? The techniques may not all be new, but the important point is to use whatever will best upgrade a man; for example, as part of the program of providing multiple skills, men are actually moved into other trades. If a man is a pipe fitter, he may spend as much as six months to a year working as an electrician. Then he may rotate into the welding department—again for six to twelve months. He knows that, by learning other skills, he increases his opportunity for advancement.

Another training approach is called engineering training. In this program, various maintenance men are brought into the engineering department, where there are about ten engineers to do the design work. The number of engineers in this group is the same as it was eleven years ago, despite the fact that the plant has doubled in size; the group is able to handle the increased engineering load by having maintenance men do more and more of the routine engineering.

The maintenance man may be assigned directly to an engineer to do the detailed drafting necessary, to order miscellaneous parts, to expedite parts into the plant, and to perform any miscellaneous calculations he can handle. Once the parts begin arriving, the maintenance man returns to the shop, where he becomes construction technician on the job and is responsible for installation, even to the point of ordering any parts he may have forgotten. In this way, a maintenance man can become aware of the problems that arise in designing a job.

In addition to the training provided by the engineer, after-hours training courses are conducted by engineers one day a week, to be taken on a man's own time. Such courses as the following are given: algebra; rotometer sizing; drafting; capital money and how it's acquired; horsepower-torque relationships; relief valve–rupture disc sizing; heating, ventilating, and air conditioning; estimating construction jobs; techniques and principles of heat transfer; corrosion; pressure-vessel design; and pump and steam-jet repair and installation.

In-house schools are utilized. Using the services of Employee Development Services, a subsidiary of International Correspondence Schools, specific after-hours training courses are set up. Courses are given in such subjects as report writing, basic chemistry, electrical-blueprint reading, basic math, measuring instruments, and unit operations—again geared to the process industry.

Another method used more and more is programmed instruction. At present, this consists mostly of courses provided by Du Pont. Some of the courses being given are mechanical seals, field sketching, steam traps, centrifugal pumps, static electricity, single-state steam jets, electrical schematic drawing, measurement and layout, vacuum pumps, positive-displacement pumps, and oxyacetylene welding.

Various correspondence schools are used. The men take the courses on their own, but the plant provides them with an opportunity to learn the practical skills. If a pipe fitter is taking electrical training, he is rotated into electrical work.

A great deal of use is made of the manufacturers' schools and the in-plant seminars given by manufacturers. When a new type of instrument or a new piece of equipment comes in, plant engineers conduct a seminar on its use and repair.

Various local colleges are also used. Finally, the plant uses foremen as trainers. A variety of training methods are used because no one method solves all problems. The training is geared to the idea that each man can solve company problems and be innovative if he has the knowledge to back him up. The individual concepts and training discussed are not new, but the concept as an entire package *is* new.

Accurate performance-oriented evaluation is an absolute essential if the program is to work.

Notes

1. This section is reprinted from "Performance-based Training," W.H. Weiss, *Plant Engineering,* May 14, 1981, pp. 90–92.

2. This section is reprinted from "Training Maintenance Supervisors," L.V. Smith, *Plant Engineering,* June 26, 1980, pp. 99–102.

3. This section is based on D.J. Barnick, "New Concepts in Multi-Craft Training," *Techniques of Plant Engineering and Maintenance* 20: 87–89. Copyright 1969, Clapp and Poliak, Inc.

11 Contract Maintenance

A plant's usage of contract maintenance can vary from very slight to complete. A plant might contract its roofing, painting, and perhaps other minor items, such as grass cutting, or 100 percent of its maintenance, including supervision and engineering. In some cases, plants use contract maintenance as a way of buying the know-how that is brought to the plant by the contractor.

For a plant to have this full range of alternatives, it must be in a favorable geographical location, where the capabilities exist to furnish these services. There are a growing number of such locations in the country today.

More rather than less involvement appears to be most attractive to new plants, since contracting all or a significant portion of their maintenance would not include changing an existing system. Under these circumstances, both the owner and the contractor can develop their procedures during the start-up phases of the operation. From a practical standpoint, contract maintenance in general is more applicable to a new plant, particularly if the owner is thinking in terms of total contract maintenance. Existing in-plant union agreements, laws, and court and NLRB rulings make it highly unlikely that an owner can find a set of circumstances in which he can realistically consider total contract maintenance in an established plant.

In a few areas of the country, the owner has a choice of using union or nonunion contractors. In general, however, with the exception of a very few specialties, the owner can consider only union contractors, because essentially all construction is done by the AFL-CIO Building Trades Crafts. Therefore, only union contractors, in most areas, have access to the skilled manpower available from the various crafts and have over the past several years been able to develop maintenance know-how.

Thus, for the purpose of further discussion, we will be thinking in terms of maintenance for new facilities, with labor being supplied by union contractors from the AFL-CIO Building Trades Crafts. Within this framework, what alternative or choices are available to plant owners?

1. The owner can have his own in-plant maintenance force performing all maintenance work—both day-to-day requirements and peak loads. Until recently, this alternative accounted for the great majority of plants.

2. The owner can use his own in-plant maintenance force for all day-to-day normal requirements but bring in contractors' forces to handle major peaks—that is, emergencies, turnarounds, and the like. In this case the contractors' forces would be building-trades craftsmen working under construction terms and conditions.
3. The owner can have his own minimum in-plant maintenance force performing only part of the day-to-day requirements and use a contractors' maintenance force to do the balance of the day-to-day work and to handle peak requirements.
4. The owner can elect to have no maintenance working personnel of his own and rely entirely on contract forces to provide all maintenance personnel for normal and peak maintenance needs.

Cases 3 and 4, which include contract personnel on a continuous basis, fall in the category by which the work can reasonably be performed under the General Presidents' Maintenance Agreement. Hereafter, we will restrict our discussion to work done under this agreement.

The Whys and Hows of Contract Maintenance[1]

In these competitive and rapidly changing times, businesses and industries are all deeply concerned with maximizing profits—make as high a return on investment as possible. When business is stripped down to its simplest concept, profit can be increased in only two ways: (1) higher production and in turn higher sales and higher average prices for products or services, in other words more income; or (2) by holding the line on or reducing costs of any and all kinds—less out-go.

Plant maintenance can have significant effects on both of these items. Plants kept in good condition can provide higher product output per unit of time and can also enjoy a higher service or use factor—both adding to increased production and in turn higher sales; therefore more income. Efficient maintenance operations can reduce direct maintenance costs and therefore reduce out-go.

Industrial managements in seeking new ideas to improve the quality of maintenance and reduce maintenance costs have developed the concept of a comprehensive contract maintenance program.

The General Presidents' Maintenance Agreement

Contracting out of part of an owners' maintenance requirement is not new. There has been contracting for specialty or turnaround work for a good many years, but usually on a supplemental or intermittent basis only. Back about 1956 or so, the concept of continuous contract maintenance began to emerge. The General Presidents (the top officers) of all the AFL-CIO Building Trades Crafts, in cooperation with national contractors, in particular one such contractor, developed a national agreement called "Project Agreement for Maintenance by Contract." This is a labor agreement setting forth the terms and conditions under which a signatory contractor can secure manpower for a specific maintenance project from the various building trades local unions.

In general, this agreement was patterned after a typical national construction agreement and provides the following same basic elements.

1. The building trades local unions will be the source of manpower, subject to the hiring provisions in existence between the local unions and the contractor.
2. Wage rates and fringes, i.e., health and welfare and pensions will be the same as those in effect for each craft under the construction agreements.
3. The contractor, as in the typical construction agreement, can discharge day by day via the reduction in force route if work is not available, without respect to seniority but rather on the basis of the qualifications and capabilities of the individual workers. On the other hand, the contractor can secure, as needed, additional personnel on relatively short notice.

Although this agreement is patterned to a substantial degree along the lines of a construction agreement, several significant advantages were provided for maintenance:

1. The contract is between the International Unions and the individual contractor rather than between the contractor and local unions. Contract administration and grievances are not handled at a local level but instead between the contractor and an International representative of the Presidents' Maintenance Committee, or ultimately, if necessary, face to face between the contractor and the Presidents' Committee.
2. This agreement provides that there will be no work stoppage of any kind, even during strikes over construction wages during negotiations by one or several crafts.
3. It provides for time and a half for all overtime except Sundays, and seven recognized holidays each year. The typical construction agreement provides for double time for all overtime. It further provides that there will be no travel pay as is sometimes the case in construction agreements.

4. It permits short-term shift work and also provides for continuous round-the-clock shift operations as necessary without excessive premiums.
5. It gives the contractor complete freedom to determine the number and type of foremen and supervisors required.
6. It permits the use of mixed crews for crossing of traditional jurisdictional lines and permits the use of craftsmen and foremen as available to get the necessary work done.
7. It leaves the owner the right to assign to the contractor only that portion of his maintenance work that he desires.

The Advantages of Contract Maintenance

1. The ability to fluctuate manpower to exactly meet day-to-day needs permits expending less man-hours per year to perform a given amount of work. This eliminates employment of full-time personnel to handle peaks and reduces the tendency for built-in featherbedding. It permits scheduling certain work such as painting to take advantage of seasonally good weather.

2. Seniority is not part of contract maintenance. Craftsmen must compete with fellow workers to retain their jobs. In normal times, a job in the permanent nucleus of a maintenance force is considered to be a choice job by most construction workers. They usually get 40 hours per week, some overtime and, of course, construction rates. They produce to hold this spot. In normal times, some managers feel that a plant can develop a more productive work force with contract labor than under average "in plant" conditions.

3. A plant can meet the extraordinary manpower needs for start-up without overmanning for normal operations and should work less overtime during start-up.

4. Since building trades craftsmen already have the basic skills of their craft, they require less training than the typical in-plant maintenance employee to reach a given level of efficiency.

5. Unlimited manpower permits scheduling of turnarounds, revamps, etc., to best fit overall plant economics. There is no need to attempt to fit these to the availability of a given number of in-plant workers. Downtime can also be shortened by using more people to the extent practical.

6. If the owner elects, he can pass on to the contractor the responsibility for essentially all or a significant part of his plant maintenance. This can leave more of the owners key personnel with more time to concentrate on process operations—the making of more and better products. This does not mean that maintenance will be forgotten or become a step child, because the maintenance contractor has a real economic incentive to do a good job—any maintenance project is part of his bread and butter.

7. The owner can draw on the contractor for additional help for services over and above normal maintenance work. The contractor—if he is the right one—can provide additional supervision as needed for turnarounds. The contractor can also provide intermittent supervision in specialty areas such as rigging, critical equipment inspection and repair, etc. The owner can use additional spot help for take-off, estimating, sketching, buying, etc., and still keep his own staff at a constant level.

8. One advantage that is often talked about is that the owner can reduce his investment in facilities, buildings, tools, and equipment. This is not a significant advantage. Over a period of time it appears that change houses, offices, etc., for contractors' personnel approach the norm for the owner's process people. Shops of various types can be kept minimal, and the larger, infrequently used equipment is available for rent by the owner from the contractor or others, regardless of "contract" versus "in-plant" maintenance. There may be some savings on investment in tools if the contractor supplies them, but this is not substantial.

9. The owner can shift essentially all responsibility for labor relations for maintenance personnel to the contractor.

10. The owner can start with contract maintenance and if not satisfied with the results, he can phase into an in-plant maintenance set-up—if he starts out with an in-plant force, he is by and large stuck with it. If he is not satisfied with his contractor, the contractor can be changed. This has been done in a few cases.

To summarize the advantages of contract maintenance in just a few words we can say that the owner has a "wide degree of flexibility" substantially more so than that enjoyed with a typical in-plant maintenance force.

The Disadvantages of Contract Maintenance

Industrial maintenance management can't look at only one side of any issue. What are the pit falls—the problems?

1. Some "in-plant" conditions permit the use of a "refinery" or "plant" mechanic who is not bound by jurisdictional lines. While under contract maintenance, historical construction craft lines, by and large, prevail except in emergencies.

(a) It is not the intention of the Presidents' Committee to lock in on jurisdictional lines to the extent of generating significant inefficiencies or feather-bedding. They clearly recognize that craft lines can not be held as tight on maintenance as on construction. On the other hand, they do not expect gross, and unnecessary disregard for craft lines. To avoid trouble, a contractor needs to use good judgment and be consistent in craft assignments.

(b) The degree of rigidity on craft lines will vary from contractor to contractor and from project to project. This is because different people are involved—different superintendents, area supervisors, craftsmen, stewards, business agents, etc. This is not the fault of the agreement or concept but results largely from a lack of thorough training, good judgment, communication, and leadership on the part of certain contractor personnel.

(c) With craft lines, good and detailed planning is an absolute necessity. If you have all of the right people (the right crafts), materials, tools and equipment in the right place at the right time, you can have craftsmen working at their specialty without delays or squabbles over jurisdiction.

2. On occasions, the "skilled" craftsmen are not so skilled—particularly when an area is faced with an extremely high level of construction. In some cases, maintenance employees have been attracted back to construction by substantial overtime, by offers of supervisory positions, and by special deals or incentives on the part of construction contractors. The owners contribute to this by not controlling overtime and their contractors.

3. A related problem has been that of manning up for substantial peaks. During periods of high construction, about the only way a contractor can attract substantial numbers of men for relatively short periods of time is to work overtime—enough to meet or beat overtime that is being worked on construction. Although the contractor and owner, for other considerations, often elect to work some overtime on turnarounds, some feel that during high construction periods, they are working more overtime than normal to attract enough capable men. Here we see a conflict between construction and contract maintenance brought on by unusual boom conditions. This situation will be improved and return to normal as the construction level tapers off and as training programs turn out more skilled people.

4. Some owners feel that they cannot go all the way on contract maintenance because the contractors and unions cannot provide skilled supervision and craftsmen with reasonable back-up for certain specialties. This has been true for instrumentation—particularly electronic. This, of course, does not preclude contract maintenance but it has been a deterrent, and a good many owners have elected to develop their own instrumentation force. Certain contractors and the unions concerned are aware of this and similar problems and are taking steps to provide these services.

5. Contract maintenance wage rates are tied to construction rates and, on the average, construction wages have risen substantially faster than in-plant maintenance wages, including fringes. Most owners watch this development carefully. They like the flexibility that goes with contract maintenance, but if the price tag on contract labor gets too high in relation to industrial rates they cannot or will not be able to afford the luxury of contracting maintenance.

Mechanics of Contract Maintenance

What are the mechanics of contract maintenance? By this we mean what steps and thought processes should an owner go through from the time he decides he wants to consider contract maintenance until he is actually underway. In logical sequence these are:

1. Most maintenance managements usually have a background and experience for in-plant maintenance but need to gather comparable knowledge on contract maintenance. The two best sources for this information are other owners using contract maintenance and maintenance contractors. After as much information as possible is gathered, a careful analysis will determine what is really applicable in the individual plant's case. The more contractors contacted, the more need to do this.

2. Next, the owner must make a comparison of the advantages offered by contract maintenance over having his own forces. However, caution and complete objectivity must be exercised because certain intangibles are being dealt with and certain assumptions must be made. Such a study should consider the start-up period, but particularly the norm for the long haul.

3. If, after an overall evaluation of dollars and cents and all other considerations, it is concluded that contract maintenance offers advantages, one should then attempt to further refine the evaluation and decide on the pattern to be followed. It should also be decided who will do and be responsible for scheduling and planning, the extent the contractor will be involved in cost accounting, whether he will be wanted to provide and staff services such as buying, drafting, etc.

This tentative pattern for operations should be established so that when taking bids or negotiating with a contractor the general scope of work can be outlined. It will also help to think about the type and size of contract organization needed.

Selection of a Contractor

Selection must be made for the long haul for a cost-plus type arrangement. The contractor and the owner must become a team. The following major points should be considered.

1. The integrity of the contractor.
2. The experience of the contractor(s)—Is his experience real? Has he performed bona fide maintenance or has he been a labor broker with owners shouldering the responsibility? What is the extent of his experience?
3. What does the contractor offer in terms of people? This is a most important consideration.

a. Does he have an adequate staff to back up his field forces? Are these people experienced on contract maintenance?
b. Who does he offer as his job superintendent? Has he been in the contractor's employ prior to this? Does the contractor really know his own capabilities? Is he experienced and skilled in supervising maintenance work?
c. Can the contractor offer enough skilled or assistant supervisors? What is their experience and background? Will he have to hire unknowns?
d. Can the contractor offer adequate supplemental and/or specialty supervisors?
e. What is the contractor's situation with regard to the several unions involved—particularly at the local level? A contractor has the protection of the Presidents' Agreement as discussed earlier but we must remember the various locals arc still his source of manpower.
    i. Does he have a good knowledge of the individual craftsmen—those capable of being foremen? Those with special skills, etc.?
    ii. Are his company and key people really respected by the unions and their members?
    iii. Is he inclined to deal in short-term expediency or does he live by the labor agreements and stand up to be counted when necessary? Does he get what he is entitled to under his labor agreements but still deal fairly with labor?
f. How complete is the contractor's service? Does he really have services to match your needs? If not, are you willing to split responsibility? Does he have necessary equipment, tools, facilities to support your operation?

It is difficult for the owner to get all the true and complete answers to these questions. Some contractors are prone to exaggerate when talking about their capabilities. The owner needs to probe deeply and take a "show me" attitude. Be sure you get the true story before drawing any conclusions.

One good measure of a contractor is his overall effectiveness. Has he been able to consistently perform repeat work for most all owners in the area? Does he get a fair share of lump-sum work? Does he make good profits on lump-sum work? Most owners should ask for financial information, not only to be sure the contractor is sound, but also to measure his progress in terms of profits. The ability to consistently make good profits is conclusive evidence of a well-managed organization.

Taking Bids or Negotiation. If the owner indicates he is considering contract maintenance he will be besieged by contractors—both local and national. By its very nature, maintenance work must be done on a cost-plus-percentage-fee basis. The owner is entering into what very likely will be a long-term agreement. The fee must be high enough to permit the contractor to provide the necessary attention and supervision to assure a quality service and still make a reasonable profit. It must be remembered that a few percentage points on fee can be lost many times over by the poor performance of field personnel.

About 90 percent of the owners go the bid route. This enables them to give several contractors a chance to present a final bid. The owner cannot be accused of favoritism. But probably the main reason is that it gives owner's representatives a set of finite figures to support their selection with their top management and auditors. They do not have to base their decisions entirely on intangibles. One word of caution—since it is unlikely that several contractors will really have equal capabilities, this can set the stage for having to award the work to someone other than the best qualified firm or have the rather difficult job of explaining to management why the contract was not awarded to the low bidder.

If bids must be taken, be sure to do the following:

1. Set a clear precise basis for bids. Do not let each contractor set his own format for bids. Make sure not to have the job of trying to compare apples and oranges.
2. Make sure that contractors understand that the owner reserves the right to award the work to the contractor of his choice—who might be other than the low bidder. And take closed bids—this leaves some flexibility in evaluating fees versus capabilities.

After the contractor is selected he must be given a letter of intent so that he can appear before the general Presidents' Maintenance Committee, explain the project, and secure the Presidents' Agreement for Maintenance by Contract. It is then time to develop and sign a formal contract covering the work.

Develop Organization and Procedures

Early in the game, when the owner decides to consider using contract maintenance, he should set a schedule to arrive at this point with adequate time left for his personnel and the contractor to do a thorough job of getting ready to go to work.

At this time, the contractor and the owner's representative must become a team—they must work close together and do a good and complete job of communicating. Some of the more obvious jobs are

1. Prepare employee information booklets: These set forth in simple layman's language, the more important contract provisions, work and safety rules, plant regulations, etc. They are prepared jointly by the owner and contractor and must be approved by the Central Presidents' Maintenance Committee before being printed and issued to workers as they are hired.

2. Owner and contractor must firm up their organizations: Key people must be assigned all the way down the line.

3. Clearly define the scope of contractor's responsibility and authority: Channels of communcation must be set up.

4. Develop and reduce to writing such things as:

Safety and plant security rules

Work procedures

Tool, equipment, and material procurement procedures

Scheduling and planning procedures

Payroll and cost accounting procedures

5. Start training and familiarization of contractor's and owner's personnel—top supervisor; other salaried supervisors; foremen (screen and select carefully): These supervisors must be given a chance to study and learn all the rules and procedures, mentioned above, and to learn their units, get training on any completely new or specialty equipment, etc.

6. Start to work: The contractor must select the craftsmen carefully and make certain they are thoroughly indoctrinated in the concept of contract maintenance. The quality of the initial group of craftsmen will exert substantial influence on the climate and productivity of the entire work force in the future.

The Contract[2]

The maintenance manager is ultimately responsible for the success—or failure—of a maintenance contract. He must avoid duplication between the staff and support functions of his own group and the functions of the contractor's organization. The maintenance manager cannot abdicate responsibilities for control or motivation, however, and he will find that the usual people problems do not leave when the contractor comes in—in fact, they may become more complex.

The maintenance manager will know he has realized the promise of contract maintenance when he can see that the contractor has not been relegated to a position of plant whipping boy but is an active, sought-after partner in a multigroup team.

The contract should be somewhat narrative, rather than "do the job as specified," as found on the purchase order form. This does not mean that every eventuality should be covered in storybook form. The contract will be well written if it clearly states what the maintenance manager and the contractor expect of each other. Points to consider in the contract are as follows:

Statement of Work. List where the contractor is expected to perform his tasks, the type of work (repair, replacement, minor new construction, and so forth), any supplemental activity (such as engineering or purchasing), and the methods used to request the work.

Staff Assignments. Specify how the owner will approve staff changes. In some detail, describe how the maintenance manager and the contractor will determine hirings and firings. This should include handling of resumés and joint interviews with prospective staff.

Term of Contract. The wording of termination clause should be tempered. It will probably have a 60- to 90-day written-notice requirement from the plant to the contractor, but the contractor should be given a feeling of security. He should be able to justify in his own mind (and to his management and employees) that he is there to stay. This will encourage investment on his part, in both material and human terms. We recommend a three-year minimum contract with an open-ended maximum. The contractor should give a minimum of six months' notice of intent to leave, and termination should take place only on the anniversary of the contract.

Either party should be able to reopen negotiations on certain sections of the contract with consent of the other. Although termination and its reasons are an unpleasant, negative aspect of a contract, it should nevertheless be well covered in precise, legal terms. Each partner should have no doubts concerning the other's understanding of the termination clause.

Records. The records the contractor is to maintain should be described in detail. The manner of their availability should also be specified. Typical records include cost reports, such as hours, wages, and equipment-maintenance records; OSHA safety and accident reports; and payrolls. If the plant uses electronic data-processing equipment, similar facilities should be available to the contractor. Combining the needs of both parties on one machine and program will usually reduce costs.

Training. If the normal contractor crew is 100 or more men plus staff, use of a training coordinator should be considered. It is always difficult to decide who pays for training. We recommend that the owner pay for the time of the people in training and be selective about who is approved to receive the instruction; and the contractor should pay the cost of instructors and supplies. The contract should clearly indicate how any training monies accruing from outside sources, such as the Veterans Administration and local and federal government reimbursements for certain types of instruction, will be distributed. If both parties do not believe in the value of training, it is best to remove training from the contract, as it will be a source of possible future conflict.

Owner Obligation. A statement of services and facilities provided by the owner is another mutual-trust builder that a contract can provide. The contractor should participate in development of this section. He will usually request shop facilities, change area, smoke area, and the like. A critical function is attention to safe-work permits to release process equipment for work; neither party should leave this responsibility entirely up to the other. Purchasing and expediting spare parts and materials are best kept under the owner's administration. He should make available a complete and current listing of all spares and commodity materials normally stocked and have the contractor check it. Any sizable operation should include a full-time materials representative-expediter from the contractor's staff within the owner's materials group.

Insurance: The plant's legal department will want to define all liability-insurance requirements deemed necessary, together with limitations and amounts. Consulting the plant insurance underwriter will probably disclose coverages that can be combined at a lower premium with those the plant already has in effect. Legal liability of property damage to third-party equipment under the contractor's control, care, and custody is an important coverage that should be included. All policies should be attached to the contract, as an appendix, within ten days after the contract is signed.

Secrecy Agreement. The owner should provide that both the contractor and company personnel will make no use of any information obtained from or pertaining to the plant or its equipment, processes, or procedures. Agreements with salaried people should include the period during and after employment for a specified time. This can cover inventions and patents developed by the contractor pertaining to the owner's property. Such inventions are usually assigned to the owner.

Compensation. It is best to state the various bases for compensation in a separate annex to the contract. Thus, the economic facets can be upgraded and modified without rewriting the entire document.

In addition to the normal statement of payment for wages and salaries with the various burdens, expenses, overheads, and other direct costs, several critical factors should be considered—for example, legal fees, tools and toolroom costs, expendable supplies, absenteeism, and special award fees based on performance.

The best way to curb excesses and motivate the contractor's local management is to have the top management in the contractor's home office exercise the necessary control. This is far more effective and usually less expensive than employing a large monitoring staff on the company payroll.

Since tool use and tool loss are generally a function of manpower, one can arrive at a tool fee based on percentage of straight-time labor billed; this cost will average 2 to 4 percent and may be higher if attendants and tool repair are included. If the contractor is diligent in proper tool use and surveillance, he will see a profit. Documentation of his tool costs should be provided for the plant's records. This will be a help in future negotiations.

Expendable items, such as rags, gloves, and brushes, may also be covered by this formula with good results.

Performance Award. A list should be developed of performance elements that the owner, his staff, and the contractor consider to be important indicators of good maintenance. Three or four basic divisions—such as planning, management, craft skills and turnaround execution—can be broken down into subelements. Weights should be assigned to each element, with a total of 100. A simplified completed rating sheet of one major division, turnaround performance, is shown in table 11-1. The four elements are

Table 11-1
Basic Division, Turnaround Execution
(percentage)

Element	*Weight*	*Rating*	*Score*
Planning and scheduling	25	80	20
Work-force balance	25	70	17.5
Supervision	25	90	22.5
Quality of work performed	25	90	22.5
Total			82.5

Source: Ronald J. Grey, "An Owner Looks at a Maintenance Contract," *Hydrocarbon Processing,* January 1973, pp. 65-67.

Value = 30% of whole

Score = (weight) × (rating); total score for turnaround execution = (82.5)(30) = 25.75%.

each weighted equally, and the owner has filled in his rating. A total score of 82.5 percent has been given to the contractor for the period covered. Since turnaround execution is merely one division, the total score is again multiplied by the value assigned that division. Here, the value is 30 percent. The resulting 25.75 percent value is added to scores obtained on rating sheets covering other divisions to obtain the whole. The final cumulative score will be something below 100 percent and can be used as a basis for an additional performance bonus.

Each owner will want to develop his own set of standards, with as many as ten elements under each major category. He can use direct labor billed as a basis for calculating the award bonus and can publish a table equating, for example, a range of bonus payments from 0 to 2 percent against a performance score of 70 to 100 percent.

A very real benefit derived from this concept is the resultant meeting of owner-contractor management teams two or three times annually to discuss performance scores. Small problems or trends are pointed up, discussed, and corrected early.

Self-regulation is paramount in establishing all fees, burdens, and overheads based on percentages of wages and salaries paid. Fair payments should be developed for excellent performance, and contractor management will control excesses or poor performance to improve earnings.

Legal Fees. The payment of legal fees arising out of hearings, claims, third-party suits, petitions, and citations is a point that is very seldom treated in contracts of this nature. They usually become a problem during the life of a contract.

Generally, litigation costs can be divided into several categories that also define payment responsibility. Fees that exclusively benefit one party or arise from negligence of either the owner or the contractor should be paid by the party responsible. Fees of a more difficult nature are cases of mutual responsibility.

Legal expenses associated with union activity, such as unfair labor practices or dismissals, organizational drives, and NLRB-supervised elections are in many ways a joint necessity. As such, their possible occurrence should be noted in the contract, and payment on the basis of amount of responsibility should be agreed upon. Both parties should thoroughly discuss and agree on each legal step before action is taken.

Nonlegal contractor-union affairs should be the sole responsibility of the contractor's management staff, with problem-handling ability a part of his overall performance rating.

Contractor Company Policies. Any company policies that affect expenses directly billable to the owner by the contractor should be an annex to the basic contract. These may then be modified as required without rewriting the contract.

Factors to Consider When Evaluating Contract Maintenance[3]

A temporary contract arrangement or one on a scheduled basis can provide the extra hands the plant needs to take care of emergencies or service breakdowns of complex equipment [see figure 11–1]. Such an arrangement permits the regular maintenance force to take care of the daily needs of the plant.

Each plant engineer must carefully evaluate the abilities of his personnel to determine whether they can handle the equipment involved under normal conditions and during possible emergencies without any serious production losses. In his evaluation, the plant engineer should also compare the costs of doing the work inhouse to the cost of having it done by an outside contractor [figure 11–2].

The plant engineer has four choices open to him regarding plant maintenance:

He can have his own maintenance force perform all daily and peak-load maintenance work. Many plants operate with such a plan.

He can use his own maintenance force for the day-to-day requirements only and bring in a contractor's force to handle peak loads such as emergencies and special operating equipment, or turnarounds.

He can have his own maintenance force perform part of the daily requirements and use a contractor's maintenance force to do the balance of that work and to handle peak requirements.

He can elect to have no maintenance personnel of his own and rely entirely on contract personnel to take care of the plant's maintenance needs.

Types of Contracts

If the final decision is to hire a contract maintenance service, various programs can be considered. Three types of contracts or agreements are in general use.

Labor and Parts Agreement (Guaranteed Contract). This type of agreement includes all labor and parts to keep the plant equipment in operating order. A list of some of the parts provided with such an agreement might include pulleys, relays, thermocouples, solenoid valves, fans, motors, belts, pressure valves, strainers, transformers, refrigerant, oil burner pumps, thermostats, and limit switches.

Inplant	Contractor
Men are more familiar with plant equipment.	Service is flexible.
Workforce is available immediately to handle most maintenance problems.	Large force can be employed to get job done faster.
Workers can be offered overtime to take care of some types of emergencies.	Workforce can be more versatile to handle many jobs.
Workers may not have skills experience on sophisticated equipment.	Workers have skills and experience to tackle complicated jobs.
Other essential maintenance jobs will be deferred when emergencies or special jobs require immediate attention.	Workforce does not have to be reduced after job is completed.
If more maintenance personnel are hired, it may be difficult to reduce workforce after project is completed.	There are no seniority problems.
Additional expense is involved if outside service tools are required to repair equipment or handle project.	Contractor's personnel supplement normal workforce, permitting plant staff to carry on regular maintenance work.
	Specialty tools and equipment are provided by contractor.
	Close supervision and control are not possible.

Source: R.L. Marinello, "Contract Maintenance Services," *Plant Engineering,* March 5, 1981, p. 50.

Figure 11-1. Comparison of Maintenance Services

	Regular Time	Overtime	Double Time
Labor	$16.75	$25.73	$33.50
Payroll taxes	1.56	2.34	3.13
Union welfare	2.52	2.52	2.52
Auto-Truck	3.00	3.00	3.00
Insurance	0.52	0.52	0.52
Serviceman's uniforms	0.17	0.17	0.17
Holiday/Vacation pay	0.67	0.67	0.67
Shop burden rate	0.18	0.18	0.18
Total cost	$25.37	$34.53	$43.69
Contractor's charge	$40.59	$53.50	$71.25

Source: R.L. Marinello, "Contract Maintenance Services," *Plant Engineering,* March 5, 1981, p. 51.

Figure 11-2. Hourly Contract-Maintenance Labor Costs

Ductwork, insulation, recording instruments, thermometers, boiler tubes, storage tanks, smokestacks, and heat exchangers might be excluded. Other items that may be excluded are

Changes, repairs, or corrections to equipment because of design, government code, or insurance requirements.

Water, drain, steam, and electrical lines beyond the equipment itself.

Problems caused by freezing, contaminated water, or atmosphere.

Electrical power failures, low voltage, burned-out main or branch fuses, low water pressure, or work of others.

Anything else both parties agree to.

This type of agreement is normally offered only on equipment under 5 years of age, preferably beginning at the end of the first year's warranty on new equipment.

The number of visits to perform inspection service should be part of this agreement. Frequency should be based on equipment use, its application, the length of heating or cooling season, and the risk of loss of product or materials that servicing of the equipment is to prevent.

Circumstances determine whether weekly, monthly, or quarterly inspections are essential. The more inspections, the higher the cost of the contract. Emergency service should be included in such a proposal.

Labor Only Agreement. This type of agreement provides all the features mentioned in the previous contract—except for parts.

Preventive Maintenance Agreement. This type of agreement provides inspection services outlined in a check list, but excludes emergency service, except at extra cost. This type of agreement is usually the most economical contract for older equipment and may include discounts on parts, labor, and emergency service based on quantity of equipment.

Preliminary Steps

Before a contract agreement is entered into, the following events should take place:

Survey. The contractor or his service manager should take an inventory of the equipment that will be under contract. The plant engineer or maintenance manager should take the contractor on a tour of the facilities, showing him all the equipment that will be involved. The contractor can gather data on the make, model, serial number, age, size (Btu, hp, capacity, throughput, etc.), and general condition of the units.

The results of this preliminary inspection enable the contractor to determine what type of contract would be suitable.

Presentation and Acceptance. The contractor should prepare a standard equipment service agreement, identifying what services are to be provided and their frequency, along with any other options that may have been agreed upon. It is important that the maintenance manager review the agreement to see that all items are included and that specific instructions as to how they should be inspected are spelled out in the contract.

The options could include water treatment for boilers and cooling towers, chemical cleaning of these items and condenser coils, overtime labor (if not included in contract), painting of equipment, and changing of air filters.

After a thorough review of all the pertinent points, it is up to the maintenance manager to accept or reject the proposal.

Service

Upon acceptance of the contract, the contractor should begin providing the required services within a few days.

When Service Starts. Normally, maintenance manager can anticipate that a fully qualified serviceman will arrive to perform a first-rate inspection during the initial visit. He will perform all necessary repairs and adjustments on each item listed in the guaranteed or "parts and labor" contract. It is in the contractor's best interest that the equipment be placed in the best operating order, to preclude a premature breakdown whose cost he will have to absorb.

If the agreement is for "labor only," the maintenance manager can anticipate the same service, except that necessary repairs will be listed, and the plant will be charged for the cost of the parts he agrees to have replaced.

If the agreement is for preventive maintenance, initial service will be performed by the contractor's servicemen. Repairing defects that are noted will cost the plant extra in labor and parts to correct. Any contract should spell out specific services the maintenance manager expects to receive.

Using Contractor as Trainer. If some members of the maintenenace staff have good working habits, are keen observers, and are seeking additional challenges, the maintenance manager may wish to consider only a 1 year contract. During this period, the maintenance manager should assign some of his more skilled craftsmen to serve as trainees to assist the contractor's serviceman during his visits. In that time, the trainee cannot be expected to become a qualified service technician, but he should learn enough of the basics so that some of the options in the contract can be performed inhouse.

Usually the contractor is not especially eager to change filters, clean and flush equipment, or paint it. He may be quite willing to adjust service costs downward if such tasks are performed inhouse, leaving his technician free to perform the more complicated work.

Cleaning condensers, compressors, permanent filters, etc., is another task the contractor may not mind relinquishing. A relatively small investment in a portable pressure cleaner and detergents will benefit the plant, and it can also be used for other cleaning tasks.

Contract Service. An outside contractor can provide services for an industrial plant in many ways. He can offer a general maintenance service that includes a review and audit of the organization of the maintenance department, the maintenance standards and procedures, general housekeeping, and the way emergencies are handled. He can evaluate procedures relating to the installation, operation, and maintenance of machinery. He may also offer inplant inspection techniques, scheduling, record keeping, and maintenance and testing procedures for boilers and power generating systems.

Inspection may extend to the complete structural examination of the facility, including rebar detection in concrete, underwater examinations, examination of timber and wood construction, load test certification, and corrosion studies. Inspection methods could include radiographic and ultrasonic, magnetic particle, and liquid penetrant testing, and visual examination.

Many contractors' service shops are equipped with all the required tools to put the plant's equipment back in service in the shortest time possible. There are, however, times when the equipment requiring repairs is so large that it would be difficult to send it out for repair. When the size of the equipment, its weight, or its stationary position prohibits such removal, on-site machining may be the only option. In such cases, the contractor will send his personnel to the plant to evaluate the problem and discuss how the repairs could be made. When the maintenance manager has selected the most practical repair option, the service shop will provide a complete range of repair tools to take care of the repairs right in the plant.

Selecting a Contractor

There are several avenues open to the maintenance manager in the selection of a good contractor. One possibility is to check with other plant engineers in the area to find out what contractors they are using and the degree of success they are having. Or, he may learn of experiences of other managers at the monthly meetings or seminars he attends.

Another possibility is the manufacturer of some of the more sophisticated equipment operating in the plant. The manufacturer may provide such a service for his equipment or may be able to suggest someone for that service. Many manufacturers of electrical equipment (motors, generators, transformers, and switchgear), boilers, air conditioning units, and temperature controls (calibration) have service shops throughout the country and can provide such a service on a contract basis or on as "as needed" arrangement. Any production or facility equipment that would cause the plant to shut down should be of primary concern to the maintenance department. If such equipment is beyond the skills and abilities of maintenance personnel, the maintenance manager should find out what types of services are provided by the manufacturer.

Many engineering and construction firms provide a maintenance contract service for all the operating equipment in a plant. Such services are offered mainly for refineries, chemical plants, and power generating stations, but such services are also available for any type of manufacturing facility. These firms can supply specialists in planning and scheduling, rotating equipment, piping, rigging, and electrical equipment. They are capable of handling all equipment maintenance, emergency repairs, capital revisions, overhauls, turnarounds, and other required services. Some of their systems incorporate sophisticated record keeping and maintenance control procedures, preventive maintenance programs, and emergency work-handling techniques to ensure that all work is performed logically and efficiently.

One other source would be to examine a directory containing the names of various contractors' associations to select those of specific interest to you. You can then contact these associations for the names of contractors in or near your area.

Another source for names of contractors is the local telephone book. After the names of contractors who offer the desired services have been listed, a few phone calls can tell if further correspondence or contact with those companies is warranted.

Judging Contractor's Abilities

How can maintenance management judge the quality and ability of the contractor? There are three important characteristics to look for in the services of a maintenance contractor: technical expertise of personnel, local availability, and service reputation. Pricing policy, experience, equipment repair facilities, and speed of response to service calls should also be considered.

Besides the maintenance manager, the maintenance supervisor, engineering supervisor, or electrical and maintenance engineers might get involved in the selection of a contractor.

Annual preventive maintenance agreements, two or more year preventive maintenance agreements, and service on an "as needed" basis are the types of contracts normally selected. Other types would be for special frequency on specific equipment and for emergencies.

Many "as needed" contracts are placed by a purchase order. It is important that the plant engineer or other responsible person spell out all the details related to the equipment being serviced. The plant and the contractor must agree to all conditions set down and to the cost.

One major consideration before signing any contract is to make sure that the contractor has insurance coverage and that such information and the amount of coverage involved become parts of the contract. Accidents that occur in the plant as a result of actions of the contractor's personnel should be covered by his insurance. Contractors should carry ample insurance in the following areas so that the plant is not held responsible for any accidents to the contractor's or company's workers or damage to plant equipment: workmen's compensation and occupational disease insurance, employer's liability insurance, comprehensive general liability insurance (bodily injury and property damage), and contractor's protective liability insurance. It may be advisable for the plant to take out an owners' protective liability insurance policy to protect itself from subcontractors that may be used but are not covered under the contractor's insurance policies.

Guarantees to take care of failures because of poor workmanship of defective materials furnished by the contractor should also be required. The maintenance manager should obtain a written guarantee for the period and services provided.

A penalty clause to protect the company would be advisable if the installation has to be completed by a certain time. It can serve as an incentive and also protect the company against costly delays. However, it may be detrimental in pushing the contractor to rush the job, resulting in poor workmanship. The final decision to include or exclude such a clause should be based on the reputation of the contractor and the criticality of the project.

No matter what type of contract service is selected, the maintenance manager should assign one person on his staff to review periodically all reports submitted by the contractor. With such an arrangement, it is possible to monitor the progress of the work being performed and to call attention quickly to costs that may be getting out of line.

Other Considerations

To avoid possible labor problems, the maintenance manager should give the plant's maintenance workforce all the work it can handle, including overtime, before using a contractor. The maintenance manager should review the contract the plant has with its maintenance force. Some plants may not use contractors except for construction of major new facilities: others may have restrictions that must be met.

The main objective is to accomplish the task on schedule, within cost limits, and without any disruption from the labor force or the contractor. Every effort should be made to eliminate any disputes by notifying plant workers and the union committee about the work to be performed by an outside contractor and why. Such communication can minimize disputes, improve the attitude of maintenance personnel, and reduce friction when the contractor's personnel move into the plant to do their job.

Can You Motivate Contract-Maintenance Workers?[4]

When plant management elects to employ an outside maintenance contractor, the final choice is usually based on reputation, experience, record of performance, scope of services, and labor relations. Two often neglected areas that should also be considered in evaluating a contractor are: methods of motivation and productivity measurement.

For contract maintenance to be profitable, the contractor must maintain a facility more efficiently and more economically than can the client by using his own in-plant work force. To accomplish this, the contractor must strive for the optimum utilization of his work force, which largely depends upon proper motivation of his craftsmen and measurement of their productivity.

Motivating the maintenance craftsmen to function smoothly can be achieved through the application of the following principles:

No Seniority. Perhaps the prime motivational factor is that, under the concept of contract maintenance, seniority is nonexistent. The maintenance contractor is free to retain the best qualified workers: complacent or habitually nonproductive workers may be terminated during a reduction in force. The employees who are forced to compete with fellow workers to retain their jobs, usually produce more per man-hour worked.

Planning and Scheduling. A detailed planning and scheduling . . . program of maintenance work will not only motivate but will also prove to be most economical.

Fundamentally, the program requires the following:

1. A disciplined work-order system defining the scope of work requested by the operating department provides that a priority be established to determine urgency and time requirements. Once this information is received by the mechanical department, a definite estimate is prepared, including job sequencing, man-hour estimate, material and tool requirements, special permits, and all other pertinent information associated with the work task.
2. A daily work scheduling system is required to control the manpower balance with relation to the backlog. This control affords maximum coordination between management and the work force and minimizes the loss in man-hours that can result from a lack of information.
3. Reporting and control of expanded man-hours by work order should be set up.
4. Management control to determine the effectiveness of the daily work schedule is required.
5. Cost reporting on a daily basis is the key to cost control.

Planning and scheduling eliminates idle time and the need to create work. Also, the lack of proper tools and equipment can easily discourage workers. Faulty organization, along with a constant improvisation for daily needs of inexpensive material items, tends not only to discourage, but also has the worker questioning management's and his supervisors' organizational abilities. . . .

Poor Housekeeping. Studies have revealed that poor housekeeping and inadequate facilities can hinder motivation. For example, an ill-furnished, shabby-looking locker room can be depressing. Such seemingly minor concerns can have serious effects upon morale and interest in the job. The employee may accept these as indications of a lack of interest in the worker by management.

Safety Programs. Safety and first-aid are other significant motivational aspects. . . . Safety programs should be comprehensive and include project inspections; frequent review of safety procedures and safety equipment; safety training; and provisions for doctors, nurses, hospitals, and ambulances.

Training Programs. A very positive approach to stimulating a worker's interest in his assignment is through training programs. . . . When contract workers come to a facility, they are already highly skilled. . . . Most contractors usually conduct training programs to acquaint craftsmen with a manufacturing facility's particular equipment. These programs can be held on- or off-site, with length dependent upon the equipment complexity.

Communications. Organization, efficiency, and motivation are contingent upon effective communications. For maintenance craftsmen, communications are divided into two classifications: work and personal or employee.

At the weekly meetings, foremen and stewards should discuss matters pertinent to these two areas. Under the category of *work* would fall safety, major projects, job progress, special job instructions, and business activities. The *personal or employee* category would include company programs, grievances, tardiness, absenteeism, etc.

Issues discussed should be conveyed to the craftsmen by their stewards. Employee complaints and grievances should flow from the aggrieved employee to the shop steward and then to management.

Who Motivates? Contract maintenance craftsmen are primarily motivated by management's team of staff supervisors. The degree of motivation is wholly dependent upon management's professional ability.

Motivation by foremen is also necessary. Qualified members of the basic maintenance crew may be temporarily promoted to foremen to direct and supervise expanded crews at peak periods or shutdowns. Once the shutdown is completed, these men revert to craftsmen's rates and status. It is therefore of paramount importance for a cadre of the basic force to have supervisory aptitude. This is achieved by thorough screening of the entire work force, with final selection based on optimum quality and skills.

Research in Contract Maintenance[5]

It was determined that there are four key areas that management should explore before deciding whether or not to use contract maintenance: (1) labor relations; (2) comparative costs; (3) quality of work; and (4) flexibility.

Labor Relations

Definite conclusions concerning labor relations were these:

There were no serious labor relations problems. The findings negated one of the prime objections to the introduction of contract maintenance—the fear that labor relations problems will follow. The companies surveyed, believed that the biggest labor relations problem was the difference in wage rates, but most did not consider it a serious one.

Contract maintenance will not cause unionization of a plant. Ten nonunion plants were represented in the survey; none has been organized because of contract maintenance. Thus the fear of unionization appears to be unfounded.

There have been very few work stoppages under contract maintenance. Only one respondent disagreed with this. However, respondents were not asked to state the length or seriousness of the few strikes encountered.

It is easier to get rid of poor workers. The survey confirms case study findings that a major advantage of contract maintenance is that it makes it easier to screen and cull the workforce.

In-plant unions will not prevent good contract maintenance. Eighteen companies, or over 60 percent of the respondents, had union-organized plants. Only one felt its union had made the proper use of contract maintenance impossible. Human relations problems do arise, and they require astute handling, but they can generally be resolved.

Comparative Costs

Surprisingly, the cost studies that would yield quantitative and unassailable conclusions have not been made—or at least have not been released—by companies using contract maintenance. From the financial analyst's view, therefore, it cannot be proved that contract maintenance is the cheaper alternative.

Respondents largely agreed, however, that there is very little "make work" in contract maintenance, and only three of twenty-eight felt that contractor's profits were too high. Moreover, they agreed that with contract maintenance, less money was spent on "frills," or the niceties of life. The majority opinion that wage rates were not too high indicates a feeling that other savings offset the wage differential.

Quality

Clear-cut conclusions on work quality under contract maintenance were as follows:

Contract maintenance offers an opportunity to secure a workforce of good quality. When management capitalized on the opportunity to weed out the misfits, work quality was excellent. Where no winnowing was done, work was less satisfactory.

Effectiveness depends on the quality of supervision. Good supervision is critical to success, yet this is an important factor rarely considered and little has been written about it. However, there was overselling on the quality of the supervision, which turned out to be less adequate than the contractors had promised it would be.

Contract maintenance is no automatic guarantee of quality work. Tradesmen hired for contract maintenance are not better per se than those recruited for in-plant maintenance forces. They lack the specific experience that the regular employees acquire in time, but they often compensate for this by bringing a greater breadth of experience to the job.

Flexibility

The biggest advantage of contract maintenance is flexibility. The ability to recruit and dissolve a large workforce quickly was most often cited as the major advantage of contract maintenance. Other advantages are that (1) contract maintenance is applicable to more than turnaround work, and (2) it is hard to cut down an in-plant maintenance force.

Recommendations

A major threat to the continuing success of contract maintenance lies in the escalating wage rates of construction workers. Excessive wage rates are not a basic need in contract maintenance because the seasonal factor is not great. Once the ever-increasing wage rate gap becomes unreasonable—despite the advantage of flexibility—contract maintenance companies will begin to lose their present clients and find it almost impossible to attract new ones.

The General Presidents' Committee, composed of international union presidents, developed the *Project Agreement for Maintenance by Contract* in 1956 and revised it in 1960. It is this agreement that a company becomes a party to when it undertakes maintenance by contract. The General Presidents' Committee should be strengthened; its direction and control are needed to maintain good labor relations. If work-force flexibility is to be maintained, project agreements must continue to be made with the international unions rather than with the locals.

There is a need, too, for an association of the employers who use contract maintenance. Interchange of knowledge and experiences would be helpful to all. Management should also be represented in deliberations between the Presidents' Committee and construction management.

Future growth of maintenance by contract will depend on open dialogue between union and company managers and, above all, on the maintenance of a fair balance between the annual incomes of contract-maintenance workers and those of their counterparts in in-plant work forces.

Notes

1. This section is from a paper presented to the seminar, "Engineering Maintenance Control," Louisiana State University, by Bert S. Turner, Nichols Construction Co., 1966.

2. This section is based on Ronald J. Grey, "An Owner Looks at a Maintenance Contract," *Hydrocarbon Processing,* January 1973, pp. 65-67.

3. This section is reprinted from R.L. Marinello, "Contract Maintenance Services," *Plant Engineering,* March 5, 1981, pp. 50-55.

4. This section is reprinted from F.P. Flesca, "Can You Motivate Contract Maintenance Workers?" *Hydrocarbon Processing,* January 1971, pp. 87-90.

5. This section is reprinted from James H. Jordan, "Research in Contract Maintenance," *Techniques of Plant Engineering and Maintenance* 19:253-254. Copyright 1968, Clapp and Poliak, Inc.

12 Managing Maintenance Craftsmen

How the Foreman Can Improve His Human-Relations Skills[1]

Since the foreman is really a first-line manager the title might well be "How can we, as managers, improve our human relations skills?"

What do we mean by human relations skills? Human relations skills are the skills that enable us to work effectively with others in a variety of situations to achieve long- and short-term objectives. Thus human relations skills comprise not only the interpersonal skills, but also the ability to integrate task needs with people needs in the work situation. I have often heard people say that good human relations means keeping people happy and maintaining harmony in the work group. Such an idea is naive at best, and it can be quite misleading. Improving human relations can involve a lot of sweat and tears but it also can lead to personal growth and better productivity.

Questions for Self-Examination

In achieving more effective human relations, a basic starting point is self-examination. Let's ask ourselves the following questions:

1. How much do I want to improve my human relations skills?
2. How effectively am I working with others?
3. What are the various ways in which I behave in the many situations I face?
4. To what extent do I keep myself open to feedback from others that can lead to my continued growth?
5. How am I growing with the help of and not at the expense of others?
6. What am I doing to measure my human relations effectiveness?
7. How do I try to identify and improve relations with those who don't understand me?
8. How frequently do we openly share ideas for improving productivity in the work group?
9. What am I doing to increase my understanding of myself and others?

We may find some of these questions tough to answer without some soul searching. However, as we dig to find the answers we are laying the foundation of improving our human relations skills.

Another phase of self-examination involves our basic assumptions about people, for these assumptions essentially represent our value systems. For example, which of the following statements do you agree with?

1. Most people dislike work and will avoid it if they can.
2. Most people must be coerced, controlled and threatened to get them to put forth adequate effort to achieve organization objectives.
3. Most people prefer to be directed and wish to avoid responsibility.
4. Most people have relatively little ambition and want security above all.
5. Expenditure of physical and mental effort in work is as natural as play or rest.
6. Man will exercise self-direction and self-control in the service of objectives to which he is committed.
7. Most people can learn, under proper conditions, not only to accept but to seek responsibility.
8. The capacity to exercise imagination, ingenuity, and creativity in the solution of problems is widely, not narrowly, distributed in the population.
9. Under conditions of modern industrial life, the intellectual potential of the average person is only partially utilized.

You may have found these statements controversial. How did you react to them? Your reactions can provide you with insight into your own attitudes and motivations. If you agree with Statements 1 through 4, your ideas are like those of managers who tend to use authoritarian methods. Statements 5 though 9 embody the ideas of the foreman or manager who believes in participative practices.

Value Systems and Employee Reactions

Value systems have an important influence on the effectiveness of a manager's human relations, for his expectations of how others will behave are often self-fulfilling prophesies. *People tend to behave toward you as you expect them to behave.* This is because your own behavior toward others is determined by the way you expect them to behave, and your behavior evokes the expected response.

For example, if we, as foremen and managers, believe that our employees are lazy, irresponsible, untrustworthy and uncommitted, we will devise strategies and standards that give employees very little leeway in working toward the goals we have set, and this may discourage any commitment on their part. On the other hand, if we believe our employees are capable, imaginative, trustworthy, and committed, then we will share our problems and our goal-setting with them and so stimulate their willing contributions.

Which kind of self-fulfilling prophecy are you working on? As a reformed autocrat I have often been pleasantly surprised when I have trusted others enough to let them use their abilities more fully.

Thus, if you are sincere in your desire to improve your human relations, you may have cause to take another look at your value system. Are you finding ways to encourage your people to use more of their intellectual potential on the job? Do you welcome their ideas for improving various aspects of the work situation? How many of their ideas have you recently used?

There is a graphic device to help us visualize our relationships with others. The *Johari window* illustrates the dynamics of one person's interactions with others:

	Known to Self	*Not Known to Self*
Known to Others	1. Area of Free Activity	2. Blind Area
Not Known to Others	3. Hidden or Avoided Area	4. Unknown Area

Improvement in human relations occurs when the area of free activity (Number 1) is expanded so that there can be freer exchange of ideas and feelings, more testing of issues and ideas, and more open sharing of problems. How can we expand this area? Of the approaches available, the one we ourselves can use most readily is to attempt to reduce the hidden or avoided area.

When you reveal something that has been troubling you and blocking your relationship with another person, that person feels free to bring out some of his concerns. Since you may have been ignorant of what he reveals, he will be helping you to reduce your blind area.

For example, one of your foremen may feel reluctant to come into your office to discuss his problems. Since he is one of your better foremen you have often wondered why he has communicated so little and so seldom with you. Finally he comes in to see you. He reveals that he has been reluctant, if not somewhat fearful, of coming to you with problems that are not directly related to getting the job done. He goes on to speak of his concern about his own self-improvement and personal growth. By revealing his fear and reluctance to discuss personal problems with you, he has reduced his hidden area in the relationship. At the same time he has helped to reduce your blind area by letting you know that you appear so involved with your tasks that you are difficult to approach with personal problems. If you are perceptive enough to see this, then you both can share the problem openly and consequently increase the area of free activity in the relationship.

Another approach is to try to increase understanding of our own behavior and the behavior of others. We should ask ourselves: (1) Who understands me well, and who doesn't understand me? (2) Whom do I understand well and whom do I understand the least? Frequently the people whose behavior puzzles you the most are those most perplexed by your behavior.

Sensitivity training is an effective means of increasing insight into our own behavior and the behavior of others. This type of training uses a laboratory approach that provides participants with the experience of creating a productive work group in an unstructured situation. In the process of developing the necessary leadership and the functions, participants get a clear view of their own ways of handling human relations problems. Sensitivity training provides an opportunity to learn, often for the first time, how our strategies for working with people *are seen by others* and how they *affect others.*

In addition, many participants learn that there is wide diversity of leadership styles—autocratic, directive, consultative, and nondirective—and that different situations may demand different attitudes.

Motivations

In attempting to improve human relations skills it is helpful to understand what motivates us and others in the work situation. One way to get at this is to ask "What do we want from our jobs?" Answers to this question usually include the following: Good pay—more money; security—job tenure; opportunity to use our own ideas; good working conditions; good working relationships—with our boss, our peers and our subordinates; an opportunity to develop skills; and opportunity for advancement; challenging work; a variety of tasks; a sense of making a worthwhile contribution; personal job satisfaction.

Each of us may not want each of these things to the same degree. However, *each of us wants some of them.* . . . From the standpoint of human relations, it is revealing that a list made out by subordinates is almost the same—they want the same things managers want.

Dr. A.H. Maslow describes man as having not only physiological needs, but needs for all the following: safety, belonging or affection, esteem, and self-actualization or self-fulfillment. He states that about 85 percent of most people's physiological needs are satisfied, but only 70 percent of their needs for safety, only 50 percent of their need to belong, only 40 percent of their need for esteem, and only 10 percent of their need for self-actualization.

When we compare the things we want from our jobs with Maslow's list of needs we find a definite relationship. . . .

So by comparative analysis we find that what we want from our jobs includes the entire list of man's needs. This is why the job situation can be such a potentially powerful source of motivation. *How can the foreman or supervisor harness these motivating forces? He can do so by getting the employee involved in the process of defining his responsibilities, identifying performance indicators, and establishing performance goals.*

. . . The employee and the supervisor have common points of reference for evaluating the employee's performance. More important, one of the supervisor's own goals is the development of his subordinates, and the process of setting performance indicators and goals helps the subordinate develop self-improvement goals. One way to test whether the supervisor is meeting his responsibility for development of subordinates is to examine the number and quality of the plans for the employee's self-improvement that have been established by the supervisor and the employee together.

A very potent approach to improving human relations skills is a participative work-improvement program. This type of program has strong motivational impact because it effectively integrates the task needs and people needs in the work situation.

When supervisors and foremen use this participative approach, employees actively seek to improve their performance, for they find that their ideas are welcome. Consultation with the people most directly concerned with the job improves communications, cooperation, coordination, and department relations. Most important, when employees are stimulated to generate improvement ideas, they find their work more challenging, develop better relationships with their boss and their peers, and gain increased job satisfaction by making worthwhile contributions. Foremen and supervisors can test the effectiveness of their participative work-improvement program by examining the number and quality of improvement ideas submitted by their employees.

What are some ways that can lead to continuing improvement in our human relations skills?

1. We must be dedicated to examining our effectiveness in working with others.

2. We need to keep ourselves open to feedback from others.
3. We must work at identifying those who don't understand us and at improving communications with them.
4. We must develop more ways of behaving effectively in a variety of situations.
5. We need to grow by helping others to grow.
6. We must be willing to set an example of participative leadership to stimulate employee improvement and development.

The path to effective relations is difficult, but it is rewarding in terms of better employees, better productivity, and long-term profit growth.

Can New Shift Schedules Motivate?[2]

The 1970s have seen an increasing number of HPI [hydocarbon processing industry] plants switching to a 12-hour-per-day shift schedule. Many others have been seriously discussing the implications. During the next few years, many more plants will confront the possibility.

The American Petroleum Institute's Refining Department conducted a survey on shift schedules and hours of work. The objective was to identify the advantages or disadvantages of the 12-hour schedules versus the traditional 8-hour schedules.

Questionnaires were sent to 961 plant managers in the United States and Canada. The 326 replies received represent a work force in excess of 66,000 employees on rotating shift schedules.

The questionnaire was designed to compare the industrial aspects of the schedules such as absenteeism, industrial accidents, grievances, overtime experience, etc. It did not cover the less tangible but equally important areas such as impact on family life or long range effects on health of the work force.

The answers to the questions support two theories:

1. That the vast majority of the work force who have been exposed to the 12-hour schedules are pleased with the results.
2. That management experiences more difficulty with overtime call-in on the 12-hour schedules.

When comparing the experience of 8-hour shifts to 12-hour shifts for the entire population [table 12–1], the 12-hour schedules look very promising. When these comparisons are made by geographic location, Canada is the only area where the 12-hour shifts do not compare favorably. Canada is also the area with the longest experience on the 12-hour schedules. Whether

Table 12-1
Eight-Hour versus Twelve-Hour Shift Data

	8-Hour Shifts	*12-Hour Shifts*
Number of employees	58,113	8,266
Average number of hours worked in the work week	40.6	40.8
Average number of hours per year of short term absences	40.4	36.5
Average number of hours per year of extended absences	84.8	67.2
Accident frequency rate	9.6	7.2
Accident severity rate	138.4	69.4
Grievance frequency rate	8.4	2.0
Percentage of spares	6.8	5.3
Percentage of overtime worked	7.9	9.2

Source: L.H. Campbell "Can New Shift Schedules Motivate?" *Hydrocarbon Processing,* April 1980, p. 252.

this is a major factor is not conclusive. However many individual locations have expressed the view that the initial favorable experience deteriorates with time. Experience alone will either confirm or disprove this theory.

When the absenteeism is broken down into districts, the increased absenteeism is almost entirely confined to Canada.

When compared with the length of time that people have been on the 12-hour schedules, it was found that there is a relationship between absenteeism and the length of time that the schedule has been in operation. The poor experience is still mainly confined to Canada where 12-hour shifts have been in use for the longest period of time.

If one makes a further comparison and considers the type of shifts (for example, average hours in the work week/or shift integrity), a more definite pattern emerges. It would appear that the type of shift has a marked effect on absenteeism experience and that the most undesirable schedules fall into the 38 to 40 hours per week range, and consist of shifts in which workers and supervisors do not rotate as a team.

The possibility of increased overtime is the most likely undesirable item that can exist on 12-hour schedules. Although more than 1/3 of the locations experience an increase, a similar number have experienced a decrease in overtime. Thus, the prospects are not completely negative.

As one would expect, the survey confirms that absenteeism and overtime are directly related. This suggests that one method to avoid increased overtime would be to institute a schedule which discourages excessive absenteeism.

Interestingly, locations that have experienced a decrease in overtime are experiencing the same difficulty with overtime call-in, as those locations with increased overtime.

On the whole, accident experience is encouraging. However, even the slightest deviation from former experience must be seriously studied.

Four of the five locations that experienced an increase in accidents also show an increase in overtime. The fifth location shows a decrease in overtime. However, their percentage of overtime worked is one of the highest of all the 12-hour schedules.

Another notable point is that seven of the eight locations who experienced a decrease in overtime are working 42-hour-per-week schedules that incorporate shift integrity. Six of these locations also show decreased absenteeism.

For each of these areas of concern (absenteeism, overtime, industrial accidents) and the area of plant efficiency, the statistics are encouraging but more detailed research and analysis is necessary.

To complete the survey, the task group later conducted an information session. The session consisted of four presentations by firms who had experienced the transition from 8 hour to 12 hour schedules. These presentations were used as a catalyst to stimulate discussion. Attendees represented organizations using both types of schedules.

Particular interest was noted when some attendees with operations using the 12-hour schedules described:

1. That Union members rather than supervisory personnel were used to "telephone" employees to replace absent workers.
2. How peer pressure policed absenteeism.
3. Supervisors and hourly workers voted together to try the new schedule.
4. That shift integrity (supervisor and employees remain as a team) increased the likelihood of success.
5. How supervisors are most likely to weed out poor performers if the supervisor rotates with the team.

The attendees agreed that changing to 12-hour-per-day schedules appeared to help morale but that it was only a part of a larger management climate needed to foster increased motivation and productivity.

Improving Relations with Unionized Craftsmen[3]

If supervisors and managers are to do an effective job of improving relations with unions and craftsmen, they need answers to some fundamental questions, such as: What is a union? Where does a union fit into the scheme of things? What is my responsibility in this area? How important am I as a manager? How and what do I communicate? What is my responsibility for employee discipline?

Each manager and supervisor should begin by looking at and recognizing the men working for him. The employees in the factory today are very different from the employee of ten or twenty years ago. Supervisors must recognize this fact. Today's employee is better educated, has more information at his fingertips, and has tastes that run to the so-called good life. In most cases, the employee is looking toward and can afford a living standard similar to that of the manager or supervisor. Managers have to consider this or they are going to have problem relationships with employees.

Politics versus Economics

A union is a group of employees who have banded together in an attempt to improve their lot with the company. Since they cannot all meet their employer at one time, they elect officers and representatives to do the job. These elected representatives must please their constituents. To be reelected, they must provide some type of service for their constituents. Therefore, when they look at a problem, they look at it from the standpoint of "How can this help me to look good to my constituents?"—that is from a political standpoint. A member of management, however, normally looks at a problem from an economic standpoint. When a man concerned with economic problems is placed in the arena with a politically motivated adversary, agreement may be difficult.

Both parties to a labor contract, although they work under the same roof, are looking at the solutions to problems from a different point of view. Many recognize that the union representatives are politically oriented; however, in analyzing a problem or attempting to improve relations with craftsmen and union representatives, how many managers have ever thought that this could be the reason for some of their problems? It is important to understand why the union and its representatives act as they do.

The Immediate Supervisor

The immediate supervisor's position in employee relations is most important. The union puts particular emphasis on the supervisor's part in handling employee relations. A labor contract is negotiated with the union for a period of time, but what happens between the negotiating periods is what determines the effectiveness of the contract. If the supervisor is not alert and does not do his job from a labor-relations standpoint, the contract will crumble. Eventually, the foreman will not be an effective member of management. Effective contract administration rests with the supervisor.

Effective contract administration builds good employee relations. If the supervisor abdicates his role to the union representative, then the employees will look to the union representative. If the supervisor allows the union representative to bully him and run roughshod over him, employees naturally will turn to the union representative. The supervisor as an effective member of management is the key to improving and establishing good employee relations. Therefore, in order to accomplish results in this area, a supervisor should know the individual employees, understand the union and how it works, and, finally, understand his role, his responsibility, and his job.

Pointers for the Supervisor

Regarding how the supervisor can improve relations with his employees, the following suggestions are offered:

Honesty with employees is of the utmost importance. Sometimes it is hard to be completely honest: It is not a crime to make a mistake—the crime is in trying to cover up the mistake and refusing to admit it. If you are wrong, admit it.

Making decisions is the supervisor's job. Cooperation does not mean comanagement. A supervisor should always make the decisions that affect his employee relations. He cannot share this responsibility with the union representative.

Periodically, a supervisor has to discipline an employee. If the supervisor goes to the union representative to get his permission or to get his ideas on the discipline, it is not the supervisor who is doing the discipline but the union. How can the union representative fully perform his function if he is a party to the decision?

We need discipline. Discipline is an art that seems to be out of vogue today. In discussions with the employees of one company where there were many nonunion operations, it was found that one of their biggest complaints was the lack of firm discipline.

The majority of the employees in the plant may be good, honest, and hardworking; however, if the majority of the employees see a minority getting away with longer breaks, more absenteeism, and so on, naturally they are going to go that way too, even though they do not think it is right.

If all employees are treated fairly but firmly, and if the rules are made known to them, the supervisor's relations with them will improve because the employees will know exactly what is expected of them.

If the employee has a legitimate complaint, admit that it is legitimate. Hearing the employee out is sometimes difficult, but if he has a legitimate complaint, he should be told so, and the complaint should be acted upon promptly. Nothing irritates an employee more or destroys a relationship more quickly than procrastination on a minor point. This is true whether the matter is a formal grievance or a single complaint. If a supervisor states that something will happen and it does not, his reliability in the eyes of the employee is destroyed.

Use good common sense. In many instances, a supervisor wins the battle but loses the war. A classic case is the one in which a supervisor refused to allow an aggrieved employee time off the job to discuss his complaint; since the man drove a forklift truck, the union representative was forced to jog along beside the truck to investigate the complaint.

There are many ways to solve a problem. Some take a little longer than others, but in the long run it may be worthwhile to take the time necessary to build and improve relations with employees. Scheduling vacations is an example; by using a little common sense and a little preplanning, a supervisor can sometimes give an employee the time he wants, even though the union contract may not require that he do so.

Enriching Maintenance Jobs[4]

One of the prime responsibilities of the maintenance manager is to do everything in his power to make the maintenance man like his job.

Job enrichment—making jobs more appealing and satisfying—can be an extremely powerful motivational force in the plant-maintenance department. Experiments conducted by experts on motivation indicate that maintenance men are more productive and more satisfied with their work when the job is enriched; and most successful companies subscribe to the theory that work habits can be greatly improved by enriching the jobs.

Not only do workers enjoy working in a professional atmosphere with adequate facilities and the right tools, they also appreciate a challenge. Although pay, benefits, and working conditions are important, they are not as crucial in motivating people as the challenge of the job itself. In most cases, the worker is interested in his wages more as a reflection of management's opinion of him than as what they will purchase.

Maintenance managers must recognize that today's employees are motivated most by the work itself and by the way management attempts to make that work more meaningful and rewarding. When workers are given more responsibility and training, they have a greater sense of achievement and a better chance for rapid advancement. Enriched jobs create an atmosphere of growth and bring well-deserved recognition to personnel.

A recent survey showed that participation approaches are winning increased acceptance as the way to improve productivity, create job satisfaction, and resolve labor-management problems. A number of successful companies are now using the work-team approach, in which teams of workers know their assignments, choose their assignments, choose their own hours, schedule their own overtime, and arrange most major details of their jobs. With this approach, companies are introducing incentive programs by which groups are rewarded rather than individuals only.

Herzberg has written and lectured extensively about some of the experiments and studies that have been conducted on motivational methods. He found that the things that were once considered motivators do not really motivate, although, if inadequate, they will produce worker dissatisfaction. He named them "hygiene factors."[5] Another authority has called them "maintenance factors."[6] Both phrases connote the importance of preventing rather than eliminating it. Some of these maintenance factors are benefits, salary, company policy, relations with supervisors, working conditions, interpersonal relationships, and security.

Herzberg discovered that, though improving these factors will not make people like their work, it will make their work a little easier to tolerate. The essential ingredient is still missing, however—a long-range satisfier. The missing ingredient can be supplied by giving a man such work motivators as

1. achievement—as he sees it;
2. recognition—associated with achievement;
3. appreciation of the work itself;
4. more responsibility;
5. advancement; and
6. growth in competence.

Basic Concepts of Job-Enrichment Programs[7]

A good maintenance-job-enrichment program can contribute in a very positive way to maintenance performance. For best results, six areas of potential work-satisfaction improvement deserve special consideration: the job itself, achievement, responsibility, growth, advancement, and recognition. Motivation and maintenance needs are listed in figure 12–1.

The Job Itself

Organization. Maintenance-department activities must be planned carefully; maintenance personnel perform best when their daily, routine work is precisely outlined and tools and material are available when needed. In addition, they must know who makes the decisions and who to ask for further instructions or answers to questions.

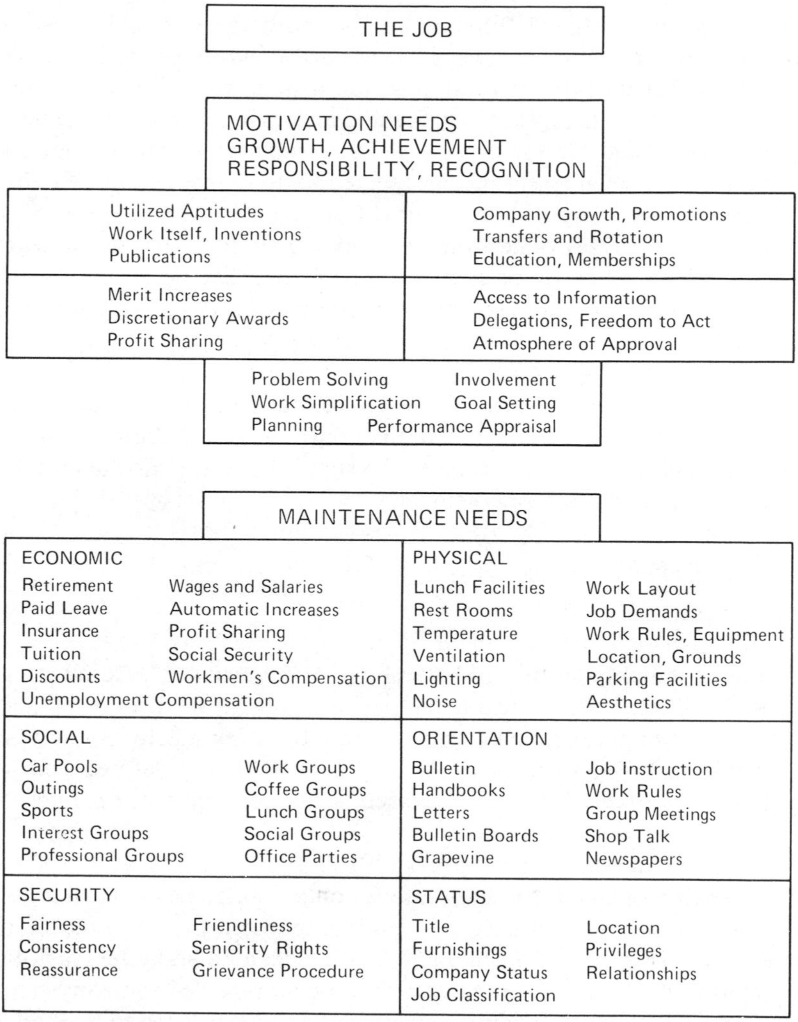

THE JOB

MOTIVATION NEEDS
GROWTH, ACHIEVEMENT
RESPONSIBILITY, RECOGNITION

Utilized Aptitudes Work Itself, Inventions Publications	Company Growth, Promotions Transfers and Rotation Education, Memberships
Merit Increases Discretionary Awards Profit Sharing	Access to Information Delegations, Freedom to Act Atmosphere of Approval

Problem Solving — Involvement
Work Simplification — Goal Setting
Planning — Performance Appraisal

MAINTENANCE NEEDS

ECONOMIC Retirement, Wages and Salaries Paid Leave, Automatic Increases Insurance, Profit Sharing Tuition, Social Security Discounts, Workmen's Compensation Unemployment Compensation	PHYSICAL Lunch Facilities, Work Layout Rest Rooms, Job Demands Temperature, Work Rules, Equipment Ventilation, Location, Grounds Lighting, Parking Facilities Noise, Aesthetics
SOCIAL Car Pools, Work Groups Outings, Coffee Groups Sports, Lunch Groups Interest Groups, Social Groups Professional Groups, Office Parties	ORIENTATION Bulletin, Job Instruction Handbooks, Work Rules Letters, Group Meetings Bulletin Boards, Shop Talk Grapevine, Newspapers
SECURITY Fairness, Friendliness Consistency, Seniority Rights Reassurance, Grievance Procedure	STATUS Title, Location Furnishings, Privileges Company Status, Relationships Job Classification

Source: E.M. Bergtraum, "Enriching Maintenance Jobs," *Plant Engineering,* July 11, 1974, p. 77.

Figure 12-1. The Job

Delegation of Authority. Management must provide each maintenance man with a direct supervisor. Employee morale suffers when personnel other than the immediate superior give orders.

Establishment of Priorities. Job priorities must be established by the maintenance manager. Men should not be pulled off half-finished jobs and reassigned.

Providing Help on Menial Jobs. Laborers and helpers should be assigned to do menial work. Often, however, large maintenance operations employ no laborers. Supervisors on these jobs are usually proud that their maintenance men do all jobs—moving furniture, digging ditches, aligning machinery, and testing electronic circuits. These supervisors fail to understand that trained craftsmen resent being expected to do laborer's work routinely.

Maintainable Facilities. The time to think about maintenance is when a facility is being redesigned or equipment is being ordered. Too often, when maintenance needs are not considered, machines are practically impossible to maintain.

Adequate Maintenance Shops. The greatest motivator of any maintenance man is a well-designed, environmentally controlled, well-illuminated maintenance ship that has modern tools and a supply of spare parts and material adequate to handle most emergencies.

Achievement

Personnel are highly motivated by feelings of achievement. Completing an assigned task successfully instills the desire to do the next job even better. Employees enjoy working on complex and interesting machinery and being responsible for their own particular segment of a well-defined and well-organized program. They are pleased when equipment operates well because of their work.

A good maintenance man enjoys his job more when he is allowed to do certain repair work and installation tasks on a regular basis. He likes to be consulted about the purchase of new equipment, the installation of a machine, or the layout of a new facility. There can be no better way of giving him a feeling of achievement than making him part of company growth.

It is very important to communicate with workers, listen to their advice, and give them a feeling of achievement when they see some of their suggestions develop into successful cost-reduction programs.

Responsibility

Many companies have achieved excellent results by making maintenance departments responsible for their work efforts—setting them up as cost centers, for example. It is desirable to give individuals responsibility for specific areas. In many companies, buildings or areas are assigned to individuals who are then responsible for equipment maintenance, cleanliness, safety, security, and so forth.

The maintenance manager should detail exactly what he expects from each individual in the effort to achieve overall departmental objectives. When expected individual contributions are known, each man will be motivated to attempt to give what is demanded.

Growth

The success of any company depends on its growth, and a maintenance department grows as the company grows. A smart maintenance manager will use company growth to enrich his department. He will promote from within the department when he has to add a foreman. A progressive company has permanent training programs to prepare men for promotions.

Maintenance-department educational programs must go beyond the necessary technical education; workers must learn more about the business their company is in. Such programs can be started with the help of other employees.

Advancement

If responsibility and growth are properly handled, individual advancement will occur naturally. A department that delegates responsibilities and prepares its personnel for growth through meaningful training will be able to advance people, both within the department and by transfer to other growing departments in the company.

A growing company will always have open positions in its maintenance department. As trained maintenance men move into lead positions, chances for advancement will open for helpers and apprentices. Some of the senior people will have an opportunity to advance into maintenance-management positions; others will move into plant or product engineering. Jobs for production foremen will also become available. Often, well-trained maintenance men will move into quality-assurance and safety jobs. Some may have the opportunity to become maintenance managers in a new plant or division.

It is important that jobs be properly defined and evaluated, and salary ranges must be realistic. If at all possible, compensation should be related closely to performance and results. A maintenance department in which performance and responsibility are rewarded is almost always able to retain its employees.

Recognition

In a well-organized company, recognition is practiced daily by the entire management team. Some of management's tools are

1. company newsletters and bulletins;
2. recognition presentations and lunches for employees who have five, ten, fifteen, or more years of service with the company;
3. quarterly or semiannual luncheons for department and company-management representatives;
4. competitive awards for safety, housekeeping, and cost savings;
5. company recognition of promotions and of employees' outside achievements; and
6. visits to shops, introduction of new managers, and the like, by top management.

In maintenance, where employees are often reprimanded by just about everybody for doing or not doing something, any kind word or gesture will enrich the job. The maintenance manager must set an example. A simple "please" or "thank you" is easy and helpful; a friendly greeting, a smile, or a personal question can go a long way. Once a manager has adopted this behavior himself, he can ask that others assume a better attitude. Some maintenance departments have even started such a trend by sending letters of praise to departments that support their efforts. When a maintenance department sets the tone, it may get praise from the departments it supports. A manager can improve the morale of his men by a pat on the back at the right time.

The general idea is to make employees feel important. Because employees in every plant have work-related problems, the foreman who is a good listener and tries to help will receive the cooperation of this workers in return.

Notes

1. This section is reprinted in part from L.G. Fred Heisman, "How the Foreman Can Improve His Human Relations Skills," *Techniques of Plant Engineering and Maintenance* 17:49–51. Copyright 1966, Clapp and Poliak, Inc.

2. This section is reprinted from L.H. Campbell, "Can New Shift Schedules Motivate?" *Hydrocarbon Processing,* April 1980, pp. 249–256.

3. This section is based on C.G. Carlson, "Improving Relations with Unionized Craftsmen," *Techniques of Plant Engineering and Maintenance* 20:84–86.

4. This section is based on Eric M. Bergtraun, "Enriching Maintenance Jobs," *Techniques of Plant Engineering,* July 11, 1974, pp. 75–77.

5. Frederick Herzberg, *The Motivation to Work* (New York: John Wiley, 1967; *Work and the Nature of Man* (Cleveland: World Publishing, 1966).

6. Robert N. Ford, *Motivation Through the Work Itself* (New York: American Management Associations, 1969).

7. This section is based on Eric M. Bergtraun, "Enriching Maintenance Jobs," *Techniques of Plant Engineering,* July 11, 1974, pp. 75–77.

13 Statistical and Operations-Research Applications in Maintenance

This chapter includes a number of maintenance applications of statistics and operations research. The relationships employed here will not be derived, as it is assumed that the reader is acquainted with statistical and operations-research techniques. The purpose of the chapter is to illustrate the application of those techniques in maintenance. The discussion is not exhaustive, in that no attempt is made to cover all applications. Furthermore, it is assumed that the reader has access to traditional statistical tables. (For continuity, see the section on statistical methods in chapter 5 and the discussion of economic order quantity (EOQ) in chapter 7, which is covered in greater detail here.)

Hypothesis Testing

As the first illustration, we know that the average length of time a mechanical seal has lasted in a weak acid pump has been 60 days, with a standard deviation of 10 days. A new type of seal was tried. If a random sample of 11 seals had an average life of 52 days, with a standard deviation of 10.8 days, using the new seals, we will test the hypothesis that the time-population mean (μ) is now less than 60, using a level of significance (α) of 0.05 and assuming the time population to be normal.

Solution.

$$
\begin{aligned}
H_0: \quad \mu &= 60 \text{ days} \\
H_1: \quad \mu &< 60 \text{ days} \\
\alpha &= 0.05
\end{aligned}
$$

Critical Region. $t < -1.812$ (from a table of t values),

$$t = \frac{\bar{x} - \mu_0}{s/\sqrt{n}}, \qquad v = n - 1 \text{ degrees of freedom (d.f.)}$$

where $\bar{x}$ = sample average life

μ_0 = large-population average life

s = sample standard deviation (s.d.)

n = sample size

Therefore,

$$t = \frac{52 - 60}{10.8/\sqrt{11}}, \qquad v = 10 \text{ d.f.}$$

$$t = -2.247$$

Conclusion. Since the calculated t value is greater than that allowed in the table, reject H_0. This means that one can conclude, with a 5 percent chance of error, that the new seals are not the same quality (using life as a criterion of quality) as those comprising the large population.

The following data are the results of two different corrosion tests on a Hastalloy C pump shaft by the standard procedure. We will determine whether the difference between the paired tests is significant at the 0.05 level, assuming that the populations are normal.

	Corrosion Rate, 0.0001 grams	
	Test 1	*Test 2*
Run 1	1.8	2.4
Run 2	1.8	1.9
Run 3	2.1	2.0
Run 4	2.2	1.7

Solution.

H_0: $\mu_1 = \mu_2$ or $\mu_d = 0$

H_1: $\mu_1 \neq \mu_2$ or $\mu_d \neq 0$

$\alpha = 0.05$

Critical Region. For two-tailed α test and 3 d.f., $t > -3.182$ and $t > 3.182$, where

$$t = \frac{\bar{d} - 0}{s_d/\sqrt{n}}$$

Test 1	*Test 2*	d	d^2
1.8	2.4	−0.6	0.36
1.8	1.9	−0.1	0.01
2.1	2.0	0.1	0.01
2.2	1.7	0.5	0.25
		−0.1	0.63

Therefore,

$$\bar{d} = \frac{-0.1}{4} = -0.025$$

$$s_d^2 = [n\Sigma d^2 - (d)^2]/n(n - 1) = \frac{4 \times 0.63 - (-0.1)^2}{4 \times 3} = .209$$

$$s_d = .457$$

$$t = \frac{-0.025 - 0}{.457/\sqrt{4}} = -0.109$$

Conclusion. Accept H_0 and conclude that the tests are not significantly different.

Estimation Theory

In choosing the best brand of equipment for a plant to purchase, it is often necessary to compare performance data (see table 13–1). Using these data, find a 96 percent confidence interval for the difference $\mu_1 - \mu_2$.

Solution.

$$(|\bar{X}_1 - \bar{X}_2|) - Z_{\alpha/2}\sqrt{(s_1^2/n_1) + (s_2^2/n_2)} < d_0 < (|\bar{X}_1 - \bar{X}_2|)$$
$$+ Z_{\alpha/2}\sqrt{(s_1^2/n) + (s_2^2/n)}$$

where $\bar{X}_1$ = average run time in hours, manufacturer 1

$\bar{X}_2$ = average run time in hours, manufacturer 2

Z = normal random variable with mean 0 and variance 1.

α = .04 [since confidence interval = $(1 - \alpha)100$—, so .96 = $(1 - \alpha)$ and $\alpha = .04$]

s_1 = standard deviation of data for manufacturer 1 valve

s_2 = standard deviation of data for manufacturer 2 valve

$\mu_1 - \mu_2 = d_0$

n = number of samples

Thus we obtain

$$(|3{,}600 - 4{,}200|) - 2.054\sqrt{(300^2/52) + (410^2/78)} < d_0$$

$$< (|3{,}600 - 4{,}200|) + 2.054\sqrt{(300^2/52 + 410^2/78)}$$

$$= 473 < d_0 < 727$$

Therefore, we can assert, with a 96 percent probability, that the standard normal variety will fall between 473 and 727.

Regression Analysis

In maintenance, as in other areas of engineering, there are situations in which some variable (dependent variable Y) depends on one or more other variables (independent variables $X_1, \ldots, X_m$). Consider the data in table 13–2. There, the dependent variable Y is the number of hours various com-

Table 13–1
Time between Failures: Pressure-Regulated Valves

	Number of Samples	*Average Run Time (hours)*	*Standard Deviation*
Manufacturer 1	52	3,600	300
Manufacturer 2	78	4,200	410

pressors ran before failure. The independent variables are X_1, the quantity of gas compressed (coded); X_2, the age of the compressor; and X_3, the average years of experience of the previous repair crew.

In collecting the data in table 13–2, it was hypothesized that the length of time a repaired compressor runs is dependent on how much work it did – that is, how much gas it compressed; the age of the compressor; and the average experience of the crew that repaired the unit.

Table 13–2
Time between Failures: Compressors

Y *Run Time (hrs)*	X_1 *Quantity of Gas Handled, Coded*	X_2 *Age of Compressor (yrs)*	X_2 *Average Experience of Previous Repair Crew*
400	200	15.1	12.4
420	300	13.2	16.5
450	380	14.0	12.4
600	450	15.1	18.0
640	480	12.9	40.5
650	500	8.2	18.0
690	520	10.1	12.4
780	890	8.4	18.0
810	960	3.2	16.5
860	980	3.2	38.2
940	1,150	15.1	18.0
945	1,210	6.8	36.1
1,005	1,300	3.2	40.5
1,020	1,470	6.8	36.1
1,080	1,490	6.8	16.5
1,210	1,620	8.2	38.2
1,260	1,690	6.8	40.5
1,310	1,800	3.2	18.0
1,600	2,500	13.2	38.2
1,670	2,600	8.4	36.1
1,800	3,000	10.1	40.5
2,400	4,000	3.2	38.2
2,490	4,800	6.8	40.5

It is realized at the outset, of course, (1) that there may be other variables affecting run life, (2) that some or all of the tabulated variables may have no effect on run life, and (3) that, even though all the variables are tabulated, the weight of each variable on the dependent variable is different.

One way to approach such problems is to use one of the various stepwise-regression techniques that present the data in such a way that the contribution of each variable can be realized. The forward-selection procedure is such a technique in that it begins with but one variable, iterating progressively until all variables have been considered. The procedure accommodates cross-product terms and terms with nonlinear exponents if the analyst so commands.

As mentioned, it is possible that there is some interaction among and between the independent variables. It is also possible that the relationships are nonlinear, so the following additional variables are identified:

$$X_4 = X_1^2$$
$$X_5 = X_2^2$$
$$X_6 = X_3^2$$
$$X_7 = X_1X_2$$
$$X_8 = X_1X_3$$
$$X_9 = X_2X_3$$
$$X_{10} = X_1X_2X_3$$

From this, the general form of the mathematical model will be

$$Y = b_0 + b_1X_1 + b_2X_2 + b_3X_3 + b_4X_1^2 + b_5X_2^2 + b_6X_3^2 + b_7X_1X_2 + b_8X_1X_3 + b_9X_2X_3 + b_{10}X_1X_2X_3 + e$$

where the b terms are the weighting coefficients of the independent variables (b_0 is the intercept term) and e is the error term. The error term includes the inherent variability in the data and any variables that might affect Y but were not identified.

The following are the iterations of the stepwise-regression procedure, phasing in each variable in decreasing importance as measured by the F value (S.S. is sum of the squares and M.S. is mean square).

Step 1. X_1 entered, $R^2 = .986$.

	d.f.	*S.S.*	*M.S.*	*F*
Regression	1	7,180,303	7,180,303	1,478
Error	21	101,977	4,856	
Total	22	7,282,280		

	b value
Intercept	375.42257
X_1	.47814

This is the model with one variable.

Step 2. X_1^2 entered, $R^2 = .991$

	d.f.	*S.S.*	*M.S.*	*F*
Regression	2	7,214,296	3,607,148	1,061
Error	20	67,984	3,399	
Total	22	7,282,280		

	b value
Intercept	301.62400
X_1	.58400
X_1^2	−.00002

This is the best model with two variables.

Step 3. $X_1 X_2$ entered, $R^2 = .992$.

	d.f.	*S.S.*	*M.S.*	*F*
Regression	3	7,222,495	2,407,498	765
Error	19	59,785	3,146	
Total	22	7,282,280		

	b value
Intercept	302.01712
X_1	0.61391
X_2	-0.00002
X_1X_2	-0.00332

This is the best model with three variables.

Step 4. X_2 entered, $R^2 = .993$.

	d.f.	*S.S.*	*M.S.*	*F.*
Regression	4	7,228,470	1,807,117	604
Error	18	53,810	2,989	
Total	22	7,282,280		

	b value
Intercept	190.56936
X_1	0.69472
X_2	8.54667
X_1	-0.00003
X_1X_2	-0.00788

This is the best model with four variables.

Step 5. X_2 replaced by X_3, $R^2 = .993$.

	d.f.	*S.S.*	*M.S.*	*F*
Regression	5	7,223,605	1,446,721	505
Error	17	48,675	2,863	
Total	22	7,282,280		

	b value
Intercept	263.19991
X_1	0.69321
X_3	−2.81218
X_1^2	−0.00003
X_1X_2	−0.01020
X_2X_3	0.43685

This is the best model with five variables.

Step 6. X_1X_2 replaced by X_1X_3, $R^2 = .994$.

	d.f.	*S.S.*	*M.S.*	*F*
Regression	6	7,236,212	1,206,035	419
Error	16	46,068	2,879	
Total	22	7,282,280		

	b value
Intercept	256.68385
X_1	0.63088
X_3	−3.67763
X_1^2	−0.00004
X_1X_2	0.00288
X_2X_3	0.46212
$X_1X_2X_3$	−0.00032

This is the best model with six variables.

Step 7. X_2 reentered, $R^2 = .994$.

	d.f.	*S.S.*	*M.S.*	*F*
Regression	7	7,240,110	1,034,301	367
Error	15	42,170	2,811	
Total	22	7,282,280		

	b value
Intercept	421.86929
X_1	0.57396
X_2	−11.83150
X_3	−9.57441
X_1^2	−0.00004
X_1X_3	0.00536
X_2X_3	0.91642
$X_1X_2X_3$	−0.00038

This is the best model with seven variables.

Step 8. X_1X_2 reentered, $R^2 = .994$.

	d.f.	*S.S.*	*M.S.*	*F*
Regression	8	7,241,814	905,227	313
Error	14	40,466	2,890	
Total	22	7,282,280		

	b value
Intercept	531.16471
X_1	0.47142
X_2	−21.69586
X_3	−12.73444
X_1^2	−0.00004
X_1X_2	0.01132
X_1X_3	0.00830
X_2X_3	1.19488
$X_1X_2X_3$	−0.00069

This is the best model with eight variables.

Step 9. X_2^2 entered, $R^2 = .995$.

	d.f.	*S.S.*	*M.S.*	*F*
Regression	9	7,243,166	804,796	267
Error	13	39,114	3,009	
Total	22	7,282,280		

	b value
Intercept	651,10297
X_1	0.42174
X_2	−39.96083
X_3	−15.19285
X_1^2	−0.00005
X_2^2	0.66422
X_1X_2	0.01487
X_1X_3	0.00987
X_2X_3	1.38876
$X_1X_2X_3$	−0.00078

This is the best model with nine variables.

Step 10. X_3^2 entered, $R^2 = .995$.

	d.f.	*S.S.*	*M.S.*	*F*
Regression	10	7,243,923	724,392	226
Error	12	38,357	3,196	
Total	22	7,282,280		

	b value
Intercept	740.16061
X_1	0.42608
X_2	−41.06771
X_3	−23.32940

X_1^2	−0.00005
X_2^2	0.65130
X_3^2	0.14797
X_1X_2	0.01667
X_1X_3	0.00985
X_2X_3	1.42930
$X_1X_2X_3$	−0.00083

This is the best model with ten variables.

One of the most important problems in the analysis of regression problems is the selection of the set of independent variables to be used in the model. It is desirable for the chosen model to contain independent variables that result in a satisfactory predictor model but minimize the variables that do not appreciably affect the dependent variable. A compromise between these objectives will serve as the best model.

Although a number of criteria exist to aid in determining the best model, the adjusted coefficient-of-determination method will be used here. Under this method, the R^2 value will be adjusted to consider the ratio of the total number of observations, *n,* and the number of terms in a specific step of the iteration process, *p*. The expression for the adjusted coefficient of determination, $\bar{R}_p^2$, is

$$\bar{R}_p^2 = 1 - \left(\frac{n-1}{n-p}\right)\left(1 - R_p^2\right)$$

These values for the ten steps are listed in table 13–3 and plotted in figure 13–1.

A simplistic definition of R^2 suggests that it states the amount of variability in the data that is accounted for by the variables considered. Therefore, considering table 13–3, 98.6 percent of the variability is accounted for by the inclusion of only X_1. A somewhat higher value, 99.1 percent, is obtained when X_1^2 is included. As additional variables are added, however, only slight improvements in both R^2 and $\bar{R}_p^2$ are noted. This indicates that, everything considered, it is perhaps best to stop at step 2. Therefore, consulting the *b* values for step 2, the predictor model will be

$$Y = 301.6 + 0.58X_1 - .00002X_1^2.$$

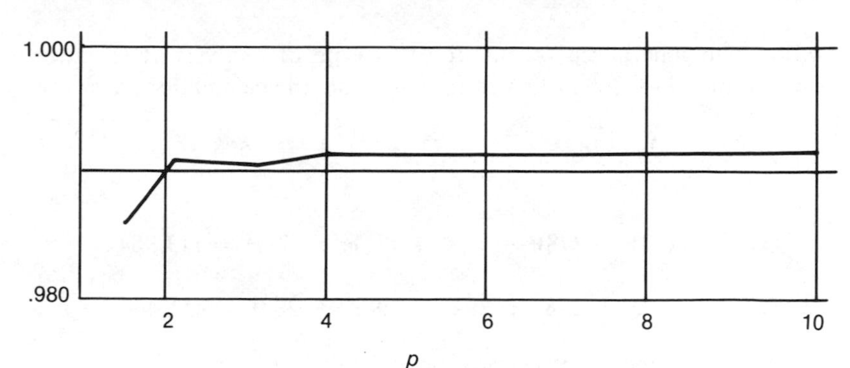

Figure 13–1. Plot of p versus $\bar{R}_p^2$

Table 13–3
List of $\bar{R}_p^2$ Values

p	R^2	$\bar{R}_p^2$
1	.986	.986
2	.991	.991
3	.992	.991
4	.993	.992
5	.993	.992
6	.994	.992
7	.994	.992
8	.994	.992
9	.995	.992
10	.995	.992

$n = 25$

Analysis of Variance: Comparison of Several Means

A maintenance supervisor wishes to determine if there is any difference among five different manufacturers of pressure-relief valves. He takes one valve from each manufacturer and mounts it on a test stand. The springs are all rated for 75 pounds. He subjects each valve to the rated pressure four times. The data he collects are shown in table 13–4. Using the analysis-of-variance procedure, determine if there is a significant difference among the valves when compared with the variation within the units.

Solution. The generalized format for this type of analysis of variance is shown in table 13-5. According to that format, the calculations are

$$\Sigma x = 73 \times 78 + \ldots + 71 + 71 + = 1508$$

$$\Sigma x^2 = (73)^2 + (78)^2 + \ldots + (71)^2 + (71)^2 = 113{,}984$$

$$\Sigma (GT)^2 = 312^2 + 318^2 + 294^2 + 296^2 + 288^2 = 455{,}464$$

$$(\Sigma x)^2 = 1508^2 = 2{,}274{,}064$$

$$N = 20,\ n = 5,\ k = 4$$

$$\frac{(\Sigma x)^2}{N} = 2{,}274{,}064/20 = 113{,}703.2$$

$$\frac{\Sigma (GT)^2}{k} = 455{,}464/4 = 113866$$

Substituting in the format, we obtain

Source	*S.S.*	*d.f.*	*M.S.*
Among groups	113,866 − 133,703.2 = 162.8	4	40.7
Within groups	(by difference) = 118.0	15	7.9
Total	113,984 − 113,703.2 = 280.8	19	

The appropriate F test for this model, using $\alpha = .01$, is

$$F = 40.7/7.9 = 5.15$$

Compared with $F_{.01,4,19} = 4.50$, therefore, one can reject the null hypothesis and can assume that there is a difference among the valves.

Analysis of Variance: Factorial Analysis

In an effort to determine which variables affected the demand for maintenance on a group of compressors, the data shown in table 13-6 were collected. The experiment is completely balanced, with data collected on four levels with three replications. The first level consists of two levels of

Table 13-4
Pressure-Relief-Valve Test Data

	Manufacturer (Group)				
Test	*1*	*2*	*3*	*4*	*5*
1	73	84	72	75	76
2	78	76	77	74	70
3	81	80	73	75	71
4	80	78	72	72	71
Group total	312	318	294	296	288

Table 13-5
Generalized Format: Single Classification for Analysis of Variance

Source	*Sum of the Squares*	*Degrees of Freedom*	*Mean Square*
Among groups	$\Sigma \dfrac{(\text{group total})^2}{\text{No. in group}} - \dfrac{(\Sigma x)^2}{N}$	$n - 1$	S.S./d.f.
Within groups	By difference	by difference	S.S./d.f.
Total	$\Sigma x^2 - \dfrac{(\Sigma x)^2}{N}$	$N - 1$	

throughput (95 percent and 75 percent of capacity); the second level consists of two levels of inhibitor (0 and 20 percent); the third level consists of identifying the shift on which the immediately previous work order was performed (main or other); and the fourth level consists of the average years of experience of the crew that last maintained the compressor. There are then three replicates of the measured variable, which is the run time between failures, in hours. Determine from the data the contribution of each variable and cross products using the sum-of-squares and F-test method.

Solution

Calculation 1 determine, from table 13-6,

$$\Sigma x = 19{,}690$$

$$\Sigma x^2 = 8{,}233{,}148$$

$$\frac{(\Sigma x)^2}{N} = 19{,}690^2/48 = 8{,}077{,}002.1$$

Table 13–6
Time between Failures of Compressors
(hours)

	95% Capacity								75% Capacity							
	20% Inhibitor				No Inhibitor				20% Inhibitor				No Inhibitor			
	Main Shift		Off Shift		Main Shift		Off Shift		Main Shift		Off Shift		Main Shift		Off Shift	
Replicates	Low Expr	High Expr	Low Expr	High Expr	Low Expr	High Expr	Low Expr	High Expr	Low Expr	High Expr	Low Expr	High Expr	Low Expr	High Expr	Low Expr	High Expr
1	355	392	415	376	473	483	514	528	387	351	409	399	433	494	509	461
2	395	350	403	357	400	488	453	435	417	329	392	363	431	404	495	404
3	323	298	346	357	420	438	489	454	373	325	386	350	400	325	477	434
Total	1,073	1,040	1,164	1,090	1,293	1,409	1,456	1,417	1,177	1,005	1,187	1,112	1,264	1,223	1,481	1,299

Calculation 2. Determine all the main-factor totals—that is, the main-factor sum of squares divided by the number of measurements minus the correction term.

Capacity totals:

95% = 9,942

75% = 9,748

Inhibitor totals:

20% = 8,848

0% = 10,842

Shift totals:

Main = 9,484

Off = 10,206

Experience totals:

Low = 10,095

High = 9,595

Capacity sum of squares:

$$\frac{9{,}942^2 + 9{,}748^2}{24} - 8{,}077{,}002.1 = 784.1$$

Inhibitor sum of squares:

$$\frac{8{,}848^2 + 10{,}842^2}{24} - 8{,}077{,}002.1 = 82{,}834.1$$

Shift sum of squares:

$$\frac{9{,}484^2 + 10{,}206^2}{24} - 8{,}077{,}002.1 = 10{,}860.1$$

Experience sum of squares

$$\frac{10{,}095^2 + 9{,}595^2}{24} - 8{,}077{,}002.1 = 5{,}208.3$$

Calculation 3. Determine first-order interactions by the total of all possible pairs of factors, calculating sum of squares and subtracting main-factor sum of squares involved and corrected factor.

	95% Capacity	*75% Capacity*
20% inhibitor	4,367	4,481
0% inhibitor	5,575	5,267

	95% Capacity	*75% Capacity*
Main shift	4,815	4,669
Off shift	5,127	5,079

	95% Capacity	*75% Capacity*
Low experience	4,986	5,109
High experience	4,956	4,639

	20% Inhibitor	*0% Inhibitor*
Main shift	4,295	5,189
Off shift	4,553	5,653

	20% Inhibitor	*0% Inhibitor*
Low experience	4,601	5,494
High experience	4,247	5,348

	Main Shift	*Off shift*
Low experience	4,807	5,288
High experience	4,677	4,918

Capacity-inhibitor sum of squares:

$$\frac{4{,}367^2 + 5{,}575^2 + 4{,}481^2 + 5{,}267^2}{12} - 8{,}077{,}002.1 - 784.1$$

$$- 82{,}834.1 = 3{,}710.0$$

Capacity-shift sum of squares:

$$\frac{4{,}815^2 + 5{,}127^2 + 4{,}669^2 + 5{,}079^2}{12} - 8{,}077{,}002.1 - 784.1$$

$$- 10{,}860.1 = 200.0$$

Capacity-experience sum of squares:

$$\frac{4{,}986^2 + 4{,}956^2 + 5{,}109^2 + 4{,}639^2}{12} - 8{,}077{,}022.1 - 784.1$$

$$- 5{,}208.3 = 4{,}033.3$$

Inhibitor-shift sum of squares:

$$\frac{4{,}295^2 + 4{,}553^2 + 5{,}189^2 + 5{,}653^2}{12} - 8{,}077{,}002.1 - 82{,}834.1$$

$$- 10{,}860.1 = 884.0$$

Inhibitor-experience sum of squares:

$$\frac{4{,}601^2 + 4{,}247^2 + 5{,}494^2 + 5{,}348^2}{12} - 8{,}077{,}022.1 - 82{,}834.1$$

$$- 5{,}208.3 = 901.3$$

Shift-experience sum of squares:

$$\frac{4{,}807^2 + 4{,}677^2 + 5{,}288^2 + 4{,}918^2}{12} - 8{,}077{,}022.1 - 10{,}860.1$$

$$- 5{,}208.3 = 1{,}200.0$$

Calculation 4. Determine the second-order interactions by the total of all possible three-factor combinations, calculating sum of squares and subtracting main-factor sum of squares, first-order-interaction sum of squares, and correction factor.

	95% Capacity		*75% Capacity*	
	20% Inhib.	*0% Inhib.*	*20% Inhib.*	*0% Inhib.*
Main shift	2,113	2,702	2,182	2,487
Off shift	2,254	2,873	2,299	2,780

	95% Capacity		*75% Capacity*	
	20% Inhib.	*0% Inhib.*	*20% Inhib.*	*0% Inhib.*
Low experience	2,237	2,749	2,364	2,745
High experience	2,130	2,826	2,117	2,522

	95% Capacity		*75% Capacity*	
	Main Shift	*Off Shift*	*Main Shift*	*Off Shift*
Low experience	2,366	2,620	2,441	2,668
High experience	2,449	2,507	2,228	2,411

	95% Capacity		*75% Capacity*	
	Main Shift	*Off Shift*	*Main Shift*	*Off Shift*
Low experience	2,250	2,351	2,557	2,937
High experience	2,045	2,202	2,632	2,716

Capacity-inhibitor-shift sum of squares:

$$\frac{2{,}113^2 + 2{,}254^2 + \ldots + 2{,}487^2 + 2{,}780^2}{6} - 8{,}077{,}002.1 - 784.1$$

$$- 82{,}834.1 - 10{,}860.1 - 3{,}710.0 - 200.0 - 884.0 = 444.3$$

Capacity-inhibitor-experience sum of squares:

$$\frac{2{,}237^2 + 2{,}130^2 + \ldots + 2{,}745^2 + 2{,}522^2}{6} - 8{,}077.002.1 - 784.1$$

$$- 82{,}834.1 - 5{,}208.3 - 3{,}710.0 - 4{,}033.3 - 901.3 = 533.5$$

Capacity-shift-experience sum of squares:

$$\frac{2{,}366^2 + 2{,}449^2 + \ldots + 2{,}668^2 + 2{,}411^2}{6} - 8{,}077.002.1 - 784.1$$

$$-10{,}860.1 - 5{,}208.3 - 200.0 - 1{,}200.0 - 4{,}033.3 = 481.4$$

Inhibitor-shift experience sum of squares:

$$\frac{2{,}250^2 + 2{,}045^2 + \ldots + 2{,}937^2 + 2{,}716^2}{6} - 8{,}077{,}002.1$$

$$- 82{,}834.1 - 10{,}860.1 - 5{,}208.3 - 884.0 - 901.3 - 1{,}200.0 = 2{,}581.4$$

Calculation 5. Determine the third-order interaction by computing the sum of squares of the replicate totals and subtracting main effects, first-order interactions, second-order interactions, and the correction term.

Capacity-inhibitor-shift-experience sum of squares:

$$\frac{1{,}073^2 + 1{,}040^2 + \ldots + 1{,}481^2 + 1{,}299^2}{3} - 8{,}077{,}002.1 - 784.1$$

$$- 82{,}834.1 - 10{,}860.1 - 5{,}208.3 - 3{,}710.0 - 200.0 - 4{,}033.3 - 884.0$$

$$- 901.3 - 1{,}200.0 - 444.3 - 533.5 - 481.4 - 2{,}581.4 = 320.1$$

Calculation 6. Determine the error sum of squares and the total of all of the foregoing sums of squares.

$$\text{Total sum of squares} = \Sigma x^2 - \frac{(\Sigma x)^2}{N}$$

$$= 8{,}233{,}148 - (19{,}690)^2/48$$

$$= 156{,}145.9$$

Table 13–7 summarizes the variables, the various cross products, and the error term (both sum of squares and degrees of freedom by difference). Since all main effects and cross products are measured at two levels, the degree of freedom is $n - 1$, or 1.

Table 13-7
Analysis of Variance for Compressor Example

Source	*Sum of Squares*	*Degrees of Freedom*	*Mean Squares*
Main effects			
Capacity	784.1	1	784.1
Inhibitor	82,834.1	1	82,834.1
Shift	10,860.1	1	10,860.1
Experience	5,208.3	1	5,208.3
First-order interactions			
Capacity × inhibitor	3,710.0	1	3,710.0
Capacity × shift	200.0	1	200.0
Capacity × experience	4,033.3	1	4,033.3
Inhibitor × shift	884.0	1	884.0
Inhibitor × experience	901.3	1	901.3
Shift × experience	1,200.0	1	1,200.0
Second-order interactions			
Capacity × inhibitor × shift	444.3	1	444.3
Capacity × inhibitor × experience	533.5	1	533.5
Capacity × shift × experience	481.4	1	481.4
Inhibitor × shift × experience	2,581.4	1	2,581.4
Third-order interaction			
Capacity × inhibitor × shift × experience	320.1	1	320.1
Error	41,170.0	32	1,286.6
Total	156,145.9	47	

To determine which main effects and cross-product terms affect the demand for maintenance on the compressors, it is necessary to choose a significance level for the statistical test. Choose $\alpha = .01$.

Since the data being analyzed here represent the entire population, the case is classified as Model 1. In order to determine which, if any, of the main effects or cross products significantly (at the .01 level) the need for maintenance, the *F* test is used, where

$$F_{\alpha,\, v1,\, v2} = \frac{\text{M.S.}_1}{\text{M.S.}_2}$$

where $\alpha = .01$

v_1 = degree of freedom for M.S._1

v_2 = degree of freedom for M.S._2

$M.S._1$ = Mean square for main effect or cross product under consideration

$M.S._2$ = Mean square for error term

In the case under consideration, if the F value exceeds the established criterion, which is

$$F_{.01,\ 1,\ 32} = 7.50$$

then that variable significantly affects the need for maintenance of the compressors. We can use this criterion because the degree of freedom is always 1 for this situation. If it were different for each variable, a different F test would have to be established for each.

Applying the test to the data in table 13–7 yields only two significant variables; both happen to be main effects—that is, inhibitor and shift. Therefore, if one wants to extend the time between failures, then one would eliminate the inhibitor. This can be determined by totaling the data for 20 percent inhibitor (8,848) and for no inhibitor (10,842). Similarily, maintenance during the off shifts is significantly better than that during the main shift.

Preventive-Maintenance Interval (Replacement)

The data in table 13–8 relate the operating costs to the weekly time periods as an item of equipment is allowed to run without replacement. Determine the minimum-cost time interval, t_r.

Solution. The curve that best represents the data in table 13–8 (from regression) is

$$C_T = 1{,}775 - 1{,}903e^{-0.3t}$$

Writing the function and differentiating the total cost expression, we obtain

$$B/K - C_r = (Bt_r + B/K)e^{-kt}r$$

where B = coefficient of the exponential in the operating-cost curve

K = curve shape factor

Table 13–8
Operating-Cost Data

Time Period (weeks)	*Operating Cost ($/week)*
2	800
4	1,100
6	1,400
8	1,600
10	1,700
12	1,800

C_r = cost of replacement or, in this case, the constant term in the operating-cost curve

t_r = time between replacements.

Therefore, substituting in the minimum-cost model, we obtain

$$1{,}903/0.3 - 1{,}775 = (1{,}903t_r + 1{,}903/0.3)e^{-0.3t_r}$$

$$e^{-0.3t_r} = \frac{4{,}568}{1{,}903t_r + 6{,}343}$$

Solving by trial and error, if $t_r = 3$, then .41 = .38; if $t_r = 4$, then .30 = .32. Thus, somewhere between $t_r = 3$ and $t_r = 4$, $t_r = 3.6$ and .34 ~ .34. Therefore, for a minimum-cost replacement policy, the equipment should be maintained every 3.6 weeks.

Weibull Distribution

The Weibull distribution was developed by Waloddi Weibull, a Swedish engineer, in 1951. Initially, it was applied to fatigue of materials. Its application to failure distributions of electronic tubes was recognized by Kao of Cornell University in 1955. He collected data from a large number of electron tubes and, from those data, proved the usefulness and applicability of Weibull's work.

Since the, it has been recognized that, although the Weibull distribution is a special case of the exponential distribution, the exponential distribution cannot be used when failures are predominantly caused by wearout rather than by chance. For this reason, the Weibull distribution is particularly useful in maintenance. A number of components of interest to maintenance management—such as pumps, compressors, bearings, and electrical components – have been shown to have Weibull time-to-failure distributions. Gamma and, particularly, normal distributions have been used to predict reliability parameters, but the flexibility of the Weibull distribution indicates that it yields more satisfactory results.

The probability-of-failure function of the Weibull distribution is

$$F(x) = 1 - \exp(-\alpha x)^{\beta}$$

and, because the probabilities are mutually exclusive, the probability-of-survival function—that is, the reliability function—is

$$R(x) = 1 - F(x) = \exp(-\alpha x)^{\beta}$$

where α = a scale parameter and

β = a shape parameter.

Shape and scale parameters must be established from the analysis of experimental data for generic items of equipment. Kao determined β for electron tubes to be approximately 1.7. As β increases, the mean of the distribution approaches $1/\alpha$, and the variance approaches zero, as shown in table 13–9. The term $1/\alpha$ is sometimes called the characteristic life and is denoted by θ.

Figure 13–2 illustrates the different patterns of the distribution with the scale parameter assumed to be unity. The scale parameter helps locate the distribution on the x axis.

Table 13–9
Mean and Variance of the Weibull Distribution as a Function of β

	Mean/θ	*Variance/θ^2*
0.5	2.0	20.0
2.0	0.89	0.22
4.0	0.91	0.07
6.0	0.93	0.03
20.0	0.97	0.01

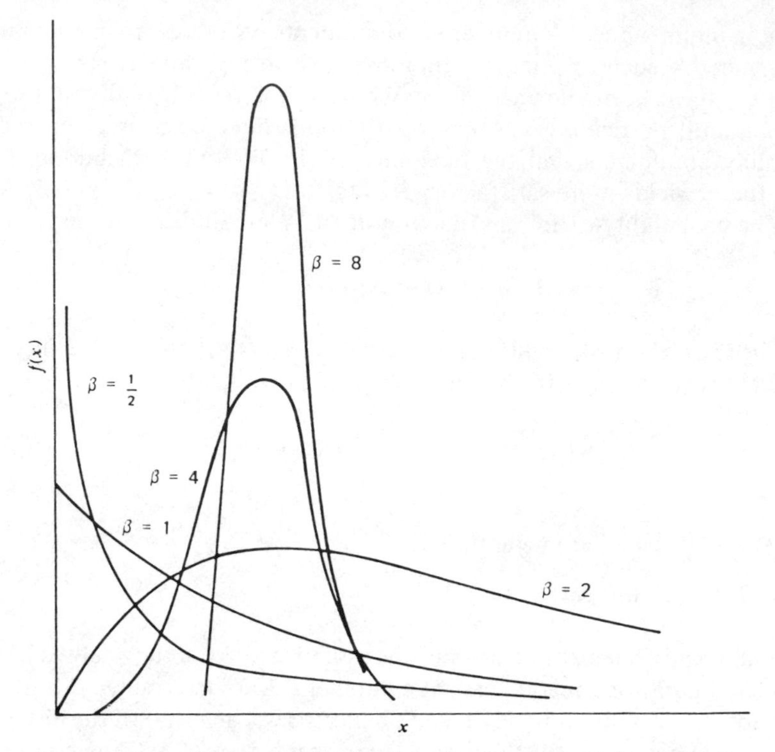

Figure 13–2. Weibull Distribution for Various Values of β

If we substitute $x = \theta$ and $\alpha = 1/\theta$ into the $F(x)$ formula, we obtain

$$F(x) = 1 - e^{-1} = 0.632 \quad \text{for} \quad x = \theta, \alpha = 1/\theta$$

This indicates that the probability of failure up to θ is 0.632; that is, for any value of β, θ will divide the area under any of the distributions in figure 13–2 into 0.632 and 0.368. This is the reason for denoting θ as the characteristic life.

Further inspection of figure 13–2 reveals that, for $\beta = 1$, the Weibull distribution becomes the exponential distribution, and, for β between 3.5 and 4.0, the Weibull distribution approximates the normal distribution.

Because of the complexity of many of the calculations having to do with the Weibull distribution, special graph paper is available. For details of the techniques, consult any comprehensive reliability textbook or handbook.

Table 13–10
Data on Pressure Regulators

Time to Failure (hours)	*Frequency*	*Cumulative Frequency*
0–250	1	1
250–500	5	6
500–750	12	18
750–1,000	16	34
1,000–1,250	30	64
1,250–1,500	21	85
1,500–1,750	10	95
1,750–2,000	5	100

Example. Previous data indicate that the shape parameter for pressure regulators is 1.85. (See table 13–10 for additional data.) What is the probability that any one regulator will perform for 1,500 hours?

Solution: The cumulative failure equivalent to 63.2 percent of failures is approximately 1,250 hours. Therefore,

$$R(1{,}500) = -e^{(x/\theta)^{\beta}}$$

$$= -e^{(-1{,}500/1{,}250)^{1.85}}$$

$$= .246$$

Another Example. Again consider pressure regulators ($\beta = 1.85$). If the characteristic life is 11 kilohours, how many hours should 90 percent of them last? (*Hint:* θ must be in kilohours.)

Solution:

$$R(x) = 1 - F(x) = e^{-(x/\theta)^{\beta}}$$

Therefore,

$$x^{\beta} = \theta \ln \frac{1}{R(x)}$$

$$x^{1.85} = 11 \ln 1/.9$$

$$x = 1{,}085 \text{ hrs.}$$

Queuing Examples

Several Service Channels

A plant has three repair crews to service all their compressors. The compressors arrive in the maintenance system (that is, fail) according to a Poisson distribution, at an average rate of three per twenty-four-hour day. The time a crew takes to repair each unit has an exponential distribution, with a mean service time of four hours. The compressors are serviced in the order that they fail. We will determine how many hours a week a crew can expect to spend maintaining compressors; and how much time, on the average, a compressor is out of service for maintenance.

Solution. According to the data given:

$$\lambda \text{ (mean arrival rate)} = 3/24 = 1/8 \text{ arrivals/hr}$$

$$\mu \text{ (mean service rate)} = 1/4 \text{ services/hr/crew}$$

First, generate the probability that a compressor that fails will not have to wait to be worked on—that is, P_0:

$$P_0 = \frac{1}{\left[\sum_{n=0}^{k-1} \frac{1}{n!}\frac{\lambda^n}{\mu}\right] + \frac{1}{k!}\left(\frac{\lambda}{\mu}\right)^k \frac{k\mu}{k\mu - \lambda}}$$

where k = number of channels—3

n = number items in system—0, 1, 2, 3, . . .

Therefore,

$$P_0 = \frac{1}{\left[\frac{13}{8}\right] + \frac{1}{6}\left(\frac{1}{2}\right)^3 \frac{3/4}{5/8}} = \frac{1}{66/40} = 0.606$$

$$P_n = \frac{1}{n!}\left(\frac{\lambda}{\mu}\right)^n P_0 \qquad \text{for } n = 0, 1, 2, \ldots, k - 1$$

$$P_1 = 1/1!(1/2)^1\, 0.606 = 0.303$$

$$P_2 = 1/2!(1/2)^2\, 0.606 = 0.076$$

Thus, the expected number of idle crews at any time is $3P_0 + 2P_1 + P_2 =$ $3(0.606) + 2(0.303) + 1(0.076) = 2.5$ crews. Therefore, the probability that any one crew will be idle at any specified time is $2.5 \times 1/3 = 0.833$, and the expected weekly time a crew works on the equipment is $1 - 0.833 = 0.167 \times (24 \times 7) = 28$ hours.

Two Separate Queues

A plant has two employees working in stores. The first handles withdrawals only; the second receives new deliveries only. It has been found that the service-time distributions for both receipts and withdrawals are exponential, with mean service time of 15 minutes. Stores receipts are found to arrive according to a Poisson distribution throughout the day, with mean arrival rate of 3 per hour. Withdrawers also arrive according to a Poisson distribution, with mean arrival rate of 2.5 per hour. We will determine (a) what the effect would be on the average waiting time for receipts and withdrawals if each employee could handle both receipts and withdrawals; and (b) what the effect would be if this could only be accomplished by increasing the mean service time to 18.5 minutes.

Solution. (a) With two separate queues, for receipts: $\lambda = 3$, $\mu = 4$ items/hr. Average waiting time is

$$E_w = \frac{\lambda}{\mu(\mu - \lambda)}$$

$$= \frac{3}{4(4 - 3)} = 0.75 \text{ hr.} = 45 \text{ min.}$$

For withdrawals: $\lambda = 2.5$, $\mu = 4$ items/hr. Average waiting time is

$$\frac{2.5}{4(4 - 2.5)} = \frac{2.5}{6} = 0.47 \text{ hr} = 25 \text{ min}$$

If both employees handle all transactions, arrivals are then by Poisson distribution, with $\lambda = 5.5$ transactions/hr, $\mu = .4$, $k = 2$, and $\lambda/\mu = 5.5/4$. Thus,

$$P_0 = \frac{1}{2.375 + \frac{1}{2}\left(\frac{5.5}{4}\right)^2 \frac{8}{5 - 5.5}} = \frac{1}{3.025}$$

and average waiting time is

$$E_w = \frac{\mu(\lambda/\mu)^k}{(k - 1)!\,(k\mu - \lambda)^2} P_0$$

$$= \frac{4(5.5/4)^2}{1!\,(8 - 5.5)^2} \times \frac{1}{3.025}$$

$$= 0.4 \times 60 = 24 \text{ min}$$

(b) If the effect of both employees handling all transactions is to reduce the service rate to $\mu = 60/18.5$,

$$\lambda/\mu = \frac{5.5}{60/18.5} = 1.7$$

$$P_0 = \frac{1}{1.7 + \frac{1}{2}\left(1.7\right)^2 \frac{6.49}{6.49 - 5.5}} = \frac{1}{11.09} = .09$$

and average waiting time is

$$E_w = \frac{\mu(\lambda/\mu)^k}{(k - 1)!\,(k\mu - \lambda)^2} P_0$$

$$= \frac{3.24(1.7)^2}{(2 \times 3.24 - 5.5)^2} (0.9)$$

$$= 0.878 \times 60 = 52.7 \text{ min}$$

Inventory-Management Examples

Basic Model

Find the economic order quantity (EOQ) for 3-inch valves, where 960 are used annually; it costs $55 to place an order; accounting tells us that it costs 33 percent per year on the average inventory value; and each value costs $125.

Solution.

$$EOQ = \left(\frac{2DS}{IC} \right)^{1/2}$$

where $D = 960$

$S = 55$

$I = .33$

$C = 125$

Therefore

$$EOQ = \sqrt{\frac{1(960)55}{.33(125)}} = 50.6$$

or approximately once a week.

Variable Demand Time and Variable Lead Time

Considering, again, the 3-inch valves, let d = daily demand and t = lead time in days to replenish stock, so that

d	*% of days that usage occurs*
0	05
1	30
2	55
3	10

t	*% of time that lead time occurs*
1	10
2	60
3	30

With these data, it is possible to develop the various combinations of use during lead time. If U is the use during lead time, its probability is $P(U)$. From the foregoing data, it can be seen that the maximum usage is $3 \times 3 = 9$—that is, usage of 3 per day when 3 days are required for replenishing.

The probability of that event is

$$P(U_9) = P(\text{lead time of 3}) \times P(\text{usage of 3 on 3 consecutive days})$$
$$= 0.3(0.10)(0.10)(0.10)$$
$$= 0.0003$$

or 3 days in every 1000 days.

$$P(U_8) = [0.25(0.1)(0.1)(0.55)]\ 3$$
$$= 0.004$$
$$P(U_7) = 3(0.3)\ [(0.1)(0.55)(0.55) + ((0.1)(0.1)(0.3)]$$
$$= 0.03$$
$$P(U_6) = 0.6\ [(0.1)(0.1)] + 0.30\ [(0.05)(0.1)(0.1)3 + (0.3)(0.55)(0.1)6 + (0.55)(0.55)(0.55)]$$
$$= 0.086$$

Summarizing, we have

U	$P(U)$
9	0.0003
8	0.004
7	0.030
6	0.086

Table 13–11
Storehouse Data: Six-Inch Pipe Tees

Units Required per Week (d)	*Frequency of Requirement (f_0)*
0	6
1	7
2	9
3	9
4	8
5	6
6	7
7	4
8	1
9	1
10	0

Assuming a stockout cost, Z, of \$250, the optimum $P(U)$ is

$$P(U) = \frac{ICQ}{ZD}$$

$$= \frac{0.33(125)50.6}{250(960)} = 0.0087$$

which falls between a usage of 7 and 8. Therefore, the reorder point might be chosen as 8.

Projecting Inventory Statistics by Distribution Identification

Poisson Distribution: This distribution is used to generate demand for storehouse items. Considering the data in table 13–11, determine whether the requirement distribution follows the Poisson law, using the chi-square test.

Solution.

$$\Sigma(d \times f_0)/\Sigma f_0 = \left[(1 \times 7) + (2 \times 9) + \ldots + (8 \times 1) + (9 \times 1)\right] 58$$
$$= 3.47 = m$$

The model for the Poisson distribution is

$$P_d = e^{-m}\frac{m^d}{d!}$$

where $m = 0, 1, 2, \ldots, 10$ in this case

$m = \text{mean}—\Sigma(d \times f_0)/\Sigma f_0$ in this case

Thus,

$$P_0 = e^{-3.47}\frac{3.47^0}{0!} = 0.31$$

and multiply by 58 to obtain the expected frequency f_e for $d = 0$:

$$f_e = .031 \times 58 = 1.8$$

For

$$P_1 = e^{-3.47} \frac{3.47^1}{1!}$$

$$f_e = .108 \times 58 = 6.2$$

For

$$P_2 = e^{-3.47} \frac{3.47^2}{2!}$$

$$f_e = .187 \times 58 = 10.8$$

Similarly,

$$P_3 = 12.5$$
$$P_5 = 10.8$$
$$P_5 = 7.5$$
$$P_6 = 4.4$$
$$P_7 = 2.1$$
$$P_8 = 0.9$$
$$P_9 = 0.4$$
$$P_{10} = 0.1$$

In order to compare the actual and observed frequencies, use the chi-square test, as illustrated in table 13–12. Thus, $\chi^2 = 15.86$. Using degrees of freedom $k - 2 = 9 - 2 = 7$ and $\alpha = .01$, The χ^2 value of 15.86 is compared with the tabulated value of 18.475 for 7 degrees of freedom at the 0.1 level. Therefore, we can conclude that there is no reason to believe that the distribution frequency is other than Poisson.

Normal Distribution to Generate Lead-Time Distribution. Consider the analysis of lead time for 6-inch tees. Let t be the lead time in weeks to obtain an order and f_0 be the observed frequency of that lead time. Using the data in table 13–13, determine whether the lead-time distribution follows the normal law, using the chi-square test.

Table 13–12
Chi-Square Goodness-of-Fit Test for Storehouse Distribution

d	f_0	f_e	$f_0 - f_e$	$(f_0 - f_e)^2$	$(f_0 - f_e)^2/f_e$
0	6	1.8	4.2	17.64	9.8
1	7	6.2	1.2	1.44	.23
2	9	10.8	1.8	3.24	.30
3	9	12.5	−3.5	12.25	.98
4	8	10.8	−2.8	7.84	.73
5	6	7.5	−1.5	2.25	.30
6	7	4.4	2.6	6.76	1.54
7	4	2.1	1.9	3.61	1.72
8	1 } 2	0.9 } 1.4	.6	.36	.26
9	1	1.4			
10	0	0.1			
					15.86

Solution. To compute the expected frequencies, it is necessary to obtain the mean and the standard deviation:

$$m = \Sigma(t \times f_0)/\Sigma f_0$$
$$= (1 \times 0) + (1 \times 2) + \ldots + (11 \times 1) + (12 \times 1) = 305/46$$
$$= 6.63$$

$$s = \Sigma(f_0 - m)^2/N$$
$$= (4.08)^{1/2}$$
$$= 2.02$$

To calculate the expected frequencies, f_e, it is necessary to graduate the normal curve, using the Z statistic and the calculated mean and standard deviation:

$$Z = \frac{t - m}{s} = \frac{t - 6.63}{2.02}$$

where t is the class boundary for the discrete values of lead time. Table 13–14 illustrates the calculation.

Table 13–13
Lead-Time Data: Six-Inch Pipe Tees

Lead Time (t) in Weeks	*Frequency of Lead Time* (f_0)
1	0
2	1
3	3
4	4
5	6
6	9
7	8
8	6
9	3
10	4
11	1
12	1

Table 13–14
Calculation of Expected Frequencies of Lead Time: Six-Inch Pipe Tees

Midvalues of t	*Class Boundary*	*Z Value*[a]	*Area below Z*[b]	*Area between Boundaries*	*Expected Frequency*[c]
1				.0055	.03
	1.5	−2.54	.0055		
2				.0152	0.7
	2.5	−2.04	.0207		
3				.0399	1.8
	3.5	−1.55	.0606		
4				.0863	4.0
	4.5	−1.05	.1469		
5				.1408	6.5
	5.5	−0.56	.2877		
6				.1884	8.7
	6.5	−0.06	.4761		
7				.1903	8.8
	7.5	0.43	.6664		
8				.1547	7.2
	8.5	0.93	.8238		
9				.0984	4.5
	9.5	1.42	.9222		
10				.0504	2.3
	10.5	1.92	.9726		
11				.0194	0.9
	11.5	2.41	.9920		
12				.0062	0.3
	12.5	2.91	.9982		

[a] $Z = (t - m)/s = (t - 6.63)/.220$; for $t = 1.5$, $Z = -2.54$, and so forth.
[b] From a Z table of areas under the normal curve.
[c] $N \times$ theoretical frequency.

To compare the observed and expected frequencies, we use the chi-square test illustrated in table 13–15. We obtain $\chi^2 = 3.77$. Using degrees of freedom $k - 3 = 9 - 3 = 6$ and $\alpha = .01$, the χ^2 value of 3.77 is compared with the tabulated value of 16.812 for 6 degrees of freedom at the .01 level. Therefore, we can conclude that there is no reason to believe that the distribution frequency is other than normal.

Control-Chart Example

A maintenance planner would like to establish a computer program to determine when his man-hour estimates for work orders differ from the actual time required in the field. To do this, he has sampled his estimate by randomly choosing five work orders per week and calculating the percentage of actual to estimated man-hours as shown in table 13–16. Since the planner has decided to monitor his work orders with the control-chart concept, the means and ranges are also shown in table 13–16. Based on these data, calculate the upper and lower control limits for a mean ($\bar{X}$) chart and a range (R) chart and plot the data. Determine whether the process is in control whether there are assignable causes that result in the estimates being out of control. Use 2σ limits in the calculations.

Table 13–15
Chi-Square Test: Lead Time of Six-Inch Pipe Tees

t	f_O	f_e	$(f_O - f_e)^2$	$(f_O - f_e)^2/f_e$
1	2	0.3		
2	1 } 4	0.7 } 2.8	3.24	1.16
3	3	1.8		
4	4	4.0	0.00	0.00
5	6	6.5	0.25	0.04
6	9	8.7	0.09	0.01
7	8	8.8	0.64	0.07
8	6	7.2	1.44	0.20
9	3	4.5	2.25	0.50
10	4	2.3	2.89	1.26
11	1 } 2	0.9 } 1.2	0.64	0.53
12	1	0.3		
				3.77

Table 13–16
Percentage of Actual versus Estimated Man-Hours

	Percentage							
Week	*W.O.1*	*W.O.2*	*W.O.3*	*W.O.4*	*W.O.5*	*Sum*	*Mean* ($\bar{X}$)	*Range (R)*
1	93	109	91	97	101	491	98.2	18
2	101	95	93	95	96	480	96.0	6
3	100	91	94	92	97	474	94.8	9
4	92	94	94	104	99	483	96.6	12
5	97	91	92	98	93	471	94.2	7
6	86	112	90	110	116	514	102.8	30
7	93	98	102	106	101	500	100.0	13
8	103	92	94	100	99	488	97.6	11
9	104	101	97	103	101	506	101.2	7
10	98	102	99	104	101	504	100.8	5

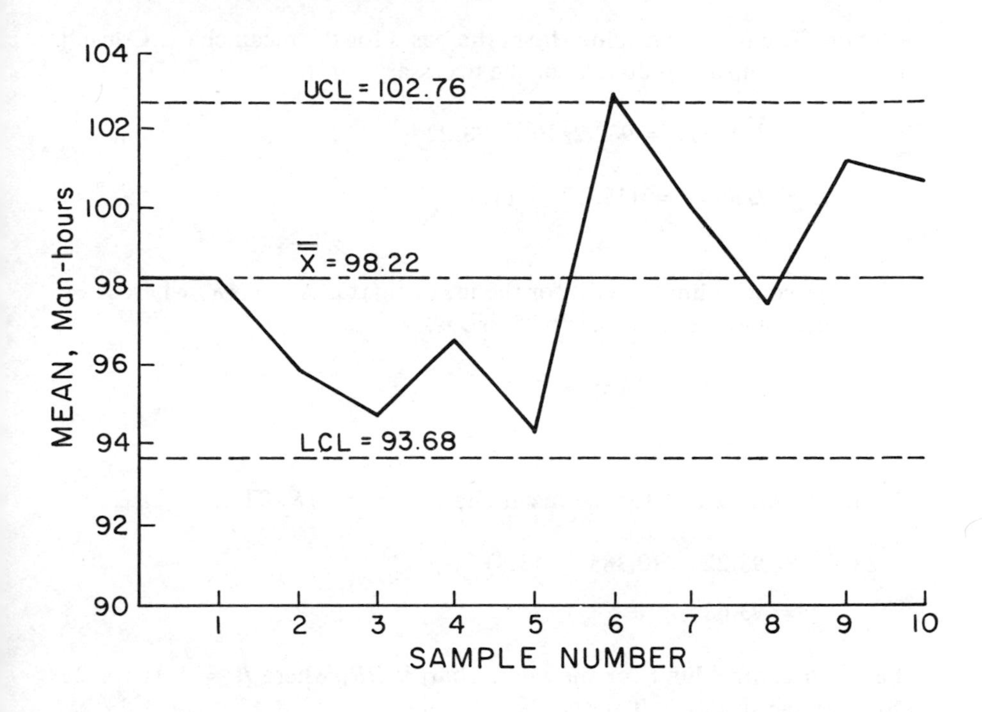

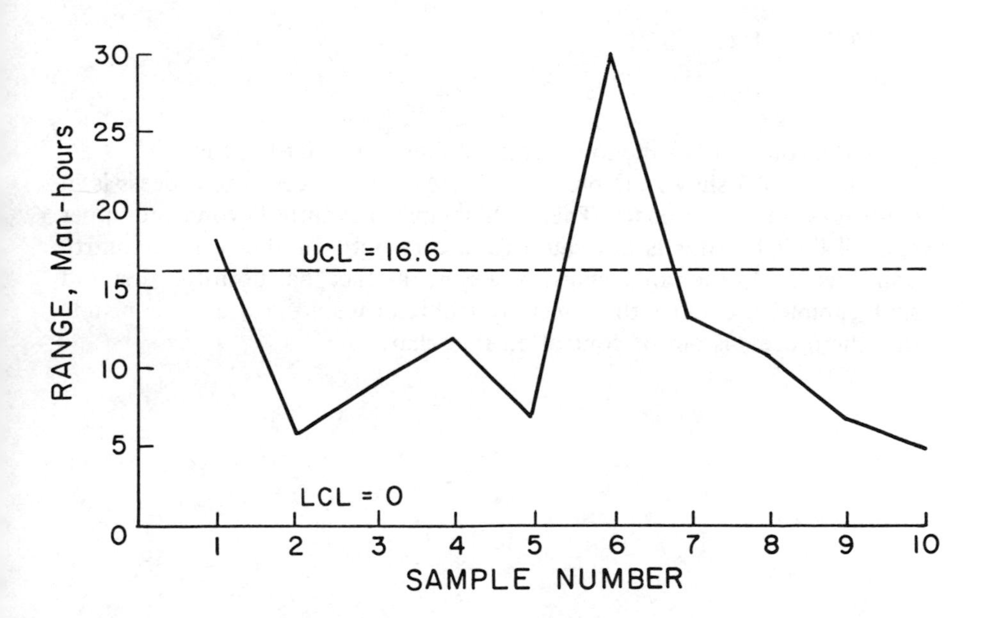

Figure 13–3. Mean ($\bar{X}$) and Range (R) Charts for Man-Hour Control

Solution. The following values form the bases for the mean chart. Consult any quality-control handbook for the necessary constants.

$$X = \Sigma \bar{\bar{X}} / N = 982.2/10 = 98.22$$

$$\bar{R} = \Sigma R / N = 118/10 = 11.8$$

The upper control limit (UCL) for the mean chart is $\bar{\bar{X}} + A\bar{R}$, where A = 0.385 for 2σ and a sample size of 5. Thus,

$$UCL = 98.22 + (0.385 \times 11.8)$$
$$= 102.76$$

The lower limit (LCL) for the mean chart is $\bar{\bar{X}} - A\bar{R}$. Thus,

$$LCL = 98.22 - (0.385 \times 11.8)$$
$$= 93.68$$

The upper control limit for the range chart is $B\bar{R}$, where B = 1.41 for 2σ and a sample size of 5. Thus,

$$UCL = 1.41 \times 11.8$$
$$= 16.6$$

Since the constant for the lower control limit is 0, that limit is 0.

Figure 13–3 shows the plotted data. As can be seen, the process is in control on the mean chart. The sixth sample is slightly beyond the upper control limit but this is not cause for worry until another out-of-control point occurs. In the range chart, however, the fact that both the first and sixth samples are above the upper control limit would lead us to conclude that the process is out of control on that chart.

Appendix A: Maintenance-Management Inventory Form

______________________________ ______________________________

PLANT DATE

____

CONTACT

ORGANIZATION CHART

NUMBER OF CRAFTSMEN ______________________________

NUMBER OF FOREMEN ______________________________

NUMBER OF PLANNERS __________, SCHEDULERS __________

COORDINATOR ______________________________

OTHER ______________________________

AREA-MAINTENANCE SUPERVISORS? ______________________________

CRAFT BREAKDOWN

DISPOSITION OF SHOPS

ANNUAL REPORT? _______________, COPY
MAINTENANCE-MANAGEMENT MANUAL? _________, COPY
AREA, CENTRALIZED, COMBINATION? ____________________
COLLECTIVE-BARGAINING AGREEMENT? _________, COPY
ANY ENGINEERING ASSISTANCE TO MAINTENANCE?

WORK-ORDER FORM, COPY
MINOR WORK REQUEST? ____________________________________
PLANNING SHEET? ____
CLOSE-OUT PROCEDURE FOR WORK ORDER

RESPONSIBILITY FOR OVERRUNS

COPIES OF WORK ORDER FILED

PRIORITY SYSTEM

APPROXIMATE PERCENTAGE IN EACH PRIORITY LEVEL

______________ % ______________

______________ % ______________

______________ % ______________

______________ % ______________

WHO DOES WORK PLANNING/SCHEDULING/ESTIMATING?

HOW DO YOU DEFINE PLANNED?

NATURE OF BACKLOG

ARE STANDARDS USED?

IS NETWORK ANALYSIS USED?

ANY WORK-MEASUREMENT ACTIVITY?

WHAT PERFORMANCE INDEXES DO YOU USE?

WORK SAMPLING? __________ IF YES, COPIES
TOTAL ANNUAL COST OF MAINTENANCE ____________________
TOTAL MAN-HOURS SPENT ON MAINTENANCE ______________
UNIT OF PRODUCTION ____
REPLACEMENT COST OF PLANT (19____) ____________________
TOTAL EMPLOYEES IN PLANT _________________________________
TOTAL OPERATIONS PERSONNEL ____________________________
TOTAL MAINTENANCE BURDEN ______________________________
TOTAL MAINTENANCE PAYROLL ____________________________
NUMBER OF MAINTENANCE PLANNERS ____________________

EXTENT OF PREVENTIVE-MAINTENANCE (PM) PROGRAM

HOW IS JOB PACKAGED?

OVERTIME HISTORY PAST YEAR

ANY EQUIPMENT-HISTORY PROGRAM?

ARE ITEMS OF EQUIPMENT GIVEN NUMBERS?

PLACE OF EQUIPMENT INSPECTION IN PM

ANY EFFORT TO IDENTIFY PROBLEM AREAS?

LOCATION OF STORES

NUMBER OF PEOPLE IN STOREHOUSE ______________________

VALUE OF STORES INVENTORY ____________________________

PURCHASING INTERFACE

USE EOQ APPROACH? _____________________________________

NUMBER OF ITEMS IN STOREHOUSE ________________________

STOCK CATALOG? ____

STOCKOUT HISTORY

WHO DELIVERS MATERIAL? HOW?

ANY OPEN STOCK? ______________________________

ANY OPEN PURCHASE ORDERS? ______________________

HOW IS COMPUTER USED IN MAINTENANCE?
STORES?

OTHER?

ANY SUBCONTRACTING?

ANY ATTEMPT TO ESTABLISH LEVELS OF MAINTENANCE?

MMIS? REPORTS

Appendix B: The Maintenance Game

Assume that you have a chemical plant with the following craftsmen: two electricians, four instrument men, five carpenters, and six maintenance mechanics. You have two cranes, two welding machines, one air compressor, and one A-frame truck.

Determination 1: Materials Control

Consider the material data in table B–1 and the work-order summary in figure B–1. (*Note:* In the work-order summary's "List Mat'l" column for work order 1030—as an example—75-4 indicates 75 feet of 8-inch pipe—item number 4—which comes in 20-foot lengths, and 8–5 indicates eight 8-inch flanges—item number 5.) Are there any work orders that you cannot work on because material is not on hand? How will this affect your schedule? Prepare a stores ledger card for each of the five items and record all transaction (*Hint:* Prepare these as you schedule). Assuming that this usage is representative of the entire year, compute the economic order quantities (EQOs). (Carrying costs are 25 percent of unit cost per year; cost of placing an order is \$35.) Economic order quantity and reorder point (ROP) decisions will not be used during this exercise, but they are to be computed. Use the following formulas:

$$EOQ = \left(\frac{2CR}{IP} \right)^{1/2}$$

$$\text{ROP} = S + RL$$

where: S = safety stock

R = annual use rate

L = lead time

C = cost of placing an order

I = carrying cost (decimal)

P = unit price of item

Table B–1
Materials Data

Item Number	*Item*	*Number on Hand 1/3*	*Time Necessary to Get New Order*	*Cost*
1	Microswitch	0	2 working days	$15 ea.
2	XKG valve	8	2 working days	$250 ea.
3	18″ exchanger gasket	10	3 working days	$120 ea.
4	8″ pipe	500′	2 working days	$3/ft
5	8″ flange	80	3 working days	$25 ea.

How many of each item should we stock and what should be the reorder point? Assume that the patterns of usage are representative of entire year and that no material is on order. Safety stock is 10 percent of EOQ for each item. If material is depleted, assume that the order is placed on the same day as the needing work order is written; for example, if your lead time is two days and you order on January 4, the material would not be available until the morning of January 7.

Determination 2: Scheduling

Schedule work orders 1001 through 1054 from the summary of work orders (figure B–1). Devise a bar-graph schedule for each craftsman and each item of equipment, with craftsmen and equipment as the ordinate and days as the abscissa. (*Note:* In the summary's "Actual MH" column, 2/6 denotes two men for 6 hours each; work orders take as long as the longest craft time. In the "List EQ" column, 2A denotes 2 hours of A-frame truck required.) (*Hints:* Make a graph for 15 working days forward from January 4—5 working days per week.) Plot "Est. MH" and "List Eq" on a bar graph against the date required; that is, try to initiate the work order on the day after it is requested. On the bar chart, use different colors or some other coding for the four priority levels. Schedule "Est. MH," assuming a 40-hour work week and an 8-hour-per-day schedule.) Crew or equipment cannot be broken up; that is, if three men are assigned, they must all be there at one time. This is a rule for the game only; in actual practice you would break them up. "Date Need" means at the beginning of the shift. You cannot split work orders, and they should be completed by the start of the date needed—at the latest. Remember to check to see if you have all materials. No overtime or contracting is permitted. Assume that January 3 is a Monday.

ABBR. ELECTRICIANS = E; INSTRUMENT MEN = I; CARPENTERS = C; MAINTENANCE MECHANICS = M; WELDING MACH = WM;
CRANE = C; A-FRAME TRUCK = A; AIR COMPRESSOR = AC; SCAFFOLDING = S

W.O. #	DESCRIPTION	PRIOR	DATE WRITTEN	DATE NEED	EST. MH				ACTUAL MH				LIST MAT'L	LIST EQ
					E	I	C	M	E	I	C	M		
1001	Replace safety valve on T-3	2	1/3	1/10				2/4				2/6	–	2A
1002	Repair valve leak on steam line	1	1/3	1/5			2/4	2/4			2/2	2/2	–	2WM
1003	Repair flowrator F-21	2	1/3	1/10		1/4		1/4		1/6		1/6	–	–
1004	Replace gasket on 18" excg.	1	1/3	1/5			2/4	2/4			2/4	2/4	1–3	–
1005	Replace pipe support	3	1/3	1/13			3/8	2/4			3/8	2/6	–	4C
1006	Tighten seal on 6" pump	2	1/3	1/10				2/2				2/1	–	–
1007	Replace microswitch on alarm	1	1/4	1/6	1/2	1/2			1/1	1/2			1–1	–
1008	Exchange bundles on 20" excg.	1	1/4	1/6			3/8	3/8			3/10	3/10	–	8C
1009	Repl. 200' 8" pipe on slpr.	2	1/4	1/11			3/40	4/40			3/36	4/36	200'-4;20-5	16C,16WM
1010	Replace XKG compr. valve	2	1/4	1/11				3/12			2/6	3/14	1–2	2A
1011	Repair flowrator F-16	2	1/4	1/11		1/4		1/4		1/4		1/5	–	–
1012	Install conc. fdn. for 8" pump	3	1/5	1/17			3/20	2/4			3/18	2/6	–	2WM
1013	Calibrate control valve	1	1/5	1/7		1/2				1/4			–	–
1014	Repl. main. #2 fuse box with break	1	1/5	1/7	2/12				2/12				–	–
1015	Isolate cont. hse. #3 from hydroc.	2	1/5	1/12	1/40	3/30	1/20		1/36	3/26	1/24		–	–
1016	Replace safety valve on T-7	2	1/5	1/12				2/5				2/4	–	2A
1017	Replace gasket on 18" excg.	1	1/5	1/7			2/4	2/4			2/4	2/4	1–3	–
1018	Replace microswitch on alarm	1	1/5	1/7	1/2	1/2			1/2	1/2			1–1	–
1019	Paint No. 2 control house	4	1/5	2/1			2/40		2/8		2/44		–	40AC
1020	Sandblast Unit #1 struct. steel	3	1/6	1/17			3/80				3/72		–	80AC
1021	Repair flowrator F-82	2	1/6	1/12		1/4		1/4		1/3		1/3	–	–
1022	Sandblast excg. bank - unit #2	3	1/6	1/17			3/60				3/52		–	60AC
1023	Tighten seal on 5" pump	1	1/6	1/8				2/2				2/2	–	–
1024	Repl. 150' 8" pipe on slpr.	2	1/7	1/12				3/30				3/28	150'-4;16-5	30C, 30WM
1025	Replace XKG compr. valve	2	1/7	1/13				3/12				3/14	1–2	2A
1026	Repl. wood steps with steel-loader	4	1/7	2/1			2/4	2/20			2/8	2/20	–	20C,20WM
1027	Replace gasket on 18" excg.	2	1/7	1/13			2/4	2/4			2/4	2/4	1–3	–

Figure B–1. Summary of Work Orders

ABBR. ELECTRICIANS=E; INSTRUMENT MEN=I; CARPENTERS=C; MAINTENANCE MECHANICS = M; WELDING MACH = WM; CRANE=C; A-FRAME TRUCK=A; AIR COMPRESSOR = AC; SCAFFOLDING = S

W.O. #	DESCRIPTION	PRIOR	DATE WRITTEN	DATE NEED	EST. MH				ACTUAL MH				LIST MAT'L	LIST EQ
					E	I	C	M	E	I	C	M		
1028	Replace safety valve on T-4	2	1/7	1/14				2/4				2/3	–	2A
1029	Replace microswitch on alarm	1	1/7	1/10	1/2	1/2			1/2	1/2			1-1	–
1030	Replace 75' of 8" O.H. pipe	2	1/7	1/14			2/20	2/20			2/24	2/20	75'-4, 8-5	20C;20WM
1031	Replace corroded struct. steel T-6	4	1/10	2/18			4/40	4/40			4/44	4/44	–	40C;40WM
1032	Tighten seal on 2-1/2" pump	1	1/10	1/12				2/2				2/2	–	–
1033	Retube 36" excg. bundle	2	1/10	1/15				2/40			2/8	2/36	–	2A
1034	Reinsulate 600' 8" O.H.	2	1/10	1/15			2/40	2/40			2/42	2/42	–	40C
1035	Replace XKG Compr. valve	2	1/10	1/15				3/12				3/12	1-2	2A
1036	Repair flowrator F-70	1	1/10	1/12		1/4		1/4		1/4		1/4	–	–
1037	Replace gasket on 18" excg.	1	1/10	1/12			2/4	2/4			3/4	2/5	1-3	–
1038	Replace microswitch on alarm	1	1/10	1/12	1/2	1/2			1/2	1/2			1-1	–
1039	Seal weld rivits - 106	4	1/10	2/30			2/60	2/60		1/4	2/58	2/58	–	60 WM
1040	Replace safety valve on T-2	2	1/11	1/17				2/5				2/5	–	2A
1041	Reroof exchanger shed	4	1/11	2/30			4/20				4/24		–	2A
1042	Replace 60' of 8" O.H. pipe	3	1/11	1/21			2/16	2/16			2/16	2/16	60'-4; 6-5	16C,16WM
1043	Sandblast T-4	3	1/11	1/21			4/40	4/40			4/42	4/42	–	40AC
1044	Replace lamps - T-5	1	1/11	1/13	2/20				2/20		2/8		–	–
1045	Rewire motor 203	1	1/12	1/13	2/30				2/28				–	–
1046	Post turnaround on unit #3	2	1/12	1/18	2/40	2/40		2/40	2/48	2/40		2/48	–	40WM,40C
1047	Tighten seal on 2" pump	1	1/12	1/14				2/2				2/2	–	–
1048	Replace microswitch on alarm	1	1/12	1/14	1/2	1/2			1/2	1/2			1-1	–
1049	Repair flowrator F-32	2	1/12	1/18		1/4		1/4		1/5		1/5	–	–
1050	Replace gasket on 18" excg.	1	1/12	1/14			2/4	2/4			2/4	2/4	1-3	–
1051	Replace XKG compr. valve	2	1/12	1/19				3/12				3/12	1-2	2A
1052	Replace safety valve on T-1	2	1/12	1/19				2/4				2/3	–	2A
1053	Replace 70' of 8" O.H. pipe	3	1/12	1/24			2/20	2/20			2/24	2/24	70'-4; 8-5	20C,20WM
1054	Pave 700 sq. ft. in area 3	4	1/12	2/30			2/24				2/24		–	–

Figure B-1 continued

ABBR. ELECTRICIANS = E; INSTRUMENT MEN = I; CARPENTERS = C; MAINTENANCE MECHANICS = M; WELDING MACH = WM; CRANE = C; A-FRAME TRUCK = A; AIR COMPRESSOR = AC; SCAFFOLDING = S

W.O. #	DESCRIPTION	PRIOR	DATE WRITTEN	DATE NEED	EST. MH				ACTUAL MH				LIST MAT'L	LIST EQ
					E	I	C	M	E	I	C	M		
1055	Rewire spare motor - 215	2	1/13	1/19	2/25								–	–
1056	Sandblast cooler box - unit #3	3	1/13	1/24			3/32						–	32 AC
1057	Replace microswitch on alarm	1	1/13	1/17	1/2	1/2							–	–
1058	Replace 20' of 8" O.H. pipe	3	1/13	1/24			2/10	2/10					20'-4; 2-5	10C, 10WM
1059	Replace strainer in 12" line	1	1/13	1/17				2/8					–	4 A
1060	Raise drains in area 2	4	1/13	2/15			3/30	2/30					–	–
1061	Replace XKG compr. valve	2	1/14	1/20				3/12					1 - 2	2A
1062	Tighten seal on 2" pump	1	1/14	1/17				2/2					–	–
1063	Repair Flowrator F061	1	1/14	1/17		1/4		1/4					–	–
1064	Replace lighting on T-6	3	1/14	1/24	2/30		2/30						–	–
1065	Replace safety valve on T-5	2	1/14	1/20				2/5					–	2A
1066	Replace gasket on 18" excg.	1	1/14	1/17			2/4	2/4					1 - 3	–
1067	Rewire computer room	2	1/14	1/21	2/40	2/20							–	–
1068	Replace 50' 8" O. H. pipe	2	1/15	1/21			2/12	2/12					50'-4; 6-5	12C, 12WM
1069	Replace microswitch on alarm	1	1/17	1/19	1/2	1/2							1 - 1	–
1070	Correct alignment on compr.	1	1/17	1/19				2/8					–	8C
1071	Clean tank 112	3	1/17	1/27			2/40	4/40					–	40AC
1072	Tighten seal on 2" pump	1	1/17	1/19				2/2					–	–
1073	Sandblast struct. steel - unit #2	3	1/17	1/27			3/30	3/30					–	30 AC
1074	Replace strainer on 18" line	1	1/17	1/19				2/8					–	4 A
1075	Replace gasket on 18" excg.	1	1/18	1/20			2/4	2/4					1 - 3	–
1076	Replace ring on tank 108	4	1/18	2/30			4/40	4/40					–	40WM, 40C
1077	Replace safety valve in T-6	2	1/18	1/24				2/5					–	2 A
1078	Replace XKG comp. valve	2	1/18	1/24				3/12					1 - 2	2A
1079	Repair flowrator F-52	1	1/18	1/20		1/4		1/4					–	–
1080	Replace microswitch on alarm	1	1/18	1/20	1/2	1/2							1 - 1	–
1081	Rewire motor 219	2	1/18	1/24	2/10								–	–

Figure B–1 continued

Determination 3: Craft-Performance Analysis

Using all the data available to you, compute a performance percentage for each craft and one for the entire work force.

Determination 4: Work-Order Close-Out Report

Assuming that priority 1 work orders should be closed out on the date needed, draw up a report listing the priority 1 work orders that make the date and those that do not. Assume that a work order has to be finished at the start of the day needed to make the date. For work orders that do not make the date, explain why they do not. Compute a percentage of work orders (all priority levels) scheduled to be closed out that actually are closed out for each day. Plot the percentage closed out against the date.

Determination 5: Backlog Report

List the backlog, by priority and craft, as of close of business on January 12, and give reasons for work orders being in the backlog. Include those written on January 12 and not started on that date. When you have finished scheduling, list the remainder of the work orders (1001 to 1054) that you did not get to, by priority levels.

Determination 6: Work-Sampling Study

You have decided to monitor the activities of the work force as a whole and broken down by crafts. Your latest results are shown in figure B–2. Compute the activity percentage for the entire work force and for each craft. Comment on the results and compare your findings with previous results, as shown in table B–2. Comment on this comparison. How do these results

Table B–2
Results of Previous Work-Sampling Studies

	Percentage		
Activity	*18 Months Ago*	*12 Months Ago*	*6 Month Ago*
Productive	56	54	53
Traveling	15	17	16
Personal time	08	07	08
Unavoidable delays	21	22	23

Work-Sampling Study

_______________ – _______________
Start Date End Date

	Activity			
Crafts	Productive	Traveling	Personal Time	Unavoidable Delays
Electricians	𝍸 𝍸 𝍸 𝍸 𝍸 𝍸 𝍸 𝍸 |||	𝍸 𝍸 ||||	𝍸	𝍸 𝍸 𝍸 𝍸 𝍸 𝍸 𝍸 |||
Instrument men	𝍸 𝍸 𝍸 𝍸 𝍸 𝍸 ||	𝍸 𝍸 |	𝍸	𝍸 𝍸 |||
Carpenters	𝍸 𝍸 𝍸 𝍸 𝍸 |||	𝍸 𝍸 𝍸 |	𝍸 𝍸 𝍸 ||	𝍸 𝍸 𝍸 ||
Maintenance mechanics	𝍸 𝍸 𝍸 𝍸 𝍸 𝍸 𝍸 𝍸 𝍸 𝍸 𝍸 𝍸 𝍸 𝍸 𝍸 𝍸 𝍸 𝍸 |	𝍸 𝍸 𝍸 𝍸 𝍸 𝍸 ||	𝍸 𝍸 𝍸 |||	𝍸 𝍸 𝍸 𝍸 ||

Figure B–2. Work-Sampling Study

compare with the productivity indexes from the comparison of estimated versus actual man hours? Explain any discrepancies.

Determination 7: Standards

Note the following groups of work orders from the summary (figure B-1):

1. Replace safety valve: 1001, 1016, 1028, 1040, 1052, 1065, 1077
2. Repair flow rator: 1003, 1011, 1021, 1036, 1049, 1063, 1079
3. Replace gasket: 1004, 1017, 1027, 1037, 1050, 1066, 1075
4. Tighten seal: 1006, 1023, 1032, 1047, 1062, 1072
5. Replace microswitch: 1007, 1018, 1029, 1038, 1048, 1057, 1069, 1080
6. Replace pipe: 1030, 1042, 1053, 1058, 1068
7. Replace compressor valve: 1010, 1025, 1035, 1051, 1061, 1078

Analyze the distribution of the actual man-hours on these groups of work orders and then recommend a standards for the repetitive work orders.

Recommendation 8: Recommendations

List the specific recommendations you would make to reduce the backlog. What do you think would be a reasonable backlog?

Bibliography

Ammer, D.S. *Materials Management,* 3rd ed. Homewood, Ill.: Irwin, 1974.

Blanchard, B.S., and Lowery, E.E. *Maintainability.* New York: McGraw-Hill, 1969.

Bleuel, W.H., and Patton, J.P. *Service Management.* Research Triangle Park, N.C.: Instrument Society of America, 1978.

Briggs, A.J., *Warehouse Operations Planning and Management.* New York: Wiley, 1960.

Cooling, W.C. *Low-Cost Maintenance Control.* New York: American Management Associations, 1973.

Corder, A. *Maintenance Management Techniques.* New York: McGraw-Hill, 1976.

Cotz, V.J. *Plant Engineer's Manual and Guide.* Englewood Cliffs, N.J.: Prentice-Hall, 1973.

———. "Engineered Performance Standards," Navdocks P-700 Series, Washington, D.C.: Department of the Navy, Bureau of Yards and Docks, 1963.

Evans, F.L. *Maintenance Supervisor's Handbook.* Houston: Gulf Publishing, 1962.

Gradon, F. *Maintenance Engineering.* New York: Halsted Press, Wiley, 1973.

Heintzelman, J.W. *The Complete Handbook of Maintenance Management.* Englewood Cliffs, N.J.: Prentice-Hall, 1976.

Hibi, S. *How to Measure Maintenance Performance.* Asian Productivity Organization, Tokyo, 1977.

Husband, T.M. *Maintenance Managment and Terotechnology.* Farnborough, England: Saxon House, 1976.

James, C.F.; Shendy, B.N.; and Stanislao, J. "The Design of a Maintenance System." Asian Productivity Organization Productivity Series 14, Tokyo, 1979.

Jardine, A.K.S. *Maintenance, Replacement and Reliability.* New York: Halsted Press, Wiley, 1973.

———. *Operational Research in Maintenance.* New York: Barnes and Noble, 1970.

Kelly, A., and Harris, M.J. *Management of Industrial Maintenance.* Woburn: Newnes-Butterworths, 1978.

Lewis, B.T. *Management Handbook for Plant Engineers.* New York: McGraw-Hill, 1977.

Lewis, B.T., and Marron, J.P. *Facilities and Plant Engineering Handbook.* New York: McGraw-Hill, 1973.

Lewis, B.T., and Pearson, W.W. *Maintenance Management.* New York: J.F. Rider, 1963.

Lewis, B.T., and Tow, L.M. *Readings in Maintenance Management.* Boston: Cahners Books, 1973.

Love, S. *Inventory Control.* New York: McGraw-Hill, 1979.

Mann, L. *Maintenance Management,* 1st ed. Lexington, Mass.: Lexington Books, D.C. Heath, 1976.

Miller, E.J., and Blood, J.W. *Modern Maintenance Management.* New York: American Management Associations, 1963.

Morrow, L.C. *Maintenance Engineering Handbook.* 2nd ed. New York: McGraw-Hill, 1966.

Morse, P.M. *Queues, Inventory and Maintenance.* New York: Wiley, 1958.

Newbrough, E.T. *Effective Maintenance Management.* New York: McGraw-Hill, 1967.

NPRA Refinery and Petrochemical Plant Maintenance Conference, Unpublished papers. Tulsa: Gerald L. Farrar.

Orlicky, J. *Material Requirements Planning.* New York: McGraw-Hill, 1975.

Page, J.S. *Estimator's Manual of Equipment and Installation Costs.* Houston: Gulf Publishing, 1978.

———. *Estimator's Equipment Installation Man-Hour Manual,* 2nd ed. Houston: Gulf Publishing, 1978.

———. *Estimator's Piping Man-Hour Manual,* 3rd ed. Houston: Gulf Publishing, 1976.

———. *Estimator's Electrical Man-Hour Manual.* Houston: Gulf Publishing, 1959.

Papers. New York: Plant Engineering and Maintenance Division, American Society of Mechanical Engineers.

Patton, J.E. *Maintainability and Maintenance Management.* Research Triangle Park, N.C.: Instrument Society of America.

Pilborough, L. *Inspection of Chemical Plant.* London: Leonard Hill Books, 1971.

Plant Engineering Management. Manufacturing Management Series Vol. 5. The Society of Manufacturing Engineers, Dearborn, Mich., 1971.

Richardson, W.J. *Cost Improvement, Work Sampling, and Short Interval Scheduling.* Reston, Va.: Reston, 1976.

Stanier, W. *Plant Engineering Handbook,* 2nd ed. New York: McGraw-Hill, 1959.

Stewart, H.J.M. *Guide to Efficient Maintenance Management.* London: Business Publications, 1963.

Index

About the Author

Lawrence Mann, Jr., received the B.S. in mechanical engineering from Louisiana State University and the M.S. and Ph.D. in industrial engineering from Purdue University. He has taught at Purdue and is currently professor of industrial engineering at LSU. He has extensive practical experience in industry and has consulted or conducted courses for Exxon, Dow, Copolymer, Mississippi Chemical, Allied Chemical, and Uniroyal. He has conducted maintenance-management seminars for twenty years. Dr. Mann is the author of a textbook, the plant-engineering chapter of a production handbook, and forty-two articles and papers on plant engineering and maintenance.